MINERVA'S FRENCH SISTERS

MINERVA'S
FRENCH SISTERS

Women of Science in Enlightenment France

NINA RATTNER GELBART

Yale UNIVERSITY PRESS/NEW HAVEN & LONDON

Published with assistance from the Annie Burr Lewis Fund.
Published with assistance from the Louis Stern Memorial Fund.

Yale University Press books may be purchased in quantity for educational,
business, or promotional use. For information, please e-mail sales.press@yale.edu
(U.S. office) or sales@yaleup.co.uk (U.K. office).

Designed by Mary Valencia.
Set in Fournier type by Integrated Publishing Solutions, Grand Rapids, Michigan.
Printed in the United States of America.

Library of Congress Control Number: 2020945845
ISBN 978-0-300-25256-9 (hardcover : alk. paper)

A catalogue record for this book is available from the British Library.

This paper meets the requirements of ANSI/NISO Z39.48-1992
(Permanence of Paper).

10 9 8 7 6 5 4 3 2 1

If it were the custom to send little girls to school and teach them the same
subjects taught to little boys, they would grasp and understand the subtleties
of all the arts and sciences just as easily as the boys do.
—CHRISTINE DE PIZAN, *Livre de la Cité des Dames* (~1405)

I seek enlightenment with an urgency that consumes me.
—MME DE GRAFFIGNY'S ZILIA in *Lettres d'une Péruvienne* (1747)

Let it be firmly *resolved* that women will henceforth be enlightened and intelligent.
. . . Courage, women, no more timidity. Let us prove that we can think, speak, study
and criticize as well as [men]. . . . I await this revolution with impatience. I will do my
utmost to be the first to precipitate it. . . . Men everywhere are being forced to recognize
that Nature made the two sexes equal.
—MME DE BEAUMER, first female editor of the *Journal des Dames* (1761–1763)

There is no female mind. The brain is not an organ of sex.
As well speak of a female liver.
—CHARLOTTE PERKINS GILMAN, *Women and Economics* (1898)

Contents

Acknowledgments

Throughout my entire career I have been enamored of eighteenth-century Frenchwomen, journalists, playwrights, midwives, revolutionary activists, and most recently those dedicated to science. I cannot count the number of wonderful colleagues and friends who have assisted me in various ways during the many phases of this scholarly love affair. Even here, where I confine myself to those who helped and inspired me with this particular project, my debts of gratitude stretch so far back that I am certain to have forgotten to include some. Conversations over decades in both the U.S. and abroad with those who shared my enthusiasm for Enlightenment France have enriched my understanding of the period, its science, and its intrepid females, and many have listened patiently, argued good-naturedly, encouraged my research, critiqued my writing, and offered constructive suggestions of all sorts.

In France, Patrice Bret, Françoise Vaysse, Françoise Launay, Elisabeth Bardez, Jean-Denis Bergasse, Gabriela Lamy, Aline Hamonou-Mahieu, Jeannine Gerbe, Gilles and Antoine Tarjus, Sophie Miquel, Nicolle Maguet, and Monsieur Fougerolle kindly assisted me in different capacities. Pascale Heurtel, formerly of the Bibliothèque Centrale du Muséum National d'Histoire Naturelle, Florence Greffe, formerly of the Archives de l'Académie des Sciences, and Madeleine Pinault-Sørensen and Catherine Voiriot, both of the Louvre, were helpful at key junctures in my research. For gracious hospitality over the years I cannot thank enough friends and relations Ida and Sydney Leach, Kathleen Wilson Chevalier, Danielle Chabaud, and Gerard Rychter in Paris, and Edith and René Vles in Périgord.

Although we have never met, two Canadian colleagues helped me greatly. Marie-Laure Girou-Swiderski corresponded with me and facilitated my access to manuscripts in the University of Ottawa archives, and Laurence Bongie graciously answered my email queries. Also from afar, Michael Sonenscher and

Neil Jeffares in England and Sergueï Karp in Russia were attentive to my questions, as was Karl Feltgen in Rouen.

Here at home in the U.S., Kathryn Norberg, Gary Kates, Laura Mason, Kathleen Wellman, Judith Zinsser, Peggy Muir, Carolyn Knobler, Liz Horst Redman, Jeff Ravel, Bonnie Smith, Karen Offen, Mary Lindemann, Tip Ragan, Candace Waltz, and Keith Baker have been keen sounding boards. Some have been readers, all have been attentive listeners and have engaged in fine talk and provided inspiration. Chairs and panel members too numerous to name at the Society for French Historical Studies and the Western Society for French History have read drafts of my presentations on this material over the years, and audiences at these meetings have offered countless helpful responses. Colleagues past and present at Occidental College—Annabelle Rea, Jim Whitney, Martha Ronk, Sharla Fett, and Robert Ellis—as well as the members of my academic co-mentoring group have been curious, sensitive, and sensible guides and models. Lynn Dumenil, retired from our history department and greatly missed, read the most recent version of the manuscript, gently but firmly pointing out excess verbiage, vagueness, and infelicities of style. Of course, any errors that remain despite the excellent advice from all readers are mine alone.

Some of the people most important and dear to me are no longer alive. I will never forget their encouragement and the ways my ideas for this book were strengthened and honed through conversations with the late Roger Boesche, Roger Hahn, Ray Birn, Mary Sheriff, Sheila Levine, and Nanci Rivière.

Archivists in Geneva, Paris, and London were very helpful, as were librarians Samantha Alfrey of Occidental College and Russell Johnson and Teresa Johnson of the UCLA Biomedical Library Special Collections.

Elizabeth Knoll read and generously advised me on my initial proposal and directed me to Senior Executive Editor for science and medicine Jean Thomson Black at Yale University Press, who warmed to this project from our very first exchange and has provided valuable insights and patient nurturing for which I am most grateful. I appreciate too the care and attention given to this manuscript by my copy editor, Eliza Childs, my in-house editor, Margaret Otzel, and my proofreader, MaryEllen Oliver. I thank also the rest of the team at Yale and the anonymous reviewers of my manuscript for their thoughtful, useful suggestions.

A Guggenheim Fellowship was a welcome early vote of confidence for this book, and subsequent Faculty Enrichment Grants from Occidental College provided additional funds which I greatly appreciate.

The support of my family never flagged even as my work on this manuscript stretched interminably, submerging me in avalanching piles of notes all over the house. Our cats helped too, as only cats can. To Bill, my sweet, wise, scientifically brilliant, and culinarily gifted husband of fifty-two luminous years, to my daughter Eva and son Matthew—computer guru and rescuer during tech meltdowns—and to their spouses, and to my grandchildren Patrik, Julie, and Niklo, I dedicate this book with love.

Chronology

1762, 1764 Lepaute's all-female eclipse map—thousands of copies distributed

1764 Barret moves with naturalist Commerson from Bourgogne to rue des Boulangers, Paris

1766 D'Arconville's *Essai pour servir à l'histoire de la putréfaction* published anonymously

1767 Barret's departure from Rochefort as valet to Commerson on Bougainville's voyage

1768 Barret and Commerson disembark on Isle de France (Mauritius) in the Indian Ocean

1770, 1771 Biheron's second and third demonstrations to the Académie

1771–72 Biheron's first trip to London with her anatomical models, to meet British scientists

1772–73 Biheron's second trip to London; consideration of a trip to Saint Petersburg

1773–74 On Isle de France Commerson dies; Barret opens billiard bar, marries a French soldier

1775 D'Arconville publishes her *Mélanges* with many additional scientific translations. Barret returns to France from Isle de France, settles in the Dordogne region

1776 Barret meets notary in Paris to claim her inheritance from Commerson's estate

1780 Lepaute requests a pension for her scientific work from Minister of Finance Necker

Basseporte dies, is buried in Église Saint Médard

1785 Barret is awarded a royal pension for her contributions to science

1786 Biheron's cabinet of models "purchased" by Marie Antoinette although Biheron is allowed to keep most of it

1788 Lepaute dies, is buried in Église Saint Roch whose façade now sports a Lepaute clock

1789 Revolution begins—disruptions ensue for the women's work

1793 *Officiers de santé* plan to obtain Biheron's secret ceroplastic formula

Biheron fights to receive her promised payments for the royal purchase of her collection

Barret struggles to get her royal pension reinstated

1794 D'Arconville is imprisoned, her son and brother-in-law decapitated at end of the Terror

1795   Biheron dies, Basseporte's self-portrait among her possessions; bones
        to catacombs

1805   D'Arconville dies, after dictating her memoirs, is buried in Notre Dame
        des Blancs-Manteaux

1807   Barret dies in Saint Aulaye, in Périgord, on the banks of the river
        Dronne

Actors in a Supporting Role

Aotourou	Tahitian prince taken by Bougainville back to Paris, befriends Barret onboard
Bachaumont	chronicler who commented on Lepaute, Barret, Biheron
Boscovich	astronomer, one of several foreign scientists squired around Paris by Lepaute
Bougainville	sea captain sympathetic to Barret despite her deception
Buffon	accuser of Condillac (and Ferrand) as plagiarists, supporter of Basseporte at the Jardin
Catherine the Great	Empress of Russia, hostess of Diderot, near recipient of Biheron's anatomical models
Clairaut	close friend of Ferrand, colleague but nemesis of Lepaute
Commerson	colleague of Barret, admirer of Lepaute for whom he named the hydrangea
Condillac	colleague of Ferrand
Cramer	correspondent of Ferrand
Dagelet	nephew of Lepaute, trained by her in astronomy, lost at sea on Lapérouse expedition
D'Alembert	gossiper about Ferrand, refuser of calculation successfully done by Lepaute
Dalibard	botanist, supporter of Franklin's electricity work, friend of Basseporte and Biheron
Daubenton	head of the Cabinet du Roi of the Jardin du Roi, recipient of Barret-Commerson specimens
Desnos	mapmaker, plagiarizer of Lepaute's eclipse chart
Diderot	admirer of Barret, close friend and promoter of Biheron

Fourcroy	chemist and admirer of d'Arconville
Franklin	friend and admirer of Basseporte and Biheron
Gosselin	geographer, late-life confidant of d'Arconville, to whom she bequeathed her library
Helvétius	close member of Ferrand's circle
Hewson	British surgeon and anatomist who worked with Biheron during her London visits
Jussieu, Antoine Laurent de	friend of Basseporte and Biheron, neglecter of Barret/Commerson collection
Jussieu, Bernard de	botanist, supporter of Basseporte at the Jardin du Roi
Lalande	colleague of Lepaute, worked with her on Halley's Comet calculation
LeRoi	clockmaker and admirer of Lepaute but fierce rival of her husband
Mably	brother of Condillac, devoted friend of Ferrand and her companion Mme de Vassé
Macquer	chemist, colleague of Mme Thiroux d'Arconville
Malesherbes	friend of Ferrand, admired by Basseporte and Biheron, friend of d'Arconville
Mentelle	geographer, devotee of Basseporte
Morand	surgeon, physician, royal censor, supporter of Biheron and d'Arconville
Nassau-Siegen	admirer of Barret on Bougainville's voyage
Poivre	supporter of Commerson's and Barret's disembarking and staying on Isle de France
Poulletier de la Salle	chemist and collaborator of d'Arconville
Pringle	doctor and admirer of Biheron in Paris and London; writer on antiseptics disputed by d'Arconville
Rouelle	chemist helped by Basseporte, teacher of d'Arconville
Rousseau	close friend of Condillac, admirer of Basseporte
Stuart, Charles Edward	Pretender to the British throne, protected by Ferrand
Thouret	founder of revolutionary École de Santé, admirer of Biheron
Turgot	member of Ferrand's circle, pursuer of Commerson's materials once alerted by Barret

Vachier	friend of Commerson's since childhood, protector of Barret's inheritance
Vivès	surgeon on Bougainville's voyage, examiner of Barret after her unmasking

Introduction

A Sextet of Firsts, Variations on a Theme

FRENCHWOMEN ARE NOW KNOWN TO HAVE PLAYED A SIGNIF-
icant role in the eighteenth century, participating in and shaping the Enlightenment and the Revolution as *salonnières,* authors, and activists.[1] Still other women,
those who daringly chose to do science, to expand our knowledge of the natural
world in diverse disciplines and who are the subjects of this book, have been less
recognized. Such intrepid females were scarce, for their path to science was
strewn with obstacles. Referring to her own difficulties, the Newtonian physicist Mme Du Châtelet wrote, "I feel the full weight of the prejudice which uni-
versally excludes us from the sciences . . . there is no place where we are trained
to think, . . . an abuse which cuts back, as it were, one half of humankind."[2] The
great feminist Mary Wollstonecraft echoed this, lamenting more generally at
the end of the century that the study of nature seemed closed to most women
and wishing that more of them would undertake serious work on this "fair book
of knowledge," would "attach themselves to a science with that steady eye that
strengthens the mind." Instead, she found such women to be a wondrous rarity.
"I have been led to imagine that the few extraordinary women who have rushed
in eccentrical [*sic*] directions out of the orbit prescribed to their sex, were male
spirited, confined by mistake in a female frame."[3]

Yet there were some "extraordinary women" of science nonetheless, a few
even achieving stardom, for example, Italy's learned ladies—Laura Bassi, the
first woman physics professor in Bologna, and her colleague the anatomical lecturer Anna Morandi—or the distinguished English astronomer Caroline Herschel.[4] These three certainly escaped beyond their normally confining spheres.
So did Mme Du Châtelet. But to find her French female contemporaries, also

working with courage and perseverance but less visible, we must adjust our lens. The pages that follow tell compelling stories about six of them.

Mme Du Châtelet herself is not among them, although she became a veritable scientific heroine of the period and deservedly so. But even she was overlooked for two centuries; as Voltaire's mistress she was mentioned by historians solely in that capacity, her own intellectual prowess not discussed, surely not celebrated. Finally, however, toward the end of the last century, she began to be studied in her own right, taken seriously as the *femme de science* that she was, brilliant, bold, and widely influential, her books on natural philosophy and physics appreciated and analyzed in depth and detail.[5] Here was a woman driven to overcome the lack of education she decried above, procuring the private tutors necessary to learn and truly master mathematics and the hard sciences. Her numerous works, especially the 1740 *Institutions de physique* and her posthumous translation of and creative commentary on Newton's *Principia*, establish her as one of the most sophisticated writers on science of her day. There is now a copious literature on her and the many facets of her oeuvre, and she has attained a kind of visibility and luster that her female contemporaries do not enjoy.[6] The six women discussed in this book, scientific peers of hers, were relegated to the shadows and are only slowly emerging into the light. Aware of Mme Du Châtelet's brilliance, Voltaire called her the "Minerva of France," invoking the Roman goddess of wisdom and the arts. He told Frederick of Prussia that she, the real brain behind his 1738 *Élémens de la philosophie de Newton,* truly deserved the authorial credit: "Minerva dictated, and I wrote."[7] I have therefore called my less-known Enlightenment women of science "Minerva's sisters." Several of them were likened to her by their male contemporaries, and the chemist of my last chapter emblazoned her own name on Minerva's shield in her commissioned bookplate. So this deity was broadly seen as the patroness of science both pure and applied.

Lest the word "science" in this context appear anachronistic, it was used frequently in eighteenth-century Europe, as by Du Châtelet and Wollstonecraft above, in addition to the more common term "natural philosophy."[8] The crowning intellectual achievement of the French Enlightenment, the multi-volume *Encyclopédie* of Diderot and d'Alembert, invoked "science" in the very subtitle, *Dictionnaire raisonné des sciences, des arts et des métiers,* and the term with its modern meaning occurred in its numerous entries concerning the study of nature. The long article in volume 8 on "Natural History," for example, remarked on the "taste for this science in the general public" and sprinkled the word throughout as it compared "the different branches of science" that deal with the mineral, animal, and vegetable realms of nature: astronomy, anatomy, botany, chemistry, and experimental physics. "Blessed is the century in which the sciences are sufficiently perfected as to . . . contribute to the happiness of man."[9] The august members of the Paris Académie des Sciences were, as the name implies, devoted to the study of the natural world, not other kinds of wisdom pursued in other academies. The men who knew the women in this book employed the term when referring to them, as did Benjamin Franklin in citing the anatomist's work and Jérôme Lalande the astronomer's. And the women themselves used the term, the chemist of my last chapter, for example, writing, "Chemistry, as well as all the sciences that have for their object the knowledge of nature, makes strides every day."[10]

It was a Cartesian, the great popularizer Fontenelle, who in 1686 first encouraged French women to participate in scientific learning, the marquise in his best-selling *Conversations on the Plurality of Worlds* refusing to be side-tracked by her teacher's repeated seductive maneuvers, instead insisting that they keep their eyes and ideas on the stars in the night sky they were discussing.[11] The Newtonians explicitly wooed women too, as did Algarotti's *Newtonianisme pour les dames* in 1738. But they did not all need such scientific simplifications. In fact Newton's 1704 *Opticks* had been translated in 1720 into French at the urging of a woman.[12] At its end this work included inviting "Queries"—"hints to be examin'd and improv'd by the farther Experiments and Observations of such as are inquisitive," speculations intended for "a farther search to be made by others."[13] Newton's overtures in these Queries stimulated new ideas, approaches, and styles of investigation not only in mathematics, astronomy, mechanics, and the physics of light but beyond—in epistemology, natural history, botany, anatomy, and chemistry, the fields of the women in this book.[14]

Elisabeth Ferrand, the subject of chapter 1, was a mathematician, an early believer in the law of attraction, and an epistemologist who studied human cognition by analyzing separately what each of the five senses contributed to it. In a portrait by Quentin de La Tour she chose to be depicted "meditating on Newton." Astronomer Nicole Reine Lepaute of chapter 2 computed the accurate prediction of the return of Halley's Comet, a triumph widely considered to prove Newton's law of universal gravitation and which required the analysis of the component forces that determined the comet's orbit. Chapter 3 deals with

3

two enthusiastic contributors to botany. Field naturalist Jeanne Barret disguised herself as a man to work with botanist Philibert Commerson collecting flora during Bougainville's round-the-world voyage. When curating these harvests they sought order behind their great profusion, discovering previously unknown plant species and always searching for subspecies, varieties, to better clarify the overarching category. Madeleine Françoise Basseporte, botanical illustrator and the king's draftsperson at the Jardin du Roi, enriched the work of Buffon and of Bernard de Jussieu by analyzing and depicting the parts of plants to discover the patterns and organizing principles of that science. Chapter 4 concerns anatomist Marie-Marguerite Biheron, who analyzed the human body, taking it apart in countless dissections to elucidate the hidden internal structures and thus perfect the wax models that she displayed and taught with for decades. And chapter 5 introduces chemist Marie Geneviève Charlotte Thiroux d'Arconville, who studied organic decomposition, echoing Newton's view, in Query #30 of the *Opticks,* that decay was a natural breakdown process in which substances were reduced to their component elements. As Newton summed up, "Nature seems delighted with transmutations," and d'Arconville confirmed, "We must therefore look at putrefaction as the wish of nature."[15]

I set out almost two decades ago to revive the stories of these women, seeking their traces in the archives, finding what contemporaries said about them, and bit by bit uncovering information that had been dropped from the triumphalist narratives of science composed by men in charge. These women had been written out of history. Well-known in their day but subsequently erased from the record, they have only recently begun to resurface.[16] Their lives are fascinating because their keen intelligence, curiosity about the workings of nature, extraordinary verve, and visionary courage led them to defy the gender conventions of their time and do the science they wanted to do. Driven to embark on what Carolyn Heilbrun has called a "quest plot," to hunt for something beyond the comfortable though stifling nest, they escaped from the traditional female script and, unafraid, became their own agents of rescue.[17] Theirs was a choice much more difficult than conforming to the expectations of their day, but for them there simply was no other option.

I have called them "firsts" because unlike girls and women today who want to pursue science, these women had no female role models to follow, and so they themselves were the pioneers. Their very presence in these fields was rare, they were un-networked, unaffiliated in any official capacity with exclusively male institutions, and not even connected with each other. Taking tentative steps on untrodden paths, then blazing the way forward as they gained confidence, Ferrand became the first woman—preceding Mme Du Châtelet—to champion Newtonianism in France, Lepaute the first Frenchwoman to be elected to a scientific academy, Barret the first woman to sail around the world, Basseporte the first (and only) woman to secure the coveted post of *dessinateur du roi* in the royal botanical gardens, Biheron the first person to teach general anatomy and sex education using models of her own invention, and d'Arconville the first person to suspect, over a century before Pasteur, that the cause of putrefaction was airborne. Theirs were full, rich lives of steely resolve, and they knew their contributions to science were worthwhile. Such women made it easier for those who followed in their footsteps, precisely because they paved the route. Love of learning about nature, perseverance in the face of obstacles, determination to be useful, will and energy to buck the norm, to break free of limits—all characteristics that distinguished them then and can be emulated today.

These are the qualities that they had in common, a shared, discernible pattern. But I want to emphasize their differences as well, for they were an eclectic mix, representing a wide range, an assortment of approaches to their scientific goals. There was no solidarity among them for they were unaware of each other except in one special instance where two of them were a couple. Overall, because their backgrounds and aspirations were so diverse, they illustrate an array of choices and trajectories that got them where they wished to go. There was not just one way to pursue science.

How did they do it? Very few possibilities existed in eighteenth-century France for women with serious intellectual curiosity to slake their thirst. Girls might at best be educated in a convent with catechism, music, needlework, some domestic skills, and if they were lucky, they might learn how to read. Then married off young to a stranger, they were immediately expected to produce children, as soon and as many as possible. Divorce was unheard of, and they were to be content with serving a husband and nurturing a family. To circumvent this, half of my six women chose female life companions and so remained unconstrained by the almost feudal marriage laws and customs that persisted in this period and denied wives any legal existence or social independence, no more rights than the criminal and the lunatic. But lesbian relations, considered a form of sodomy, were then punishable by death if "lewdness," "depravity," and "unnatural acts" could be documented; there was, for example, an execution of this sentence on 5 June 1750 when two individuals were burned alive in the Place de Grève.[18] Whether the intimacies shared by the women in this story were sexual as well as emotional and scholarly we cannot know; in any case their intense female friendships seemed respectable enough to not raise alarms.

The remaining three women did marry, but they did not compromise their work to do so; one delayed marriage until after her scientific activities ended, and the other two tolerated oddly decentered husbands who were far less significant than the male colleagues with whom they found fulfilling cerebral relationships and who assisted them in their work. These three thus managed, astonishingly, to carve out freedoms for themselves despite the confines of matrimony. One even arranged a sort of chaste ménage à trois, living for decades with both her complacent spouse and her male science colleague in an unusual configuration. The boldness of these women manifested itself in still more ways, one hiding a famous political fugitive for years at great personal risk, another masquerading as a man to achieve her purpose, another refusing to retire in her old age and preventing powerful men of science, and royal ministers, from forcing her to do so.

Tenacious, independent, and resolute, these six boundary-breaking women did not flare out, were not dilettantes, but instead dedicated researchers who kept at their work over many years, the shortest for about fifteen, the longest for more than half a century. We could reasonably argue that their commitment enhanced their longevity, that they exemplified creative aging, as their lifespans far exceeded the thirty-eight-year average for their day, the youngest dying at fifty-two, the others living exceptionally to be sixty-five, sixty-seven, seventy-six, seventy-eight, and eighty-five. They were motivated not by fame, glory, or fortune—what Dorinda Outram has called "dirty" power—but by genuine thirst to satisfy their deep curiosity about the natural world and determination to make useful contributions to scientific knowledge. Fierce competition among men swirled all around them, priority disputes, intellectual meanness, plagiarism accusations, feuds of all kinds, grasping claims for entitlement to patrons, pensions, prizes. The Enlightenment was a time of burgeoning public appetite for science; there was a flourishing market for it and also the inevitable jockeying for position to enjoy the spoils. The women had to negotiate this without damage to themselves, find a way through it that would not seem threatening and so would not derail them from pursuing the science they loved. Spanning several fields, the women strengthened their century's understanding of the mind, the cosmos, the wild plants of the wide world, the cultivated botanical garden, the human body, and the border between vitality and decay.

Were they feminists? The term "feminism" has evolved over the centuries, and even today when it sounds familiar and is used loosely it means many different things.[19] In the eighteenth century the word did not exist. These women wanted to do science, which was not conventional, and they thus figured out how to prevail, but they did not necessarily picture other women doing the same. None of them believed they were inferior to their male colleagues, nor did they resent those men. They did not rail against inequity or voice grievances against male prerogatives.[20] Their actions, however, spoke louder than words. They made original life choices for themselves, and they maneuvered to accomplish exactly what they wanted yet without disruption to the fixed order.

While sensitivity to gender is of course central to my book, the sociologist Dianne Millen's work on female scientists has shown how tricky it is to "do feminist research on non-feminist women." Millen argues that "a concern for power relationships is *the* defining feature of feminist research," but my six women did not necessarily consider themselves downtrodden or oppressed, did not all have "full awareness of the systems which surround and constrain them." Mme Du Châtelet did and was vocal about it as we saw, but my six subjects were less outspoken and functioned differently. They never berated the establishment, never imagined that the world would change to accommodate them.[21] Instead, without overt protest, they learned to work the system, to recognize and exploit opportunities, believing that their own efforts, and not external factors, would be responsible for successful outcomes. Inner drive propelled them to pursue their scientific plans. When encountering structural barriers they may have lacked the conceptual vocabulary, or the desire, to speak about them; instead they negotiated smart means to do what they wished without causing a stir. Nimble and purposeful, they navigated the male-dominated turf without leaving it.

These women deployed what Michel de Certeau pointedly calls *tactics* as opposed to *strategies,* coping mechanisms for operating in difficult imposed terrain, improvised measures to circumvent the established scheme, to encroach cleverly without being detected as transgressors. In fact, to do anything other than pairing and bearing in their day, women had to develop all manner of methods to steer through obstacles and get where they wanted to go. Certeau makes a meaningful distinction between *strategy,* which requires power and having one's own proper place from which to go out and master more places, and *tactics*, which must be played by the less powerful on the existing field and which require maneuvering in uncongenial space, using the constraining order unconventionally, artfully. In his formulation, strategies are repressive, tactics are oppositional and involve the seizure of chances that arise to gain one's objectives, creative resistance, manipulation of prevailing rules, and the imagining of an alternative vision to attain desired ends.[22] There is, as Certeau says, always some play in the machine, ways to map out room within constraints to make the situation habitable. The women in this book contrived idiosyncratic routes for their lives through and out of traditional expectations.

Although they had in common their unorthodox choice to do science, in many other respects they were dissimilar, hailing from different classes and backgrounds and working in unrelated fields. The first three were atheists, the last three were Jansenists, critical of official Catholicism but devout believers in pure biblical teachings and thus enemies of despotism, hierarchies, tyranny, and arbitrary powers of church and state.[23] The women were socially diverse as well. Barret was a domestic, Biheron the daughter of a pharmacist, d'Arconville a high society grande dame. Their motives and aspirations were not the same. Except for Basseporte and Biheron, who were partners, we cannot be sure the others were acquainted, although they should have been. They had no access to the scientific sociability enjoyed by their male contemporaries, who formed bonds through numerous institutional affiliations, academies, university faculties, observatories, the periodical press, and the joint effort of the magisterial *Encyclopédie*. All of these provided for men a sense of belonging to lofty, historically significant endeavors. There was no analogous arena for smart women to come and work together, no shared forum for discussing or exchanging ideas of mutual scientific interest. My six women had to craft their own stimulating communities, enlist teams of men who could give them the help they needed to get started. Once launched, however, they were pretty much on their own, outside of formal institutions, persevering in original ways in salons, studios, workshops, makeshift observatories, dissecting enclosures, on boats, in gardens and home laboratories, at dinner tables.[24] That they did this without learned female camaraderie is an object of wonder and an overarching point of my book. Each had to find her own path, and their stories illustrate the diversity and creativity of the choices they made in order to live scientific lives.

Known in their day, they were later deemed insignificant, rendered invisible, literally obscured. As biographical theorist Judy Long plainly puts it, such socially produced "obscurity" accounts for the "shortage" of notable female subjects, for they are "caused to disappear" by later male gatekeepers in an active process of submersion.[25] And while my women have in recent years begun to attract some scattered attention, they have been discussed in isolation, never together, and so remain apparently singular. Treated separately, disconnected, they seem anomalous, exceptional and rare birds, *admired* in the old sense that, as Germaine Greer writes, "carries an undertone of amazement."[26] When they are set in context and discussed together in depth and breadth, however, a pattern emerges that illuminates a kind of female courage and investigative energy that ran throughout the Enlightenment. There were quite a few other scientific women in this period of whom I caught glimpses, but too much evidence had been lost to time. I concentrate here on the ones I was able to flesh out. And the taxonomic urge is strong, as historian Jenny Uglow says. Whereas their individual lives are interesting in and of themselves, their cumulative experience is fuller and more empowering.[27] "A collective biography," Alison Booth explains, "requires an additional rhetorical frame besides that of any biography: the definition of the category or principle of selection . . . and the *encouraging* view that noteworthy lives differ enough from each other to leave space for others to join them."[28] Of course I believe that such examples of path-blazing women of science are "encouraging" for female scientists today, as Booth would hope them to be. Assembling their stories of fervor and stamina results in a picture and a message that is greater than the sum of the parts.

These six women, once their resolve and staying power became clear, found support from some male contemporaries. This fact reveals an important tolerance and even welcome among the men of science and philosophers that they selectively gathered around them, a surprising elasticity of mind in a world too easily dismissed as narrowly patriarchal. It enriches our current understanding of the French Enlightenment to observe how a considerable number of men wanted these women to succeed and be properly credited for their contributions. This approach reveals unexpected hospitalities in the scientific milieux from which the women were officially excluded yet with which they were so intimately involved. My more gender-inclusive narrative shows not only the content of these women's works and days but the fact that they were, due to their own efforts and tenacity, accepted rather than thwarted. To probe this dynamic meaningfully it does not suffice to concentrate on their male colleagues and then just "add women and stir." Such a technique does nothing to alter the traditional picture.[29] Instead we must flip it around, adjust our lens, zoom in on the women at the center, and *then* consider the orbiting men who become visible in their lives as we pan out.[30]

So the focus of this book is on the women's stories, with male supporting actors added as needed. Yet they are needed. Most of the letters, scientific papers, and artifacts of these women were not preserved by the custodians of culture and are lost, through neglect, unceremonious discarding, or deliberate material destruction—the net effect was erasure.[31] We do have Basseporte's botanical illustrations and d'Arconville's manuscript memoirs dictated in her old age. But archivists, librarians, and museum curators failed to acknowledge the value of, and therefore did not keep, Ferrand's mathematical communications, or Lepaute's astronomical papers on the transit of Venus of 1761, or Biheron's anatomical models. And they did not attempt to distinguish, while they still might have been able to, Barret's herbarium labels from Commerson's. We know that all six women corresponded with savants, but we have a mere handful of letters by Ferrand, Lepaute, Biheron, and d'Arconville, one by Basseporte, and phantoms of a couple from Barret to which only the replies remain.

These women, therefore, are known to us mostly through the words of their famous male contemporaries, who luckily for me had a lot to say about them, so there is rich material here. I make extensive use of such sources—letters defending them, sometimes vehemently, ship logs, obituaries and eulogies (the men were particularly eloquent in mourning), newspapers, memoirs, dedications, portraits, homages, mini-biographical entries, chronicles, and gossip sheets. Diderot, d'Alembert, Rousseau, Buffon, Mercier, Bougainville, Lalande, Commerson, Condillac, Bachaumont, Clairaut, Jussieu, Grimm, Mentelle, the artists Quentin de La Tour and Voiriot in France, and foreigners such as Ben Franklin, John Pringle, John Wilkes, Linnaeus, members of the Bernoulli family, the monarchs of Sweden and Denmark, and various German princes, to name only some, make clear that the subjects of this book were forces to be reckoned with and scientifically significant in their day.

I write their stories with as much texture as possible. Determined to set their science in the context of their ways of life, in what Steven Shapin calls their *habitus,* I agree with him that science should be studied "as if it was produced by people with bodies, situated in time, space, culture and society."[32] Their science bore the marks of where and when it was produced, of their quotidian situations. While mining the sources mentioned above, what others said about them,

I also test those reports, holding them up against the exertions of the women themselves, their behavior, what they actually *did*. Deeds do not lie. Kathleen Barry argues that "feminist-critical biography must assume a self that is knowable through its doing and actions, that is, through intentionality." Even for women who did not leave accounts of themselves, we can study the highly significant choices they made, their interactions, the empirical world of their experiences, and in this way we can retrieve some of their lost subjectivity.[33] These six women were not strident about their problems or the difficulties they encountered, but, as Barry reminds us, women who did not say much still had a battle to fight, suggesting that because women in all ages "usually know more about domination than they speak," we may assume such consciousness even as we recognize the reasons it might have been submerged.[34] These women were quietly but concertedly persistent and, while they surely encountered deterrents, took deliberate steps forward and stayed the course.

Although they did not leave much evidence of their private thoughts, we have a bit more in their works meant for public consumption, whether textual or figural: Lepaute's star charts, eclipse maps, and computational tables; Basseporte's hundreds of botanical illustrations, some drawn with pencil and pastels, others painted on vellum, these last still part of the well-preserved permanent collection called *Les Vélins du Roi;* Biheron's four-page pamphlet advertising her first *Anatomie artificielle* exhibit in 1761; and d'Arconville's (always anonymous) printed volumes of scientific translations plus of course her original work, a treatise on putrefaction. These are relatively meager traces, however, and the women seem almost mute compared to their contemporary Mme du Coudray, the "King's midwife" and subject of my last book. Because du Coudray's was an official royal mission to arrest infant mortality and she therefore a celebrity, I found hundreds of letters in numerous departmental archives by, to, or about her.[35] But such a paper trail was exceptional for a woman at that time. Moving from my *sage-femme* who enjoyed publicity to these much less conspicuous yet equally strong *femmes sages* has had its challenges.

These women refused to accept exclusion from the march of progress. The need to be useful was a trait they shared and absolutely central to their motivation, and they repeatedly articulated it as a moral imperative for all their work. Their life choices gave them purpose and self-fulfillment, taking control of their destinies, learning courage by being courageous. And they upheld the highest standards: intellectual precision and logical rigor, computational accuracy, dis-

ciplined collecting in the field, "truth-to-nature" botanical images and anatomical models, and scrupulously recorded laboratory experiments. They expressly
considered it "criminal" to be careless when presenting scientific results.[36]

The independence of these women resonated with some writing by their
nonscientific contemporaries. Mme de Graffigny's 1747 novel, the best-selling
Lettres d'une Péruvienne, featured the heroine Zilia who gains strength, resists
marriage, finds and frees herself through learning and chooses a single life that
satisfies her mental thirst and fulfills her. Graffigny's critics were constantly
trying to change the ending of the novel, to marry off her heroine, but the author stood firm. Female journalists also proclaimed the Enlightenment the *Siècle
des dames,* the Century of Women. Some of my six in this book were explicitly
mentioned and lauded in the *Journal des Dames,* a periodical whose very existence signaled something new and whose first female editor, Mme de Beaumer,
championed women adamantly.[37] There was excitement in the air, a sense of
possibilities, of openness for women to accomplish things and to do it relatively
unscathed if they knew how, as these did, to choreograph the requisite dance.

Subsequent historical accounts, however, damped down all this energy, re-
ducing the players to silence. The erasure of women from the history of science
in particular has been noted and rued by many scholars, historians, sociologists,
and biography theorists. Margaret Rossiter early stressed their "systematic under-
recognition" in previous centuries, and Naomi Oreskes their "invisibility," how
they end up in the "ellipses" instead of the limelight because they do not buy
into the male rhetoric of "heroism" and conquering the unknown. Hilary Rose
explored the shared experience of oppression among women scientists.[38] To re-
dress this imbalance it is necessary to highlight and foreground what women
have accomplished, to embark on a new and different kind of recounting, a
counterpoint narrative. The "objective" cradle-to-grave, womb-to-tomb presentation is, after all, a patriarchal invention that needs some unpacking. Here
feminist biographers have chimed in: Phyllis Rose on the "deliciously wicked
absence of impartiality"; Carolyn Heilbrun and Judy Long on telling women's
lives in original ways based on connection with the subjects and the validating
of empathy and affinity; Paula Backsheider on how feminism, through reciproc-
ity, has brought the biographer back into the frame; and Amy Richlin on the
need to argue with the archival silence that continues to engulf women and the
importance of explaining the author's use of the personal voice, "not everyone's
cup of tea" but honestly imperative for social change.[39] It is not only feminists

who uphold innovation and making the teller part of the tale. James Clifford has criticized the traditional approaches by old-school omniscient biographers bent on making lives seem tidy, what he calls the "myth of personal coherence," Robert Rosenstone has promoted unorthodox historical writing of all kinds, and Thomas Söderqvist has argued for risk-taking, including specifically "open collaboration" between biographer and scientific subject.[40]

Such scholars have inspired me to experiment with both storytelling and structure in this book. Thus it has a hybrid format, combining the age-old yet newly relevant genre of scientific biography with more personal elements.[41] The tales of these six women are told in separate chapters, excepting a central fulcrum chapter that combines two dedicated but very different contributors to botany, which provides a certain symmetry. While the chapters incorporate the women's geographies of inquiry and various theoretical lenses through which they can be viewed, they are biographies, tales of lives lived whole. In addition, these discrete sections are braided together through short links, interludes in a different voice that bridge the individual acts and join them. The chapters cover, the links uncover, investigate, try things—in the spirit of the original meaning of the "essay."[42] There is much about the lives of these women that we cannot know. In the interludes I meander through the neighborhoods where they lived and died, allowing myself to pry, to entertain possibilities, even probabilities for which no evidence remains. For historians, hard facts are not all we need. We also wonder, wander, ponder, speculate, and my links reveal such processes; they are more musing and playful. Increasingly, historians have been using an active rather than a neutral voice, showing their unique relationships with their subjects, their involvement, borrowing devices from other genres, including fiction.[43] My interludes take the form of imaginary letters from me to my protagonists, and for a reason. Enlightenment archives are full of correspondence between men, much of it entirely inconsequential, but far fewer exchanges between women have been preserved. After close to twenty years of living with this project and thinking of these people as "my" women I take the liberty of writing to them, asking questions, probing what might have caused their awakening. Because of course they do not answer, readers get a chance to pause, to pose their own questions, and to contemplate what those replies might have been. Most importantly, I try to inform my six women of their modern relevance, bringing them up to date on things I think they have a right to know.

In stressing the individual agency of these women I have tried to make theirs a "usable past" for us in the twenty-first century.[44] Women have been deleted from histories of literature and art as well, but the effect of their absence from the scientific record is especially damaging.[45] There is still nothing close to parity for women in science today; their numbers in the STEM fields remain distressingly low. Many decry this situation and seek persuasive ways to attract women.[46] I believe that inspiring examples of females in science from the past are much more than interesting; they encourage girls to enter and women to *stay in* this field. Such accounts of female perseverance do advocacy and crucial social work.

My object is not to lionize or make exaggerated claims for the scientific contributions of the six women in this book, although they were certainly knowledge producers, enriching epistemology, astronomy, field exploration, botany, anatomy, and chemistry. Men respected them, made room for them and relied on them in their day. Combining and juxtaposing their surprising lives reveals a vital though unsung female presence in eighteenth-century French science. They are not broadly known, even less so in the English-speaking world, and their stories have not been told together. It will be enough to give credit where it is (over)due.

Dear Elisabeth, Reine, Jeanne, Madeleine Françoise, Marie-Marguerite, and Geneviève,

There. Now I've told the reader what I am trying to do in this book. Next I will tell all of you. I address you collectively here because part of my plan is to introduce you to each other, for although two of you were intimately linked there is no evidence that the rest of you were acquainted. But you should have been.

I believe that your different stories will inspire others who read them today. Why? Because you had grit and stamina, did unprecedented things, and designed satisfying lives for yourselves through science. Many more women do research now than in your time, but still nowhere near enough. We need to hear how you, the trailblazers, gained confidence to work independently in original ways. As your colorful lives prove, there were multiple pathways to such achievement.

You knew the worth of your contributions, seemed not to feel subordinate at all. Your actions made clear what driven women can do. Two of you criticized female frivolity, urging women to think rather than flirt and to make themselves "useful citizens." One of you insisted that women be granted official scientific recognition and academic membership. One of you literally did a man's job as well as any male, indeed better. One of you set up an art school for impoverished orphaned girls, teaching them a skill that launched their economic freedom as flower painters. One of you even educated adolescents and women, graphically, about their own bodies.

Without making a fuss about it you simply refused limits, transcended stereotypes to become yourselves and, fueled by your passion for the sciences and by your mental and practical skills, fashioned full, strong lives. The imaginative ways you employed your talents and fulfilled your ambitions are told in the chapters that follow. The examples you set matter even more today.

ONE

Mathematician and Philosopher

The "Celebrated Mlle Ferrand" (1700–1752)

This demoiselle knows her mathematics.
—JOHANN II BERNOULLI, 1733

My character is not supple.
—ELISABETH FERRAND, 1751

MADEMOISELLE ELISABETH FERRAND WAS NEAR DEATH AS SHE sat to have her portrait done by the great pastelist Maurice Quentin de La Tour, choosing to be depicted "meditating on Newton," a title hardly necessary given the very visible name on the giant volume behind her. The stunning image reveals that she literally rose to the occasion, getting up, dressing up, and sitting up gamely in a conspiracy with her artist to keep her fatal illness secret from beholders. But Ferrand had another secret to which La Tour was not privy, for at that moment the Young Pretender to the British throne, Charles Edward Stuart, nicknamed Bonnie Prince Charlie by the Scots and the object of a massive, continent-wide man-hunt, was hiding in her home. Contemporaries declared "we have absolutely no idea in what part of Europe he lives. They say he has run through the whole North."[1] Keeping the mercurial fugitive here in what he called his "right nest," concealed not only from her artist but from the group of savants who regularly gathered at her salon, may have contributed to the strain of Ferrand's illness, but it also surely added a frisson to the day-to-day life she shared with her intimate partner, Madame la Comtesse de Vassé, the former Antoinette-Louise Gabrielle des Gentils Du Bessay.

La Tour, an admirer of his accomplished contemporaries, was an astronomy enthusiast himself and the owner of at least three precious telescopes, so he might have been especially pleased to honor and immortalize this female Newtonian.[2] Aware of the urgency to complete his pastel while his subject was alive—one of his reviewers discretely commented that the execution of this beautiful work was fraught with difficulties—he was still putting finishing touches on it in his studio when Ferrand wrote her last will and testament, in which she asked that it be bequeathed to Vassé and that the artist make a copy for one of her close friends.[3] Would it have pleased her to know that La Tour exhibited her portrait in the Salon of 1753, a year after her death, alongside his others of mathematician and explorer La Condamine, engineer Montalembert, physicist Nollet, and philosophes d'Alembert and Rousseau? With this public display of Ferrand as a *savante among savants* the artist seemed to be making the point that she belonged on a par with the better-known male luminaries.[4] Now people could finally behold the brilliant but reclusive invalid, "la célèbre Mlle Ferrand" (figure 1).[5]

Born in 1700 in Champagne and coming of age under the Regency, Ferrand was accomplished enough as a mathematician to win the respect of Alexis Clairaut in Paris, Gabriel Cramer in Geneva, and various members of the Bernoulli dynasty in Basel. A professed Newtonian before Mme Du Châtelet, she had probably been taught by her longtime friend Lévesque de Pouilly, who was readily acknowledged by Voltaire as the man who introduced Newton's thoughts into France and remained a member of Ferrand's circle until his death shortly before hers.[6] Her interest in the sciences was documented from an early age, another likely tutor being Pierre Rémond de Montmort, her neighbor in Champagne and an expert on probability, chance, and game theory who, according to Clairaut, esteemed mathematics "above all else."[7] Ferrand chose not to marry, moved to Paris, and shared the rest of her life with Vassé, a widowed countess ten years her junior who admired and adored her; the mutual devotion of these two "sisters" was legendary among those who knew them. Whether or not they were lovers, they were surely most intimate, and the sweetness of their friendship was tinged with the sad realization that, as Derrida discusses in his numerous writings on mourning, inevitably one of them would see the other die first, a fear of loss articulated by both women explicitly.[8] They embraced many causes together. Behind the scenes of the Académie des Sciences the women lobbied for the election of mathematicians they favored, in particular seeking foreign associate membership for Cramer. Politically engaged as well, they were com-

Fig. 1. Maurice Quentin de La Tour, *Mademoiselle Ferrand Méditiant sur Newton,* pastel (1753). [bpk Bildagentur/Alta Pinakothek, Bayerische Staatsgemaeldesamm-lungen/Collection HypoVereinsbank, Member of UniCredit/Art Resource, N.Y.]

mitted and daring Jacobites eager to see the Stuarts recapture the British throne and agreed to provide asylum and multiple kinds of assistance for "le Prince Edouard," as the Parisians adoringly called the royal outlaw.

Ferrand's "bonne compagnie," Cramer's name for the members of her select salon, included, among others, Pouilly, Clairaut, the medievalist La Curne de Sainte-Palaye, members of the Protestant Jaucourt family, the hedonist Helvétius, along with the increasingly ardent republican theorist Mably and his brother the philosophe Condillac, with whom Ferrand developed an especially close cerebral bond in the last five years of her life. She became his inspiration, sounding board, keenest critic, and collaborator as he formulated his epistemologically radical *Traité des sensations,* considered by many to be the foundational enlightenment work in cognitive psychology. Published two years after Ferrand died, Condillac's *Traité* was dedicated to Vassé in big letters right on the title page, and it contained an effusive sixteen-page homage to the departed Ferrand, without whose sharp analytical mind and crucial assistance, the author claimed, he could not have done the work. Although she described herself as "not supple" and "incurably frank," Condillac found Ferrand to be brilliant, self-effacing, and unfailingly generous. Fourteen years older than he, she took him under her wing. In fact, Ferrand made Condillac and his success her last intellectual project.

La Tour's portrait and Condillac's tribute in the *Traité,* appearing as they did after Ferrand's 1752 demise, brought her only posthumous renown, and some modern scholars have deemed her too "obscure" to learn much about.[9] But obscurity is socially produced by the neglect or deliberate erasure of later generations. If we pay heed instead to Ferrand's contemporaries, read the letters of her distinguished colleagues, look for references to her in periodicals, travel diaries, eulogies, and testaments, we can piece together a rich picture of her life as a mathematician and epistemologist around whom male scholars gravitated.

We first encounter Ferrand in the written record in 1733 when the young math prodigy Clairaut included a visit to her home, not once but twice, when he was escorting Johann II and Daniel Bernoulli on a scientific tour of Paris. The brothers were taken to meet the venerable La Condamine, Maupertuis and Fontenelle, attended a session of the Académie des Sciences, then spent time at the Royal Observatory and at the Jardin du Roi where they conversed with the world-famous botanist Bernard de Jussieu. At Ferrand's home the discussion centered on Newtonian forces and capillary action. Mme Du Châtelet and Voltaire, who had just met, were not yet started on such studies, whereas Ferrand was already known to be well-versed in them. One of the visitors reported in his journal: "This demoiselle knows her mathematics; for us she performed physics experiments on attraction with a glass tube."[10] The Bernoullis mentioned that their late brother Nicolaus II had met Ferrand long ago at the home of Pierre Rémond, seigneur of Montmort.[11] This would have been during the years when Montmort was mediating as diplomatically as possible the dispute between Newton and Leibnitz over the invention of the calculus and actively collaborating and publishing with Nicolaus I, the Bernoullis' first cousin, on game theory.[12] Montmort's expertise in probability would have provided Ferrand with an excellent mathematical foundation. Another early influence on her, Pouilly, twice met Newton in London and then presented the great Englishman's ideas about the mechanics of the universe to the Académie des Inscriptions et Belles Lettres in Paris. Ferrand might well have met Cramer the first time he came from Geneva to Paris in the years 1727–29 and worked with the teenage Clairaut, after which the two men maintained a continuous mathematical correspondence.[13]

The salient point here is that Ferrand was connected to math circles from girlhood, was trained by some of the best, and was taken seriously in Paris by the early 1730s, perhaps even before. Supporting Newton at this early juncture, in 1733 when she entertained the Bernoullis, was both daring and unusual. She was starting a fashion, not following one, because although this would soon change, many on the continent distrusted the Englishman's ideas. At the Académie it was still difficult to declare oneself a Newtonian, as Maupertuis was tentatively beginning to do, with the help of Clairaut who had recently been elected as the youngest member in that institution's history. They inside of the academy, and Cramer and Ferrand outside, were really the vanguard.[14] For Ferrand, as we will see, being a Newtonian meant appreciating elegant reasoning, understanding math and maybe even calculus (although this is not certain), accepting the law of attraction, and embracing an orderly, lawful view of nature.

It may seem odd not to make more of the brilliant Emilie de Breteuil, Marquise Du Châtelet, about whom a great deal has already been written, in this context. She is not mentioned in the few letters by Ferrand that have survived, although she was to embark on concerted scientific work of her own, in fact hiring Clairaut to tutor her. But Du Châtelet got started later in the 1730s and was at first somewhat equivocal in her acceptance of Newton, which is why Ferrand stands out as having embraced physical and mathematical research earlier, even doing experiments as the Bernoullis noted. Well-deserved attention has been paid to Du Châtelet, who soon began to publish extensively, writing her *Institution de physique* in 1740, translating and commenting extensively on Newton's *Principia* until her death in 1749. She was the most visible woman doing science in this period, and much of her story has been told. It would of course be good to know if there was any contact, directly or through their mutual friend Clairaut, between these two women whose interests overlapped in numerous ways. So far no trace of a connection has appeared.[15]

When exactly Ferrand and Vassé became a pair remains a mystery, but they were probably living together by the end of the 1730s, the countess escaping from an unhappy, even tragic and threatening past into the peaceful, healing seclusion of a new life with Ferrand in the Couvent Saint Joseph, where secular sections with discrete private entrances on the rue Saint Dominique also provided refuge for Mme du Deffand, for the mother of Mme de Genlis, and for many other ladies of rank seeking calm and tranquility. Vassé's mother had died when she was a child, the husband she married at fourteen had died in 1733, her grandmother and aunt in 1735 and 1736, and her two sons in 1734 and 1739; before she turned thirty in 1740 she found herself grief-stricken and very much alone.[16] It appears she also had some despicable relatives on her husband's side; the murderous exploits of a vile and violent sister-in-law during those same years publicly shamed the family in the pages of various dispatches and in the *Journal de Bruxelles,* and another sister-in-law was doing her best to deny the young widow Vassé her inheritance.[17] Profoundly aware of a variety of human cruelties, she had in 1738 helped provide asylum for a Swedish woman fleeing her dangerous, abusive husband.[18]

Little wonder, then, that Ferrand's civilized milieu of learned friends and useful conversation seemed a nurturing, congenial atmosphere in which Vassé could mend, grow, and thrive. The two women became intimate and at some point moved to their shared adjoining quarters at Saint Joseph. Ferrand had an income, a pension of 3,000 livres, and a once-large though now-dwindling sum from the sale of some land in Bouleaux inherited from her father.[19] Vassé, although she had little if any income, owned the grand house and grounds opposite the Chateau of Marly-le-Roi to which the two women sometimes retired when their already reserved life in the city seemed too demanding.[20] In 1751, perhaps to pay for the dying Ferrand's medical care, Vassé finally took legal action to receive the money long overdue from her husband's estate.[21] The intense attachment of these two women was well known. One author who wrote of them in the nineteenth century delicately suggested the "afterglow of a romance"; the few authors today who mention them say simply that they were lesbians.[22] As we will see in later chapters, the botanist Mlle Basseporte and the anatomist Mlle Biheron chose female partners as well, and appear for many years to have chosen each other. Such relationships had erotic potential, surely, but the sustaining bond between these women was as much emotional as physical. Historian and gay activist Alan Bray suggests in his work *The Friend* that "the inability to conceive of relationships in other than sexual terms says something of contemporary poverty, or, to put the point more precisely, the effect of a shaping concern with sexuality is precisely to obscure that wider frame."[23] Suffice it to say that without the encumbrances of husband, family, and the conventional gendered expectations that accompanied the married state, these women could and did exercise significant and unusual kinds of freedom.

Ferrand was an atheist, and among the men who frequented her salon, those she chose to host and spend precious time with, we can discern some patterns. To remark that this was not a religious group would be an understatement. Even the medievalist La Curne, a distant relative of chemist Mme Thiroux d'Arconville (see chapter 5) and like her a Jansenist, was not observant and saw Jansenism, as he told his brother, to be an expression of independence from the establishment, more political than anything else.[24] Jaucourt was a Protestant and committed to religious toleration. The brothers Mably and Condillac had both studied theology and taken the title "abbé," but Mably became increasingly pagan through his interest in the republican governments of pre-Christian antiquity, and neither brother had continued in the church.[25] Voltaire would later approve of Condillac as "a good enemy of superstition."[26] Pouilly's most recent book, *Théorie des sentiments agréables*, although it put forth a view of a benevolent creator who inspired virtue in people, had started as a letter to the famous libertine Bolingbroke and eventually became the basis for the entry "Plaisir" in the *Encyclopedie* attributed to Diderot. And whereas Pouilly wrote about life's pleasures, the hedonist Helvétius notoriously lived for those pleasures, and Clairaut, who also loved his wine, women, and song, was said to have died prematurely of excess and dissipation.[27] Turgot probably joined her circle around 1749 or 1750, just when he was deciding not to take holy orders because it would be hypocritical, requiring him to "wear a mask" and hide his true feelings.

Ferrand was not judgmental about the personal life styles of these friends, as all were engaged, lively scholars and excellent company. Some were a bit older than she, for example, Pouilly and La Curne, born in the last decades of the seventeenth century; Cramer and Jaucourt were close to her in age; Mably slightly younger; Clairaut, Helvétius, and Condillac (like his friends Rousseau, Diderot, and d'Alembert) were all born between 1712 and 1717; and, finally, Turgot and his occasional guest Malesherbes were young enough to be her sons. Because Ferrand deliberately chose and gathered these particular men around her, a more detailed examination of them is warranted. Seeing how intellectually and personally intertwined they were with each other and with her helps to conjure the affinities and stimulating atmosphere at her salon.

Jean Baptiste de La Curne de Sainte-Palaye, an expert on the deeply Christian Middle Ages, was nonetheless indifferent to Catholicism and in fact something of a skeptic. Scrupulous historian, friend of the philosophes, he was a progressive thinker who later in the century would sympathize with the *patriote* movement, which I have elsewhere called *frondeur,* and was eager to uphold the fundamental laws of the realm against ministerial despotism.[28] He studied troubadours, savoring the study of old French literature, a choice that some found odd but others daring and original.[29] Reviewers admired his erudite and graceful works.[30] He counted among his close friends Fontenelle, La Condamine, Voltaire, Buffon, and Malesherbes, and within Ferrand's circle he was particularly close to Pouilly.

Louis-Jean Lévesque de Pouilly lived mostly in Reims but was often in Paris. He and his two brothers, Champeaux and Burigny, had begun in 1718 directing *L'Europe Savante,* a journal that sometimes propagated the ideas of Newton.[31] Seven years earlier Pouilly had first made known the ideas of Newton's *Principia* in France.[32] He frequented Bolingbroke who had been minister for the so-called Old Pretender, le Prince Edouard's father—who incidentally referred to his own son as "a continual heartbreak"—and had fled to France. At his chateau "La Source," near Orléans, Bolingbroke gathered around him other French anglophiles, including Ferrand's mathematics tutor Rémond de Montmort, many of whom early accepted Newton's notion of attraction and did experiments together. The Stuart party in France had supporters in the Saint Sulpice area of Paris where Ferrand lived and also ties to the Jansenists.[33] Pouilly's main book, *Théorie des sentiments agréables,* assumed its final form in 1747. It argued that all the senses seek what is agreeable and shun what is repugnant, providing in this seemingly simple pleasure/pain binary a seminal idea for the work that Ferrand and Condillac were beginning to do together.[34]

The Protestant chevalier Louis de Jaucourt, who would become Diderot's loyal, prolific, indispensable collaborator on the *Encyclopédie,* could have joined

Ferrand's salon any time after first coming to Paris in 1736. Former medical student of Boerhaave, classmate and great friend of the doctor Tronchin who would later treat (or not treat, in his famously Hippocratic, noninterventionist style) Ferrand during her last illness, Jaucourt was neither atheist nor materialist, but he was deeply committed to tolerance. For the *Encyclopédie* he wrote this about religious zeal: "If every zealot carefully examined his conscience, it would often teach him that what one calls *ʒeal* for one's religion is in the end pride, self-interest, blindness, or malice. . . . I like to see a man zealous for the advancement of good morals, and the common interest of mankind, but when he employs his *ʒeal* to persecute those whom he likes to call heterodox, I say of the good opinion that he has of his belief and his piety that the one is vain and the other criminal."[35] Vassé was related through her mother's first marriage to the large Jaucourt family, and a marquis de Jaucourt would become one of her executors when she died in 1768.[36] The chevalier de Jaucourt frequently praised Condillac in the *Encyclopédie,* usually without naming him, as for example in the article "Sens (metaphysique)," where he featured an "excellent modern author who provides an ingenious notion of the sensations."[37]

Swiss mathematician Gabriel Cramer was of course not in Paris most of the time, but he was a vicarious participant in Ferrand's salon and devoted to her, as his letters to Condillac and Clairaut show. A prodigy, he received his doctorate at eighteen and became co-chair of his department at the University of Geneva at twenty. He was a skilled, innovative professor who taught in French rather than the customary Latin, and who used a Newtonian cubic curve model for his own writings on algebraic curves in 1750. Johann Bernoulli, father of Ferrand's early visitors, thought so highly of Cramer that he would accept no one else to edit his collected works.

Alexis Clairaut, the only one of his numerous siblings to survive to adulthood, was even more precocious than Cramer, delivering his first paper to the Paris Académie des Sciences when he was thirteen years old and being elected at eighteen as the youngest member ever, an exception to their minimum age requirement of twenty. He had been a convinced Newtonian since the 1730s. While attending Ferrand's salon, he received a prize from the Saint Petersburg Academy of Sciences for his *Théorie de la lune* (1750) where he showed that the moon's orbit is affected by the pull of both the earth and the sun. He also became involved with Voltaire and (amorously for a time) with Mme Du Châtelet when the pair began their Newtonian work. And in collaboration with Mme

Lepaute, the astronomer who did most of the calculations (see chapter 2), he would predict the return of Halley's Comet in 1759, which was considered proof of Newton's law of universal gravitation.

Claude Adrien Helvétius's father was physician to the queen, herself a staunch supporter of the Jacobite cause. Hume reported that the younger Helvétius also helped le Prince Edouard—who referred to him as "the philosopher"—picking up his mail for him, taking risks by meeting with his partisans, but eventually regretting doing so, having "found at last that I had incurred all this danger and trouble for the most unworthy of all mortals."[38] Helvétius was adored by Voltaire, and he had especially strong ties to Turgot, Mably, Condillac, and Vassé, whom he called his "chère amie" and with whom he remained close until her death in 1768.[39] Helvétius was fiercely anti-clerical, notoriously debauched, devastatingly charming. In 1751 while attending Ferrand's salon he married the author Mme de Graffigny's niece, Mlle de Ligniville, nicknamed Minette, although he believed marriage to be unnatural, a "crime against philosophy." Police records during this time describe his orgies with prostitutes and chambermaids in Paris and at his estate in Voré, but Graffigny had been determined to bring about this union of her ward with "the genius" and amazingly the couple stuck together, establishing a famous salon of their own.[40] Ferrand's artist Quentin de La Tour, part of this circle too, knew Graffigny well, probably did her portrait, and socialized with Helvétius.[41] When he moved from Auteuil (now the 16th arrondissement but then a village) where he lived and had his art studio from 1750 until 1772, he sold his house to the newly widowed Mme Helvétius.

The young Anne Robert Jacques Turgot was also occasionally present chez Ferrand.[42] Only in his mid-twenties, he would deliver on 11 December 1750 a speech at the Sorbonne on progress, "A Philosophical Review of the Successive Advances of the Human Mind," in which he stated plainly, in agreement with Locke and Condillac, that knowledge is grounded in experience, and expressed the conviction that rigorous science moves humanity forward. Abstract speculation about faculties and essences, in contrast, leads the sciences astray.[43] Had he read Condillac's 1749 *Traité des systèmes* which makes that very point? Almost surely. Turgot's close friend Guillaume-Chrétien de Lamoignon de Malesherbes, who was from a distinguished legal family and also attended Ferrand's gatherings now and then, assumed the important post of director of the book trade in 1750, a job Clairaut seemed to feel he had helped him secure. In his new capacity he soon appointed Condillac as one of the royal censors.[44] Malesherbes would become an avid botanist, so we will meet him again in later chapters.

The fiery socialist of the French Revolution, Gracchus Babeuf, would invoke as his inspiration a group of thinkers he called the "four levellers": Rousseau, Diderot, Helvétius, and Mably. Gabriel Bonnot de Mably, Condillac's older brother, believed that studying the pagan ancients was a fine alternative to Christianity for modeling morality and politics. His writings were a major turning point toward the eventual victory of republicanism in France. Despite his noble birth he promoted an "emancipation philosophy" similar to Rousseau's, shaping the radical democracy of the Revolution and, beyond that, the utopian socialism of the nineteenth century with his egalitarian views.[45] Like Helvétius, Mably remained devoted to Vassé after Ferrand's death; years later she would die in the home he and Condillac shared.

Diderot, Rousseau, and d'Alembert, Condillac's fast friends at this time, might have popped in chez Ferrand on occasion. D'Alembert definitely knew Ferrand, and in the mid-1740s these men were all aware of and involved with each other's work, meeting regularly to dine together in the Palais Royal as Rousseau described in his *Confessions:*

> I had also become intimate with the Abbé de Condillac, who, like myself, cut no figure in the literary world, but who was born to be what he has become today. I was the first, perhaps, to see his stature, and to estimate him at his true worth. He seemed also to have taken a liking to me. . . . He was then engaged in his *Essai sur l'origine des connaissances humaines* which was his first work. . . . I spoke to Diderot about Condillac and his writing; and introduced them to one another. They were born to agree, and they did so. Diderot induced Durand the bookseller to take the abbé's manuscript, . . . as we all lived in widely different quarters the three of us met once a week at the Hôtel du Panier-Fleuri. These little weekly dinners must greatly have pleased Diderot; for though he almost always failed to keep his appointments, even with women, he never missed one of them.[46]

Rousseau was deeply committed to the fledgling *Encyclopédie*, promising d'Alembert the music articles which they would eventually work on together. And Condillac was actually related to d'Alembert, whose natural mother, Mme de Tencin, was a distant cousin. The two took trips together and were probably at each other's elbows when d'Alembert wrote the *Preliminary Discourse* to the *Encyclopédie.*[47]

These, then, were the people Ferrand invited to her salon for more than a decade. La Tour even grouped some of this extended company together, exhibiting his portraits of d'Alembert, Rousseau, and Ferrand in the same Salon of 1753.

But her strongest connection in the last years of her life was with Etienne Bonnot de Condillac. Born in 1714 into a noble but modest family, he was a weak and sickly child with bad eyesight, a slow learner who did not read until age twelve, and whose parents feared he was not very bright. Both he and his brother Mably wished to get out of Grenoble—later he would describe it as a stultifyingly boring provincial backwater—so after spending some time in the home of their eldest brother in Lyon, where Rousseau was briefly a tutor to their young nephews, Condillac followed Mably to Paris to undertake seminary studies. By 1740 he had acquired a thorough training in theology but decided against pursuing the priesthood. By May 1744 Condillac had finished his first book, *Essai sur l'origine des connaissances humaines,* but the manuscript was held up at the censor. Meanwhile Condillac socialized, with an interest in theater and literature and a sensitivity that made him welcome in many circles. The Encyclopédistes enjoyed his company, Turgot emulated him, and Voltaire loved him.[48] Condillac was conventionally sexist, not even praising women with the suspect gallantry of some of his friends. In his *Essai,* which finally saw publication in 1746, he lamented how easily the heads of girls were turned by reading romantic novels or devotional works that can make them believe they have visions and converse with angels. "Since their mind is often too little engaged in education, it eagerly seizes fictions that flatter the natural passions at their age."[49] These words were written before he knew Ferrand, and in all fairness he laid blame on their lack of schooling rather than on the girls and women themselves.

A quick look at Condillac's first book is in order here, for it is what brought him into Ferrand's life. She was one of its earliest and sternest critics. When it appeared in 1746, the work of a fresh new voice, the *Essai* made a splash. The influence of Newton, the search for "one single principle," is suggested in the very title, *Essai sur l'origine des connaissances humaines: Ouvrage où l'on réduit à un seul principe tout ce qui concerne l'entendement humain.* In it Condillac touted his method of analysis as "the true secret of discoveries because it always makes us go back to the origin of things. . . . It is the enemy of vague principles and of everything that is contrary to exactness and precision."[50] The new author was seeking a unifying idea of his own, but in this book he accomplished nothing of the kind, as Ferrand was quick to point out. Condillac attributed the acquisition of knowledge to the senses, but *also* to reflection as had Locke, and *also* to the expression of thoughts in gesture, and *also* to speaking and writing which he referred to as "signs." So while Descartes's innate ideas were rejected, inborn mental processes like Lockean reflection were not. Condillac argued that "it is reflection that makes us begin to discern the capability of the mind."[51] He did not examine precisely what the senses can teach. Citing thinkers like Molyneux, Locke, and Berkeley, who had explored the sudden restoration of sight in a formerly blind person, he stated that we do not need experience to perceive depth, taking for granted that just by opening our eyes we would automatically understand spatial relations. He also devoted half of the *Essai* to signs, which he deemed absolutely essential for the development of reasoning and knowledge. So despite the work's title and intent, there was no simple, lone principle unifying his approach at all.

Although the reviews of the *Essai* were mostly favorable—in the *Mercure*, the *Journal de Verdun*, the *Journal des sçavans*, and the *Mémoires de Trévoux*—both Cramer and Ferrand objected to it forcefully, recognizing immediately the fuzziness of the fledgling author's thinking, his too hesitant departure from previous authors, his failure to go back to the true origins of thought.[52] They may have attributed it to his poor comprehension of mathematics, for in the *Essai* Condillac acknowledged "arithméticiens" but condemned "geometriciens" because, he said obscurely, they never help us arrive at any new truths.[53] Ferrand took issue with several things, but she was above all focused on inconsistencies in his method, where he had gone astray in his attempt to find a single, overarching Newtonian principle. She argued that understanding *precedes* language and that there is initially no need for signs. She also maintained that sight alone cannot apprehend or teach us about space or even about two-dimensional surfaces because vision does not take us beyond our own limits. Sight unaided cannot make sense of the outside world or know it is there. Only when it is combined with touch (and the ability of the hand to move and encounter external objects) do we get knowledge of things apart and separate from our own bodies. So Ferrand struck at two fundamental points in Condillac's *Essai*, and he would eventually reverse his position on both. Together now he and Ferrand would begin to untangle (*démêler*), to analyze rigorously the contributions of the five senses one by one.

How exactly did this collaboration function? The details are lost because any letters Condillac and Ferrand might have exchanged have not survived and because they often worked together in person in Paris, "in conversation" as Condillac repeatedly said.[54] But letters between their colleagues and friends allow us to reconstruct what kind of interactions were taking place from 1747, the time they first undertook an original approach to epistemology, until Ferrand's death in 1752.[55] We can see how she, proceeding systematically, got him to admit mistakes, curbed his enthusiasm, questioned the logic of his positions, but then always challenged her own arguments to see if they held up under repeated scrutiny.

By early 1747 Condillac was already in discussions with Ferrand, according to a long "Mémoire" to his other exacting critic Cramer, in which he reviewed problems in his *Essai* that he needed to revisit. "Well-made objections are, I think, of a greater worth than the most flattering praise. The care with which I will respond to these critiques proposed to me will prove how grateful I am for them."[56] He stated earnestly how very much he wished his works would help others reason well. "It is not simple curiosity that made me undertake research on the origin of human knowledge, but rather in order to discover among the principles of which it is composed, those we can make use of to facilitate the progress of our mind." Rejecting Leibniz's system of monads, he wrote that he saw no practical use in supposing that we have perceptions of which we are not conscious.[57] "Analysis alone is both the source and the touchstone of ideas, definitions and principles."[58] Ferrand, he reported to Cramer, was already pushing him to analyze each of the five senses separately, zeroing in first on sight, on the problem Molyneux had posed to Locke about just how much the unaided eyes could apprehend. "Locke, Berkelai [*sic*], and I are all three wrong. Ask the reason of a *demoiselle* who made me see it. She will not grant you or Berkelai that sight alone can give the idea [even] of two-dimensional surfaces, and I think that she is correct."[59] Then, in another note that suggests Ferrand's influence, "Only analysis yields good definitions. It is to this [method] that mathematicians owe theirs."[60]

Clearly, Cramer had encouraged Condillac to meet and think through his ideas with Ferrand. She specifically objected to the overweening and misplaced faith the author of the *Essai* and other epistemologists had put in the sense of sight.[61] Cramer came to Paris in the spring and worked with Condillac on various points while Ferrand was briefly away with Vassé in Marly, but she was still testing Condillac even when out of town. As he told Cramer, she was keeping him on his toes, making him rethink his positions. "Mlle Ferrand did me the honor of communicating her observations on my work. I don't need to tell you how fine and solid they are. I know nobody who understands these matters better. In a conversation we had about signs, we did, I think, reduce the question to the point where the difficulty lies. But I have not yet come around to her position. It is not that I persist in sticking to the view in my book. I see the thing as not yet decided. It is not so bad to reduce an author to admitting that he does not know if he is right or wrong." Ferrand had also raised her other profound criticism of the *Essai,* that signs were not needed as the mind initially learns— babies obviously learn a great deal before they can talk—and that therefore the author of the *Essai* had not gone back far enough to the origin of understanding, had not *analyzed* with sufficient rigor.[62] Now Condillac referred back to Clairaut's 1741 *Éléments de géométrie,* startling evidence of his new appreciation for a subject he had previously dismissed so harshly.[63] In keeping with his spirit of eagerness for ongoing critiques, he chided Cramer's "indulgence toward me, in not having additional quarrels with my work."[64] Both Ferrand and Cramer, the two mathematicians, seemed to be enhancing his appreciation of precision, simplicity, and logical elegance.

Before leaving Condillac's *Essai,* it is important to note his surprising enthusiasm for the imagination. Given the value he placed on analysis this perhaps appears contradictory, but neither of his critics had any objection to this seeming incongruity, for Condillac confronted it squarely: "Analysis and imagination are two operations that are so different that they usually raise obstacles to each other's progress. It takes a special temperament for the advantages to lend each other mutual assistance without at the same time doing harm, and this temperament is the middle between the two extremes."[65] Regarding literary imagination particularly, "we see fiction, which would always be silly without truth, decorate the true that would often be cold without fiction."[66] This appreciation for the complementarity of creativity and rigor is interesting, and it helps to explain the awe with which Condillac regarded Ferrand, who seemed uniquely to possess both qualities in equal measure, as he would state explicitly on later occasions.

In 1748 a curious event occurred—the publication of an anonymous piece titled *Les Monades* in a collection of the Royal Prussian Academy of Berlin, a work written secretly by none other than Condillac in response to a prize proposed in June 1746 for the best essay evaluating Leibnitz's doctrine of monads.[67] Although Condillac did not win the competition, he worked on his entry in 1747 when he was already involved with Ferrand. They chose an epigraph from Cicero to suggest the modesty and uncertainty of his effort, or to underscore that it was really just an exercise, in the spirit of Montaigne's explorations of various subjects in his *Essais,* where he was so famously fair to opposing views that both sides of a controversy could feel vindicated, or both defeated. The epigraph translated roughly as "It would have been better, Velleius, to admit that you don't know what you don't know, than to regale us with nonsense, declaiming like a man who doesn't believe what he is saying and so even disgusts himself."[68] The Berlin Academy's prize announcement read like this: "We ask that, in starting by exposing in an exact and clear manner [Leibnitz's] doctrine of monads, one examine if on the one hand it can be solidly refuted and destroyed by indisputable arguments, or if on the other hand one can, after proving monads, deduce from them an intelligible explanation of the principle phenomena of the Universe, and in particular of the origin and movement of bodies."[69] Part 1 of Condillac's *Les Monades* attacked Leibnitz, but part 2 ended by using Leibnitz's errors to arrive at a new and different system of monads, which the author claimed he was able to demonstrate.

What are we to make of this? Arguing two sides of a question was a long-standing rhetorical strategy in the history of science, Galileo famously using it to play the Ptolemaic and Copernican cosmologies against each other, so Condillac was following a venerable tradition. He disagreed with but also had a fascination with Leibnitz, whose views appeared throughout his epistolary exchanges with Cramer. And Ferrand would also have been well acquainted with Leibnitz from her days with her tutor Rémond de Montmort, the mediator in the German's quarrel with Newton. Only years later, in his 1755 *Traité des animaux* where he was refuting Buffon and wanted to show that he was not a materialist, did Condillac openly claim ownership of *Les Monades,* transposing a large chunk of it into the new work.[70] But when it first appeared in 1748 Ferrand was probably the only person who knew the true author of this anonymous essay, and she surely valued the mental exercise they had gotten arguing both sides of the matter.

But she was not well, experiencing one of her frequent bouts of serious illness. And the salon had meanwhile lost Cramer, who returned to Geneva after a year in Paris and was immediately missing the group. He wrote to Clairaut as soon as he heard that Ferrand was gravely ill: "Don't forget, please, Mlle Ferrand and her amiable society. The sweet moments I spent at her home will stay in my memory forever. I send for her health my most tender and ardent wishes."[71] He also wrote directly to Ferrand and to her "bonne compagnie," recounting that after leaving Paris he had traveled with Mably as far as Lyon where they parted, and sharing his relief that the grizzly eight-year war of the Austrian Succession with "all of its horrors" was finally over.[72] Despite her illness she replied to him a few months later, writing of her delight in including him in discussions with the others around her hearth and hoping he would return soon.

This 1748 letter bears further comment as it shows the extent to which Ferrand was aware of academic politics and goings on. She was using her connections to get Cramer elected to the Académie des Sciences to fill a vacancy as a foreign member, but she needed to acknowledge that her faction had not succeeded, writing to him in a fluent hand: "Clairaut, who is extremely attached to you, did his very best to show you proof of this in the last election at the Académie. Fontenelle and Mairan and several others were also very zealous [in your behalf]. But Bradley's party was stronger. I think the absence of some of your friends may have been the reason." She was sorry to report this disappointment, of course, but frankly confident that an elderly foreign member was sure to die soon, giving Cramer another chance. In the same letter, writing about the infamous mathematical and astronomical disputes between Newton and Robert Hooke, Ferrand argued that Newton's discovery of the calculus and his "beautiful and grand" laws of planetary motion made him the greater thinker, but she found it a shame that such brilliant men had been fierce rivals and unable to put aside their differences because "human virtue is too weak for such an effort."[73] This November letter makes us wish that more of her correspondence had been preserved, but we work with what we have (figure 2). On Christmas Eve she and Condillac together sent Cramer their good wishes for the new year.[74]

Condillac's next book, *Traité des systèmes,* which appeared early in 1749, clearly showed Ferrand's influence in honing his thinking and impressing him still more with Newton.[75] He pushed the need to decompose our thoughts to find the germ, the seed, the origin, of our ideas.[76] In general he argued that form-

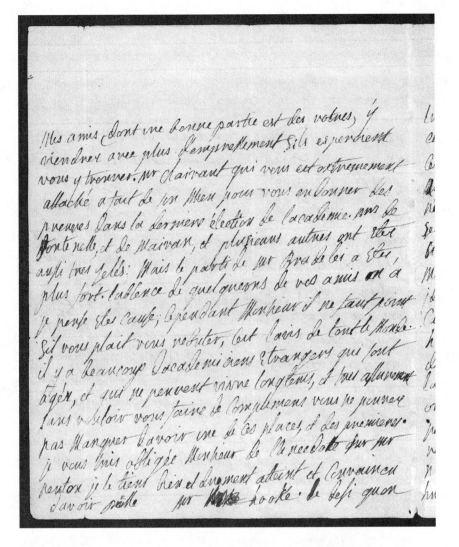

Fig. 2. Page from Ferrand's letter to Gabriel Cramer regarding elections at the Paris Académie des Sciences and her views on the feud between Newton and Hooke (1 November 1748). [Bibliothèque de Genève, MS suppl. 384, fols. 212–213]

ing a strict system of the world, as Descartes did, leads us astray, and that it is far better, as Newton understood, to observe phenomena and explain them simply. The Englishman's idea of attraction, even though he could not explain *why* it worked, seemed to accord well with the movements of celestial bodies. Astronomy, progressing from Ptolemy through Copernicus to Newton with hypotheses suggested by observations and the subsequent correcting of them based on newer and better evidence, seemed to Condillac the model science. "Located as we are on an atom that rolls around in a corner of the universe . . . there will always remain phenomena to be discovered. . . . Newton found too many problems with Descartes's system . . . [and] without undertaking to re-create the world he was satisfied to observe it . . . less bold than Descartes, but wiser."[77]

In the *Traité des systèmes* Condillac once again hailed the imagination as refreshing, revitalizing, indeed essential to the process of thinking. Perhaps he was now preparing for the great leap of imagination he and Ferrand would propose in the *Traité des sensations*, a work based entirely on a thought experiment. The imagination, by literally making *images* of our thoughts as the word implies, invigorates our ideas, supplying "finesse, fertility and breadth." It must, however, be subdued, reined in, kept in check by analysis.[78] A mind with too much imagination is superficial and unruly, but one with too little is cold, slow, and ponderous, examining everything but without inspiration, unable to find truth, and incapable of presenting it gracefully.[79] We will see Condillac press this formulation into service in the *Traité des sensations*, where he very explicitly describes Ferrand as possessing and mastering both of these qualities, precision and inventiveness, in a seemingly paradoxical mix that made her the quintessential ideal thinker.

It was while Condillac's *Traité des systèmes* circulated, again reviewed quite favorably in the spring and summer of 1749, that Ferrand was first contacted directly by Charles Edward Stuart.[80] A year earlier, when the flamboyant prince was still welcome in France, La Tour had made and exhibited his portrait,[81] but he was now on the run and able to return only in secret. Arrogant and volatile, the grandson of the deposed Roman Catholic King James II of England, Ireland, and Scotland had been in exile in France since his defeat on 16 April 1746 at the Battle of Culloden, where supported by the Highland clans he had fought to reclaim the British throne. Initially, because France was at war with England, he was allowed to circulate freely, but the Young Pretender remained a threat to the Hanoverian King George, and in exchange for peace in the Treaty of Aix-la-Chapelle, the French had agreed to surrender him to the British in 1748. Because of his enormous popularity in France he was allowed to leave on his own recognizance. But in late February 1749 the restless prince stole furtively back into France, beginning a long period of masquerade and wandering, during the first three years of which he was provided a clandestine base chez Ferrand and Vassé.

In his communications Charles Edward necessarily used code names, and he now introduced himself as "Cartouche," addressing Ferrand as "Mlle Luci" and acknowledging that he was bold to write out of the blue, explaining cryptically that "people in our profession are usually impudent."[82] Soon afterwards he snuck into Ferrand's residence, and from then until her death she gave him shelter and other kinds of assistance as he flitted in and out of Paris with incessant demands. One of his lovers, a relative of the queen named the Princesse de Talmont, lived in the same complex of apartments in the Couvent Saint Joseph. How convenient.

Ferrand's salon colleagues knew nothing of this new arrival, of course, and they continued to correspond back and forth in their customary vein, discussing math, philosophy, and theater. In one letter Condillac scorned Mme Du Bocage's play *Les Amazones*. After describing the five acts as respectively good enough, less good, very bad, very good at the end, and awful, he concluded that her play "would be beneath mediocre if written by a man of letters, but being by a woman it deserves indulgence and does her honor."[83] Voltaire called Du Bocage the "Sappho of Normandy," but many were offended by her play, an exploration of female power in a successful matriarchal society. Condillac's patronizing comment about the author and her subversive subject matter makes clear that he was no feminist. He did not humor Ferrand out of gallantry; rather, he regarded her from the start as someone on a superior scientific footing from whom he could learn in profound conversations and whose advice and criticism he welcomed. Indeed, as he had written in an earlier letter, her thinking outshone that of everyone else he consulted.

In the same letter Condillac asked Cramer if he knew of anyone born blind who might be helped (what he meant was experimented upon) by an "oculiste" heading to Geneva who had made quite a stir in Lyon and most recently Paris with his successful cataract operations. This question further revealed Condillac's continuing interest in the subject of sight, which Ferrand had raised, and in the Molyneux problem that was to occupy both of them as the *Traité des sensations* took shape. Condillac had personal as well as intellectual reasons for studying sight; his own vision had been extremely weak since birth—one scholar referred to him as "half-blind" as an adult—and it might have pleased him that his work was naturally leading to a devaluing of that particular sense as unable to convey knowledge without the help of touch.[84] Vision, in Condillac's studies with Ferrand, was being dethroned in cognitive significance. In connection with this he asked if Cramer knew of Diderot's just-published *Lettre sur les aveugles à l'usage de ceux qui voient.* Condillac was so lavishly praised in this book that he refused out of modesty to say any more about it, although here prudence may have been the real reason for his silence as Diderot's *Lettre,* ostensibly on blindness, had led to its author's incarceration in Vincennes prison, where he would remain for more than three months. The work discussed the remarkable results of this same oculist in restoring sight to the blind, but it had raised the red flag of materialism by arguing that not only our vision but also our morality, and even our religious faith, depend entirely on the sense impressions our bodies receive.[85] Spelled out so baldly, Diderot's message upset the authorities enough to call for his arrest. Clairaut visited him in jail just a week later and no doubt reported back to the Ferrand circle. He encouraged the despondent prisoner to write to the chancellor and police lieutenant requesting clemency, which he did but to no avail.[86]

A month later, from Grenoble where he was visiting his mother, Condillac told Cramer, "Mlle Ferrand writes to me of the marvels of a Genevan doctor who forbids her to take all kinds of remedies," this being the famous Théodore Tronchin, Jaucourt's dear friend from medical student days, who adhered to the wise Hippocratic belief that we must respect nature's healing powers and that heroic measures like bleeding and purging served only to make the patient worse. He recommended instead nutrition, fresh air, and exercise.[87] Recognizing that Ferrand's illness was terminal, Tronchin thus spared her all manner of indignities. Condillac's last letter in this series to Cramer, dated 22 November 1749, ended by announcing the new idea Ferrand encouraged him to embark upon: "I am working on style and a treatise on the origin and generation of *sentiment.* I formed the project of this last book in conversations with a person of whom you think most highly and who deserves [your esteem] for her qualities of heart and of mind. Thus [the book] will not be entirely mine."[88] As Condillac uses the word, "sentiment" does not mean emotional feelings but most often "opinions," "beliefs," "views," or "thoughts."[89] Here we see that although Condillac and Ferrand were and had been miles apart for the last half year, and although she had been gravely ill, their joint work on the *Traité des sensations* had begun in earnest.

And much else was happening. At the very end of December 1749, Condillac learned to his surprise and joy that he had been elected to the Berlin Academy of Science, news delivered to him in Grenoble by d'Alembert.[90] He wrote letters of profuse thanks to Maupertuis, the academy's president, and to Samuel Formey its secretary, for this honor which he protested he little deserved.[91] The next year his dear friend Rousseau became an overnight star with his provocative "First Discourse," prepared for a 1750 prize competition offered by the Academy of Dijon, in which he argued perversely that the arts and sciences do not advance but rather impair human progress by giving us airs, affectations, and pretensions and thus alienating us from our true, honest, healthy natures. Rousseau had gotten the idea of writing this on his way to visit Diderot in jail, and he famously won the Dijon prize. At the same moment Turgot, Condillac's younger friend, gave a speech at the Sorbonne saying exactly the opposite, that history demonstrated the continual improvement of humanity, that science and moral excellence developed hand in hand, that to be smart was to be good, and that there were in principle no limits to human perfectibility. Who was right? Meetings chez Ferrand on this controversy must have been lively, as Turgot himself was sometimes present and Rousseau might have made occasional appearances. Had there really been progress? Was civilization harmful or beneficial?

Meanwhile le Prince Edouard had temporarily slipped out of Paris, and Ferrand, fearing that his impulsive behavior would betray him, was eager to get him back to safety, provide whatever intelligence she could gather, and accommodate his other demands, as we see from a coded letter in which she promised that as soon as he returned to headquarters in her home the "nuns" would make him even more comfortable and secure than he was before.[92] It was nonetheless a fraught situation, and a later author gave a colorful description of what it must have been like for the fugitive in the Ferrand household:

> The unfortunate prince Edouard, after getting out of the Bastille, stayed hidden for three years in Paris, at the home of the marquise [*sic*] de Vassé who lived then with her friend, the celebrated Mlle Ferrand, at Saint Joseph, in the Faubourg Saint Germain. The Princesse de Talmont, a Polish-born cousin of the queen's with whom he was still very much in love, lived in the same house. He holed up during the day in a little garde-robe of Mme Vassé, where there was a hidden staircase by which he descended at night to the princess, and spent the evenings behind an alcove of the *cabinet* of Mlle Ferrand. He enjoyed there every

day, without being noticed, the conversations of a very distinguished society. They spoke often of him, saying much that was good and much bad, never suspecting the hidden witness in front of whom they were talking. The existence of the prince in this asylum, and the profound secret that hid him from the eyes of the entire universe for so long among three women, and in a house that received the elite of the city and the court, smacked of the miraculous. M Choiseul, several years after the prince's departure, heard of this singular anecdote and could not believe it. Minister of foreign affairs at the time, he wrote himself to Mme de Vassé asking for details. She admitted everything to him, and added that she had needed to kick the prince out of her house because of the too lively scenes between him and Mme de Talmont, scenes that always started tenderly but ended often in quarrels and even blows. We have this fact from a very particular friend of Mme de Vassé.[93]

Entirely oblivious to what was going on literally behind the scene, Ferrand's circle continued to gather and to exchange letters about their work, and it is interesting to contrast Cramer's attitude toward criticism with Condillac's, for where the philosophe genuinely craved it and saw it as a way to refine his thought processes, the mathematician was more thin-skinned and concerned about his reputation. He requested the group's frank opinion of his *Analyse des courbes,* still in manuscript, but followed up that he wanted Clairaut's critique to be delivered only privately: "I beseech you to not be so outspoken with the public, and to gain for my book the favor it will get from your advantageous review." At stake here was Cramer's "amour-propre," his fragile self-esteem. In another letter he reiterated that he counted on Clairaut's "politeness of character. I'm sure you understand that your remarks should be for me alone. Forgive me, my very dear friend, all this petty timidity of a new author. . . . This exaggerated delicacy comes I think from my situation since my birth which has been of a laughable mediocrity." Clairaut in response was thoughtful enough to underscore the "great elegance, clarity and originality" of the work, for which Cramer would thank him lavishly.[94] Yet Cramer was perfectly capable of encouraging others who received the criticism he so feared for himself.[95] Ferrand must have understood that Cramer was less sturdy than Condillac and tempered her critiques of the work he showed her, because he referred to her always as "kind." She wrote mathematical letters to him that have not survived, but we know she was involved in discussions of his book on curves.[96] She also discussed many other subjects with him and with Clairaut, for example, her friend Pouilly's *Théorie des sentiments agréables*, which would provide food for her later thought.[97]

When Condillac finally did come back to Paris in 1750 after eight or nine months away, he and Ferrand began an intense collaboration at her home to make up for their long separation. We can picture Condillac leaving his Right Bank house on rue Notre-Dame-des-Victoires,[98] from which it would take him about half an hour, walking briskly southwest, to arrive at Ferrand's on the other side of the river. Perhaps he zig-zagged through the familiar Palais Royal, where every week he met his friends Diderot and Rousseau for dinner at the Panier-Feuri. From there he may have chosen the rue Saint Honoré, crossed the Place du Carrousel to the Quai des Galeries du Louvre, or he might have skirted the Palais des Tuileries with its lovely gardens on his right. There were not many bridges across the Seine then, and the only one on the way to Ferrand was the stone Pont Royal, distinguished from the others by its gracefully curved lines with higher arches in the middle and one of the few that was not lined with houses. Once on the Left Bank, straight ahead was the rue du Bac, so named for the ferry that used to shuttle constantly to and fro before the bridge was built. He would take it until it met the rue Saint Dominique—the Boulevard Saint Germain did not yet exist and so had not pierced and demolished big parts of these streets—where he would turn right and reach his destination just before the esplanade of the Invalides.

Condillac must have been shocked by Ferrand's condition, as her always precarious health had taken yet another alarming turn earlier that year. She seems even to have suffered some kind of attack or stroke, but she forged on with their collaboration. In May Cramer revealed to Pouilly's brother Champeaux the large and challenging part Ferrand played as the *Traité des sensations* took shape: "Condillac's work on the senses cannot help but please (since it is inspired in large part by a person whose thoughts are as noble as they are subtle)."[99] In June 1750 Condillac gave Cramer specific details about his project "on the origin and generation of thought. . . . I animate a statue by degrees, first touch, then hearing alone, then the two together, what comes from each. Understanding has its principle in the senses. . . . Mlle Ferrand knew I'd be writing to you so sends her compliments. . . . She is better but not completely recovered from her accident."[100] We have here "proof" that Condillac could not have stolen the statue idea from Diderot, whose *Lettre sur les sourds et muets* would not appear until eight months later but who nonetheless accused him of plagiarism.

On 12 August 1750 Condillac wrote to Maupertuis from Segrez, twenty miles south of Paris, where he and d'Alembert were visiting the beautiful estate and famous gardens of the Marquis d'Argenson "where life is the best in the world." He reiterated his plans. "For myself I have several projects that I work on in turns: the one occupying me presently deals with the origin and generation of sentiment. It is a statue that I animate by degrees. I've encountered some difficulties but I think I have surmounted them. I will put it aside to rest a few months as is my custom."[101] So once again the collaborators were separated— maybe the interruption of regular conversations chez Ferrand was the reason Condillac was having "difficulties"?—but he was wrestling with their statue idea. Cramer meanwhile, waxing nostalgic for the company of his hostesses, wrote to Clairaut about the two women as if they were one single being with a single soul: "I am so glad I am remembered by my Paris friends. Don't overlook especially Mme de Vassé and Mlle Ferrand. I will never forget their courtesy toward me or the idea they gave me of a mind as large as it is beautiful, as respectable as it is kind."[102] Mably assured Cramer that "Mlle Ferrand speaks often of you and would be upset if you didn't think of her. She enlisted me to tell you a thousand things that only she knows how to say and I have not the temerity to spoil them."[103] Too sick to write herself at that point, she may have been eager to communicate some mathematical ideas that Mably was simply unable to accurately transmit.

How did Ferrand manage to hold it all together during this time, intensely intellectual, severely ill, politically committed, and all the while responding to the furtive prince's eccentric demands? Away from Paris yet again, he wrote asking "Mlle Luci" to send him books, some razors, and a shaving mirror and demanding that before he came back his post chaise be repaired.[104] She honored his requests for certain novels and for Montesquieu's *L'Esprit des lois* hot off the press, but she refused to send other works by a quack mathematician that she called amateurish rubbish, "worse has never been printed."[105] More coded missives followed, reprimanding Ferrand for showing his letters to the Princesse de Talmont (whom he called his "tante" while now calling himself Mlle Chevalier), explaining that he had broken off relations with his lover. Ferrand warned the prince at one point the next year that their friend Helvétius overheard a M. Le Fort boast of knowing his hiding place but that deftly "the philosopher turned the conversation." Meanwhile Talmont accused Ferrand, briefly out of town herself in Marly, of interfering in her relationship with the prince, attacking her in shameful terms which she would later regret. Disdainful of pettiness and frivolity, Ferrand responded sternly to this hysteria that her own ties to the man were nothing more than political support and friendship, and that she had no interest in meddling in his love life: "I am strongly attached to Madame your friend"—the prince was then using female code names—"[and] for her I would suffer and do everything short of stooping to an act of baseness. . . . My character is not supple. The one thing that makes my frankness endurable is that it renders me incapable of conduct for which I should have to blush." With characteristic sangfroid, Ferrand maintained her steadiness despite the chaos all around.[106]

Diderot's follow-up to his *Lettre sur les aveugles,* the *Lettre sur les sourds et muets,* appeared in February 1751. In it he presented not a statue but a progressively animated man, and at this point Condillac could well have accused *him* of plagiarism, for Diderot certainly knew about the joint project with Ferrand that had already been in the works for years. But instead it was Diderot who cried foul, at some point before the *Traité des sensations* appeared in 1754, despite the fact that there were enormous differences between the styles and approaches of the two books. Diderot's did explore the five senses but playfully, doing "anatomie metaphysique" by labeling the eye superficial, the ear too proud, the nose voluptuous, the mouth superstitious and inconstant, the hand profound and philosophical. His literary method bore little resemblance to the serious analysis that Condillac and Ferrand were doing, but strains between these once close friends began to be felt at this time, although the bruising priority dispute would not erupt publicly until Condillac's *Traité* was almost finished. This ugly accusation caused a permanent rupture between them. The pain of this was perhaps partly mitigated by some good news that Condillac received, yet another honor to celebrate with Ferrand, when he learned at the very end of 1751 that their friend Malesherbes, now director of the book trade, had named him a royal censor.[107]

During 1752, the last year of Ferrand's life, Cramer died in January, and Condillac contracted a debilitating illness, explaining to Maupertuis: "I had a 'fluxion' which made it impossible for me to read or write for over eight months. Since then I must be very careful, because even today I can only apply myself an hour at a time or I feel pain in my eyeball. . . . I was working on perfecting a treatise that was almost done and I intended to publish a year ago, but now I do not know when it will be ready."[108] Because both Ferrand and Condillac were sick, there was a long delay before their work actually got into print; she did not live to see it happen.

Ferrand dictated her testament in February. Nicolas Baille, a man seventeen years her senior and a trusted advisor, was to be the executor of her will, and Vassé of course was to be her heir. She had 18,000 livres remaining from her long-ago sale of properties in Bouleaux, and some possessions which she believed to be of little value. She bequeathed (in this order) her yellow damask furniture to Clairaut; 6,000 livres to Condillac "to enable him to buy books"; 300 livres to her male servant; some money, her bed, and many things in her closet and *garde robe* to her chambermaid; 200 livres to Vassé's chambermaid; her little commode and lacquered writing desk to Mably. Everything else went to Vassé, her *legatrice universelle*, "whom she has always loved with the greatest tenderness and whose virtue and friendship made the joy of her existence, and [whom] she begs . . . to not refuse these feeble marks of her tender devotion which she will feel until the last moment of her life." She wished Vassé to have the portrait she had commissioned, meditating on Newton, which was still with La Tour who was adding finishing touches, and asked her to have the artist make a copy for Baille, who had always been so loyal. The furniture bequests must have had some special significance that we cannot know, perhaps written on, or sat on, or admired by Clairaut and Mably when attending her gatherings.[109]

An avowed atheist, Ferrand insisted that there be no sermon, no ceremony at all when she died, just a burial of the greatest simplicity. D'Alembert commented on her atheism to his friend the gossipy and shrewd Marquise de Crequi. In an irreverent letter meant for her eyes only, after thanking her for her holiday gift of partridges and woodcocks and reporting that he was busy with many tasks, writing errata for volume 2 of the *Encyclopédie*, thrashing the Jesuits on one side, lambasting the Jansenists on the other, d'Alembert then wrote that Ferrand's death was imminent. He quipped that the ever-practical Condillac would surely take care of her body, but as he had had to fight Helvétius for her soul in life, maybe he would have to vie with the devil for her soul after she died. Referring to the recent deaths of two other well-known atheists, d'Alembert continued: "Woe to the priest, the sacraments, extreme unction. . . . For one can die like Boindin [30 November 1751] or like La Mettrie [11 November 1751]."[110] Nicholas Boindin was so blatantly antagonistic to religion that the curé of Saint

Nicolas des Champs refused to bury him in the church grounds, a rejection that would have pleased the dead man, but his much more traditional family managed to sneak his corpse into a grave there in the wee hours of dawn. La Mettrie, the famous sensualist, materialist, and unapologetic hedonist, died after overindulging on pheasant with truffles. He had insisted there was no afterlife and hoped to be buried at the French embassy in Berlin, but perhaps for spite he was interred in a Catholic church cemetery, which would have outraged him. D'Alembert's implication here was that atheists have a hard time being disposed of as they wish. But Mlle Ferrand, a less noisy nonbeliever, was buried without fuss in Saint Sulpice on 5 September 1752, two days after she died.[111]

Six days later an inventory was done of her possessions. Furniture may have been moved around in her final agony, for indications of her illness cropped up everywhere: a folding "fauteuil de malade," a wheeling stretcher, toilet chairs, numerous chamber pots of faience or of tin with leather covers, two commodes. The rooms were full but not cluttered, her belongings high quality but not luxurious, her taste eclectic, some furniture in the Regency style, many pieces made of exotic cherry and rosewood, and her preferred colors seemed to be red and yellow. The walls were covered with damask panels. Faded porcelain sets, crystal flasks, and snuffboxes were listed, along with a large engraving of Prince Edouard in a sculpted gold frame and a chart used for measuring the meridian. Many tables were placed around her home with mirrors and candlesticks for illumination, some with drawers that locked, one containing a safe. A room adjoining Ferrand's bedroom and overlooking the inner Cour de l'Horloge contained an armoire with clothes, including an outfit similar to that worn in La Tour's picture, a few dresses, among them a black one but also some of a lighter, more playful variety in apple green cotton or striped taffeta, and a country hat with a rose pink head-scarf, these last probably more fitting for excursions to Marly.

There was also another safe in which documents were found, well-organized papers showing Ferrand's tax obligations and her financial situation. Perhaps because she was an invalid she had made many prudent arrangements earlier in her life, securing during the 1730s several annuities, some from private parties and some from the government, to guarantee a steady income stream. These funds, with other belongings, she had bequeathed to Vassé, and it was left to this grieving companion to keep, repair, sell, or otherwise dispose of these personal effects. The rooms technically belonging to Vassé—the kitchen, dining area, library, and common room spacious enough to host Ferrand's salon meetings—

were not listed, but clearly there was no meaningful divide between the two women's quarters. They truly shared their space and their lives.[112]

With Ferrand gone Vassé was entirely devastated, as was Condillac, about and from whom we hear nothing in 1753. Much as he longed to make known to the world the work he and Ferrand had done together, at that moment discretion seemed the better part of valor given the scandal over the Abbé de Prades, a theology doctoral student who had defended a dissertation condemned by the university faculty, the *parlement* of Paris, and even by the pope as irreligious. De Prades had argued, among other things, that all of our ideas come from sensations. Although negotiations were underway for his reconciliation with the church, it must have seemed politic for Condillac to lie low. He was also in mourning, indeed immobilized by Ferrand's death, desperately trying to keep her memory and the memory of their project on the imagined statue alive, as became obvious in his tribute to her when the *Traité* finally did appear in 1754. He dedicated it to Vassé on the front page, in letters almost as large as the title itself: *A Madame la Comtesse de Vassé, par M. l'Abbé de Condillac, de l'Académie Royale de Berlin* (figure 3).

The book opened abruptly with an "Avis important au lecteur," a device meant to highlight something and put his readers on active alert. Condillac wrote:

> I forgot to warn about something I should have said, and even repeated, at several places within this work, but I believe that the admission of this oversight is worth more than repetitions, and has no disadvantage. I caution, then, that it is very important to put oneself exactly in the place of the statue which we will observe. One must start to exist with it, have only one sense when it has but one, acquire only the ideas it does, contract only the habits which it contracts: in a word, one must be only what it is. [The statue] will only judge things as we do once it has all our senses and all of our experience; and we will judge as it does only when we suppose ourselves deprived of all that it is missing. I believe that readers who put themselves exactly in its place, will have no difficulty understanding this work; the others will have countless objections for me.[113]

It would thus be the readers' fault if they did not follow and understand the statue as it awakened, one sense at a time. They would need to be virtual wit-

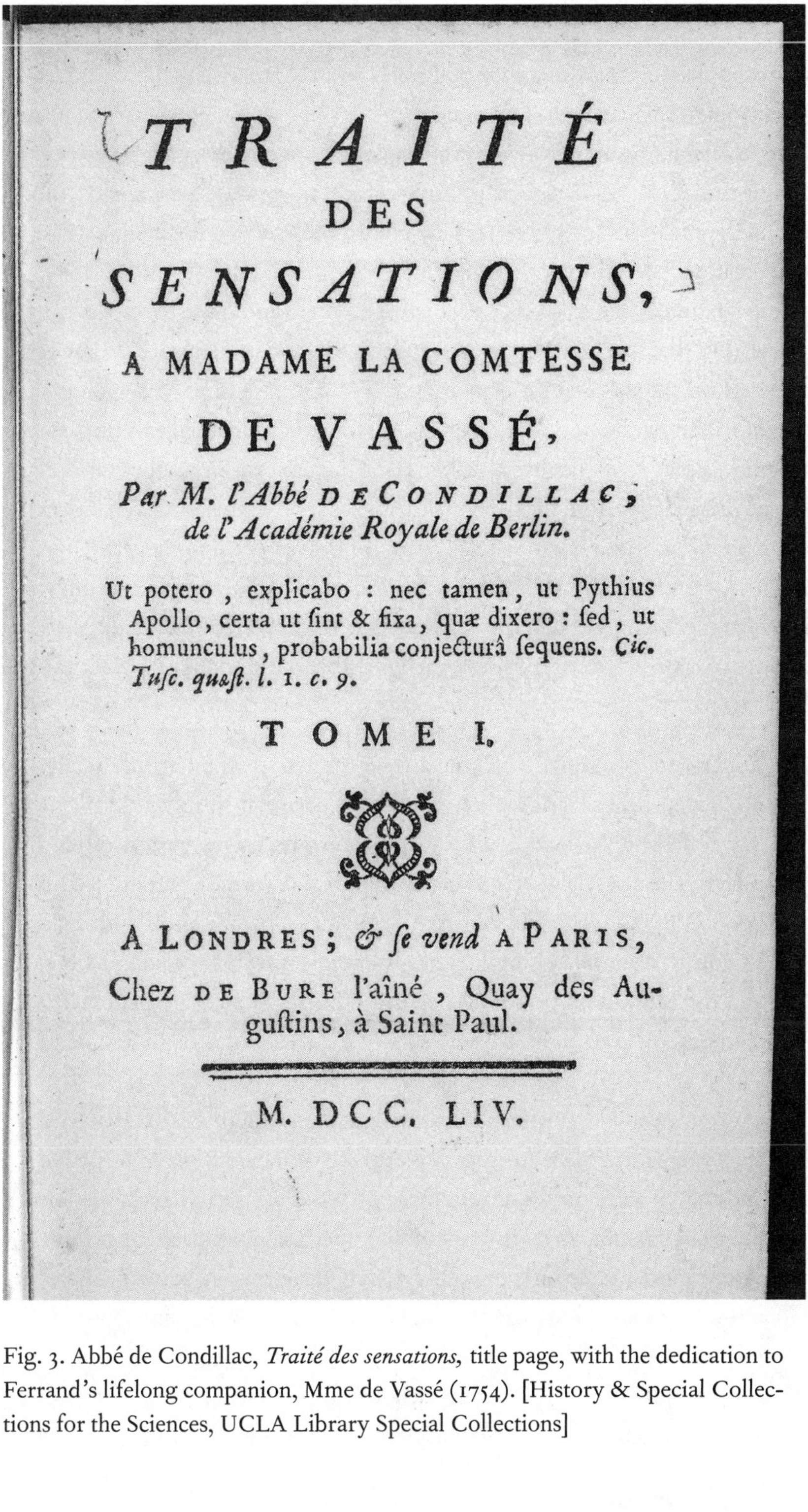

TRAITÉ

DES

SENSATIONS,

A MADAME LA COMTESSE

DE VASSÉ,

Par M. l'Abbé DE CONDILLAC,
de l'Académie Royale de Berlin.

Ut potero , explicabo : nec tamen , ut Pythius
Apollo, certa ut fint & fixa, quæ dixero : fed , ut
homunculus, probabilia conjeſturâ fequens. *Cic.*
Tuſc. quaſt. l. 1. *c.* 9.

TOME I.

A LONDRES ; *& ſe vend* A PARIS,
Chez DE BURE l'aîné , Quay des Au-
guſtins, à Saint Paul.

M. DCC. LIV.

Fig. 3. Abbé de Condillac, *Traité des sensations*, title page, with the dedication to
Ferrand's lifelong companion, Mme de Vassé (1754). [History & Special Collections for the Sciences, UCLA Library Special Collections]

nesses to the ensuing thought experiment, so they better be on board from the start. Sometimes called the "laboratory of the mind," thought experiments were used in science, even by Newton who explained for example that space does not "come under the observation of our senses." When an experiment cannot be run in the real world, we can learn by picturing, by using the imagination, one of Ferrand's special talents. Condillac ended his "Avis" to the reader: "The statue I propose to observe is unknown as of now; and this alert will seem misplaced; but that is yet another reason to notice it, and remember it." His presentation of this "Avis" as an afterthought was of course entirely disingenuous.[114]

Notice how the statue that he and Ferrand conceived of steals the scene, so to speak, without yet having been properly introduced. From the start Condillac surprised, involved, and sought rhetorically to manage his readers, to get them on his side immediately or to scare off those not up to the challenge. His "Avis" was followed by a many-page dedication called "Dessein de cet ouvrage," in which the intention of the work was explained for all readers but directed personally to Vassé (figure 4). It was a lengthy, heartfelt homage to their beloved Ferrand:

> You know, Madame, to whom I owe the enlightenment that finally dissipated my prejudices; you know the part played in this work by someone who was so very dear to you, and who was so worthy of your esteem and friendship. (It was she who proposed the epigraph to this book.) It is to her memory that I dedicate this work and I address it to you to experience simultaneously the pleasure of speaking of her, and the chagrin of mourning her. May this monument perpetuate the memory of your mutual friendship, which I have had the honor of sharing. . . .
>
> How can I not expect success when so much of this treatise is hers? Its clearest and finest insights are due to the penetration of her mind and the vivacity of her imagination; qualities seemingly incompatible but united in her. She felt the necessity of considering each sense separately, of distinguishing precisely which ideas we owe to each, and of observing with what progress they instruct one another and lend mutual support. . . .
>
> For this purpose, we imagined a statue organized internally like us, and animated by a mind empty of ideas. We supposed in addition

that the marble exterior allowed none of its senses to be used, and we reserved for ourselves the liberty of opening them as we chose, to the different impressions to which they were susceptible. . . . We thought to start with smell, because of all the senses it seems to contribute the least to the knowledge of the human mind. The others followed as objects of our research, and after considering them separately and together, we saw the statue become an animal capable of looking after itself. . . . The principle that determines the development of its faculties is simple; it is within the sensations themselves; because being either agreeable or disagreeable, the statue is interested in enjoying the first and avoiding the others. . . . Judgment, reflection, desires, passions, etcetera are nothing but differently transformed sensations. . . .

This approach is new, and it shows the simple ways of the author of nature. Isn't it admirable that he only had to make man sensitive to pleasure and pain to engender in him ideas, wishes, habits, and talents of all kinds?

There are of course many difficulties to surmount in order to develop this entire system, and I often realized how much such an undertaking was beyond my powers. Mademoiselle Ferrand enlightened me on these principles, on the plan, and on the smallest details; and I must be all the more appreciative to her that her intention was neither to instruct me nor to make a book. She did not see that she was becoming an author, and she had no other design than to engage with me on subjects that interested me. Also, she never stuck stubbornly to her own views. And if I almost always preferred them to those that I held previously, I had only the pleasure of becoming more enlightened. I esteemed her too much to agree with her for any other motive, and doing so would have offended her. However, I so often recognized the superiority of her views that she suspected me of complacency. She scolded me sometimes for this, she feared, she said, that she would spoil my work; and examining scrupulously the opinions that I was abandoning, she tried to convince herself that her criticisms had not been well-founded. . . .

Had she taken up the pen herself, her work would have been the best proof of her talents. But she had a delicacy that prohibited her from even considering it. Obliged though I was to approve, I felt sorry, because I could see from her advice what she could have accomplished so well on her own. This treatise is thus unfortunately only the result of the conversations I had with her, and I fear that I haven't always presented her ideas in the best light. It is frustrating that she could not guide me up to the moment of publication. I regret especially that there were two or three questions on which we were not entirely in agreement.[115]

The tribute that I give to Mademoiselle Ferrand I would not have dared do had she still been alive. Eager only for the glory of her friends and freely giving her contributions to them to claim as theirs, she would not have recognized her own part in this work. She would not have let me acknowledge it, and I would have respected her wish. But today should I deny myself the pleasure of rendering her this homage? It is all that is left to me in my loss of a wise counsel, an enlightened critic, and a sure friend.

You share with me this satisfaction, Madame, you who will miss her all your life, and thus it is with you that I love to speak of her. Both of you equally admirable, you had that discernment which reveals the entire value of a lovable object and without which one cannot really love. You knew the reason, truth, and courage that fitted you for one another. These qualities were the seals of your friendship, and you always found in your relations that enjoyment which is the character of virtuous and sensitive souls.

This happiness was fated to end. When she knew that her death was drawing near, your friend's one consolation was that she would not need to endure outliving you. I saw how glad she was for that. It was enough for her to live on in your memory. She liked to dwell on this idea, but regretted your suffering at her loss. "Speak of me sometimes with Madame de Vassé," she would say to me, "and may it be with a kind of pleasure." She knew in fact that sorrow is not the only form of regret, and in such cases, the more delight we find in thinking of a friend, the more vividly we experience the loss we have sustained.

It consoles me, Madame, that she thought me deserving of sharing both this pain and this pleasure with you! I am honored that you agree with her judgment. Could there be a greater proof of your shared esteem and friendship for me?[116]

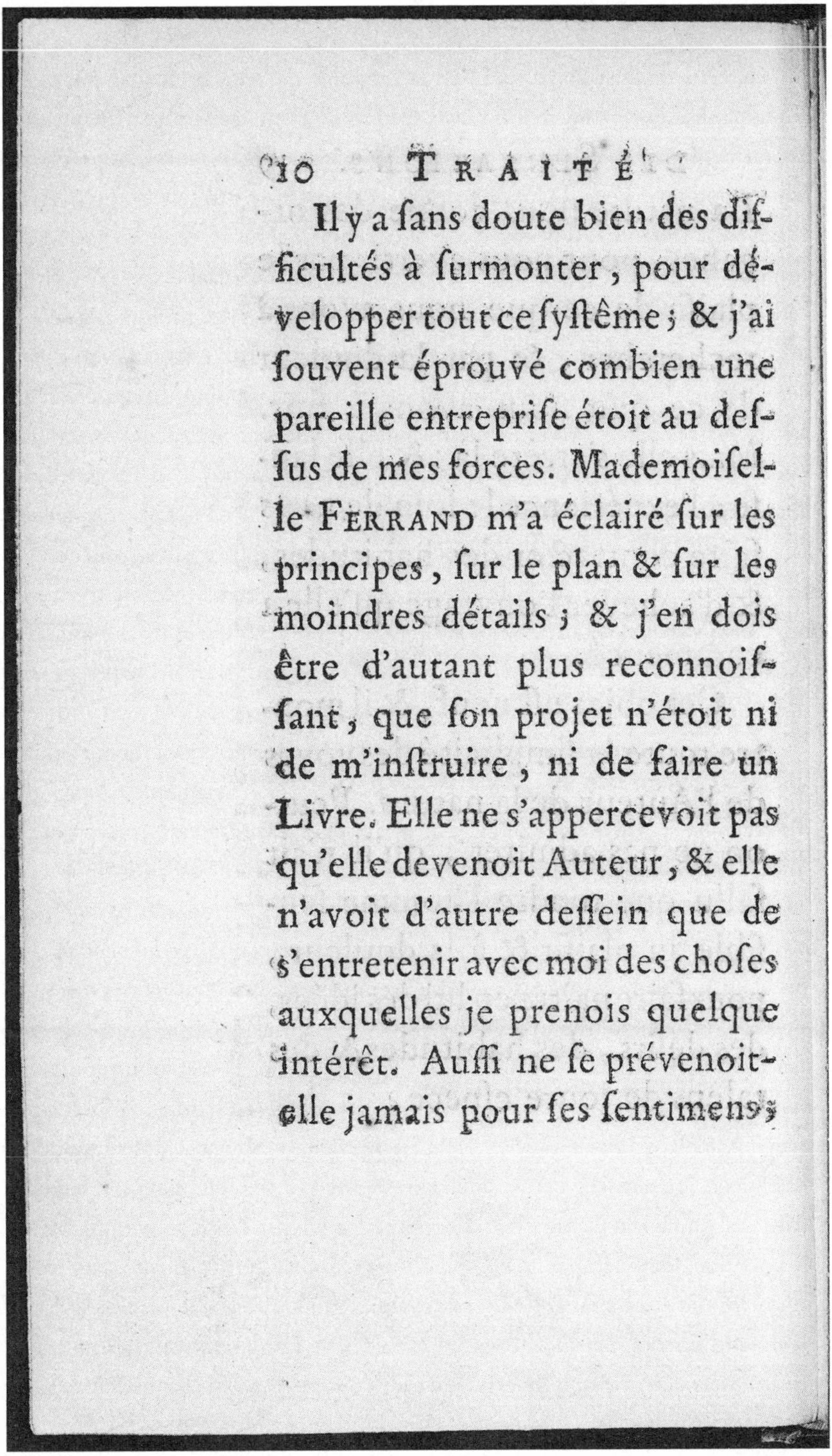

Il y a fans doute bien des dif-
ficultés à furmonter, pour dé-
velopper tout ce fyftême ; & j'ai
fouvent éprouvé combien une
pareille entreprife étoit au def-
fus de mes forces. Mademoifel-
le FERRAND m'a éclairé fur les
principes, fur le plan & fur les
moindres détails ; & j'en dois
être d'autant plus reconnoif-
fant, que fon projet n'étoit ni
de m'inftruire, ni de faire un
Livre. Elle ne s'appercevoit pas
qu'elle devenoit Auteur, & elle
n'avoit d'autre deffein que de
s'entretenir avec moi des chofes
auxquelles je prenois quelque
intérêt. Auffi ne fe prévenoit-
elle jamais pour fes fentimens ;

Fig. 4. Page from the *Traité des sensations* where Condillac explicitly notes his debt to Ferrand. [History & Special Collections for the Sciences, UCLA Library Special Collections]

We see from this tribute that Ferrand, though determined to stay out of the limelight, thought in a scientific way, critiquing their work together but then also reexamining and testing the discarded ideas to make sure they truly no longer stood up. She and Condillac proceeded by trying to falsify their results at each step; there was nothing facile about this systematic examination of the cognitive process. In the *Traité* they attempted to find simplicity, lawfulness, order, and, most importantly, a single principle for human cognition. Like Newton's gravity, sensation was to be the overarching, subsuming, universalizing phenomenon. But because the pursuit of truth was ongoing, because science was always revealing more, Ferrand suggested that their ideas be introduced not dogmatically but with modesty, choosing for an epigraph, as she and Condillac had for *Les Monades,* a quotation from Cicero: "I will explain as far as I can, yet not as if I were Pythias Apollo saying things that are fixed and certain. But rather like a mere man, following probable conjectures." Another translation: "Let us speak as one man to another and it will suffice if his reason be as probable as another man's. For exact reasons are neither in his hands, nor in those of any mortal man."[117]

The statue, then, as imagined by the two collaborators, was enclosed in a layer that isolated it from its environment until the experimenters opened it up selectively in a sequence of exposures, to learn what we owe to each of the senses. Such manipulations and sensory experiences were recorded, in a setup that was almost clinical.[118] The *Traité* focused on the period before language, the pre-linguistic learning of a solitary creature who would have no need to talk, who would, as Hans Aarsleff says, be "radically silent," and it was meant to demonstrate in detail the process by which sensations, coming gradually and in succession to the statue, would naturally lead to knowledge of itself and the outside world. The statue was presented with various sensations. For example, given the smell of a rose, it believed that it *was* that rose scent. Unpleasant odors would make it desire the better smell experienced earlier, and it would begin to notice that some smells satisfied, others frustrated that desire. It would start to pay attention, to remember, to compare. After its nose, its hearing was activated, then taste, and so on, but still the statue would think it actually *was* these smells, noises, and flavors, because it would as yet have no understanding that there was an external world and that these were properties of objects outside it. Sight alone was no better. Even touch by itself would know only the feeling of warmth or cold, or wet or dry, and would think it *was* those qualities.

But with the introduction of movement, and the touching of its own body, the statue would learn about itself. The moving hand would encounter other body parts and so would be both the toucher and the touched, receiving an answering sensation each time it touched another part of itself, realizing "c'est moi," "c'est moi," "c'est encore moi," and thus learning of its solidity, extension, and limits. Only the encounter with something solid, a tree for example, that did not give an answering sensation when touched, would result in the statue's first notion of something outside of itself and distinct from it. Then and only then, with the eye watching the hand feel the different properties of these external objects, would the statue be able to tell, for example, a globe from a disk, a cube from a square. Vision unassisted could not possibly do this. The perception of depth, of three dimensions in other words, was not automatic and had to be learned through the experience of touch. Condillac was proposing that he and his contemporaries learn how to attend, observe, and notice better, that they employ their faculties with more focused awareness.

How original was Ferrand's idea of the statue? When the *Traité* appeared in 1754, it already included a defense against accusations of plagiarism by Diderot which had circulated before its publication and which Condillac addressed immediately in print in a "Réponse a un reproche qui m'a été fait sur le projet exécuté dans le *Traité des sensations*." "This project is not new, they've said to me, it was proposed in the *Lettre sur les sourds et muets* of 1751. I grant that the author of this *Lettre* proposes to decompose a man, but long before that Mademoiselle Ferrand gave me the idea. Several people knew well that this was the very object of a Treatise I was working on, and the author of the *Lettre sur les sourds et muets* knew too." Condillac then devoted numerous pages to quoting from the *Lettre* in which Diderot described five different men, each deprived of a different sense, all this to demonstrate the glaring differences in their two approaches.[119] A society of five people, wrote Diderot, each of whom has only one sense, would break into five mutually intolerant sects, each thinking the others crazy. His tone was light, his book anything but methodical. Although Diderot protested that he had been copied, his work could not have been a direct source of Ferrand's idea.

Might she have gotten the statue idea from Buffon, who would also accuse Condillac of literary theft, very nastily in an outburst right to the author's face at a dinner chez Helvétius?[120] Buffon's third volume of his monumental *Histoire Naturelle* in 1749 had indeed proposed a being who would awaken gradually through his senses. "How do our first ideas get into our minds? . . . How can we find again the first traces of our thoughts?" There followed an almost poetic treatment of an awakening man's growing awareness of himself and nature around him, and his discovery of a female person, the story ending with the sexual union of these two beings.[121] Buffon's treatment was impressionistic, romantic, even rhapsodic in an effort to win and charm readers. Condillac footnoted Buffon and disagreed with him in the *Traité*, certainly not hiding that he knew of his discussion but arguing with it.[122] The idea of the awakening senses was in the air, and some similarities did exist between the two treatments. It is even possible that this idea was shared and exchanged long before the publication of the *Histoire Naturelle*. We know that Buffon and Helvétius socialized with Cramer in April 1747, just when Ferrand and Condillac were initiating their project.[123] So either Helvétius or Cramer could easily have relayed this idea from Ferrand's salon to Buffon, or vice versa.

In any case, it was the team of Condillac and Ferrand who approached this subject with disciplined argumentation, though the end result was far less fun to read. Grimm would call Condillac a flat-footed pedant who had drowned Buffon's statue in a barrel of cold water, adding gratuitously that "if you want to be read, you must know how to write."[124] Of course Ferrand, dead two years by the time the *Traité des sensations* was published, was not responsible for the actual writing, and the two were not trying to be crowd-pleasers. In the *Opticks* Newton had argued that the "investigation of difficult things" required analysis, deep study, and intense application. Their exploration of the mind's workings was an analytical, radically empiricist attempt to make cognition into a science. As Condorcet would later recognize in an obituary of Condillac, he had been trying to "make Locke Newtonian."[125]

The eighteenth century was fascinated with the Pygmalion myth; writers, playwrights, composers, and artists all invoked the story of the female statue coming to life.[126] André-François Boureau-Deslandes's popular, tiny little 1741 book *Pygmalion, ou la statue animée* had numerous subsequent editions, and the Ferrand circle might well have known its author, because he had earlier popularized Newton's scientific method.[127] His statue also began with a blank slate, and he did write about children learning through their senses.[128] But beyond that there was no serious explanation of the process, the work was rather lightweight, and it is unlikely that it influenced Ferrand or Condillac strongly, if at all.[129] Other authors dealt with the subject romantically, and of course artists did too. In the Salon of 1763 Diderot would praise a beautiful sculpture of "Pygmalion at the feet of his statue" by his good friend Falconet, a statue soon to be owned by Mme Thiroux d'Arconville, the chemist of chapter 5.

Was Condillac's statue a woman? A Galatea? Its sex is never mentioned but there are scholars who believe it was, completing the female triangle with Ferrand and Vassé.[130] Lending some credence to this idea is the possibility that the creators of the statue were inspired by the rhetorical strategies and by the heroine in Mme de Graffigny's 1747 novel *Lettres d'une Péruvienne*, a best seller surely known to Ferrand and Condillac, and there were even tighter connections through their close friend Helvétius and through his wife who was Graffigny's niece. Graffigny's novel used a framing device similar to Condillac's "Avis" in his *Traité* to introduce her book, an opening "Avertissement" followed by an "Introduction historique," both of which summoned, indeed dared her readers to read well, intelligently, and generously, to suspend disbelief and get into the spirit of what was being presented.[131] The captured heroine of the story, Zilia, wrested from her native land of Peru, bore some resemblance to the awakening statue in that she had to come alive in another culture/language/setting and had to form completely new ideas from scratch. Zilia's need to unravel and decode what was going on around her in her strange new world was similar to the gradual awakening of Condillac's tabula rasa statue. This novel was "epistemological fiction," for as Foucault noted, the struggles of a stranger in a strange land and of a being deprived of a sense had much in common.[132] Graffigny gave as the fictitious imprint of her book not the name of a place but "À Peine," perhaps implying that the tasks of putting philosophy into a story and depicting a strong, independent female protagonist were painfully hard, or punning that she and her heroine only just, only barely, managed it.[133]

Most reviews of Condillac's *Traité des sensations* singled out the author's homage to Ferrand, and while a few brought it up tongue-in-cheek, most were genuinely touched by this sincere acknowledgment of her help, whatever they thought of the work itself. One journalist called Condillac "one of the best minds of our century who blazes a path no one else has taken," describing the *Traité* as a "monument erected to Mlle Ferrand's glory, a monument more precious and doubtless more durable than all those that magnificence and the fine arts could have executed."[134] The *Journal de Trévoux* was particularly impressed: "Gratitude takes [Condilllac] to the grave of Mlle Ferrand; it is to her memory that he devotes this treatise. . . . Struck by an observation that was a critique of

his ideas as extensive as it was important, this celebrated metaphysician immediately picked up his pen, and chose for the guide of his new work the censor of
his former one."[135]

The *Correspondance Littéraire,* written as it was by several authors, had two
quite different reactions to the *Traité*. Raynal in his "Nouvelles Littéraires"
rather liked the book but dimissed the *savante* Ferrand as "a person of little wit
and disagreeable manner who nonetheless knew mathematics and bequeathed
things to Condillac in her will." But Grimm, in the same newsletter, hailed Ferrand as "a person of rare merit, philosopher and mathematician, dead two or
three years ago and sorely missed by our author whose intimate friend she was,
and by all who knew her." Grimm continued that because, unlike Orpheus, we
cannot descend to the underworld to bring others back to life, we must try to
revive our lost dear ones by "the force of our thinking and by speaking warmly
of them." The work itself, however, was not to Grimm's liking, for he stood
loyally by his friends Diderot and Buffon who claimed the *Traité*'s author had
stolen their ideas.[136]

Ferrand showed up again in Condillac's next book, part of what one recent
author calls his "publicity campaign," his determination to stay in the limelight
and justify his position.[137] This *Traité des animaux* of 1755, motivated by Buffon's
attack on him, ended with an "Extrait raisonné," a summary and clarification of
his earlier *Traité des sensations,* admitting some of its shortcomings. "Mademoiselle Ferrand would no doubt have caught [these]. Although she had a greater
part in this work than I did, she was not satisfied with it when I lost her, and she
believed there was much that needed to be fixed. I finished it all alone, and I
reasoned poorly, because I could not yet establish the state of the question.
What is more astonishing is that all those who claimed to criticize me directly or
indirectly did not know how to establish it any better than I, and reasoned badly
also." In a last nod to mathematician Ferrand, whose reasoning *was* keen, he
conceded that "the language of this science [epistemology] lacks the simplicity
of algebra."[138]

And what of Mme de Vassé? She quickly disentangled herself from the
troublesome and increasingly erratic Prince Edouard, who in 1765, bizarrely,
sent a turnip seed to her and Helvétius.[139] Vassé's loyalty to this author of the
scandalous *De l'Esprit* had never wavered, although that book cost him many of
his other former friends. In general, she stayed close with most from Ferrand's
circle. Helvétius's sister-in-law died in Vassé's house in 1760, and in 1768 Vassé

herself would die chez Mably, in the house he shared with Condillac when he returned from Parma, on the rue des Francbourgeois-Saint-Michel—today rue Monsieur le Prince.[140] Just as Ferrand had bequeathed 6,000 livres to Condillac to buy books, Vassé left him a diamond worth 6,000 livres, and she named one of the Jaucourts an executor of her will.[141] Mably wrote movingly of the brothers' chagrin during her final illness and devastation after her death from breast cancer. Vassé was, as Ferrand had known better than anyone, "not a person who inspired superficial attachments."[142]

When Condillac died in 1780, even *his* obituaries invoked the memory of Ferrand. Citing earlier thinkers who had realized the dependence of our knowledge on the senses—Aristotle, Bacon, Locke—one eulogist wrote that Condillac took this old idea and used it to develop new ones. Among his many works, it was the *Traité des sensations* that made him most famous, not least because of Ferrand's contributions.[143] But Condillac never seems to have become comfortable with mathematics, telling Cramer in 1747 that he failed to grasp the infinitesimal calculus, telling Maupertuis the same in 1750. Condorcet, perpetual secretary of the Académie des Sciences and a mathematician, wrote an anonymous obituary in the form of a mean-spirited "notice" on Condillac in the *Journal de Paris* that was scathingly critical of the dead man's lack of comprehension in this area.[144] But perhaps to honor Ferrand's memory, Condillac had continued to try; at his death he left unfinished a work on *La Langue des calculs*.[145]

The most lasting reminder we have of Mlle Ferrand is La Tour's portrait of her "meditating on Newton." It is a bold self-presentation of an intellectual woman, commissioned by her. This picture is now in the Munich Alte Pinakothek, where Ferrand is no longer surrounded by the men of science with whom she shared wall space at the 1753 Paris Salon: the Abbé Nollet, d'Alembert, and La Condamine, who spent years on an expedition to measure the meridian at the equator confirming Newton's theory of the flattened earth, and to whom Voltaire quipped, "Vous avez trouvé par de longs ennuis / Ce que Newton trouva sans sortir de chez lui."[146] In the La Tour, Ferrand sits amidst the worn charm of her soft blue velvet desk cloth and her simple, well-used furniture without any gilded ornamentation. She is unadorned, the hair under her bonnet brown and un-powdered, the eyebrows un-doctored. Most art historians argue about the clothing in which she is portrayed.[147] But Mary Sheriff inverts this question, pivoting to the work of Iris Marion Young, a feminist philosopher who "shifts attention away from how garments look to a viewer, to how they feel to the wearer" by asking how the subject inhabits her body, how fabrics touch that body, suggesting tactility, a perspective that challenges the "specular logic of passive object and mastering subject, gendered respectively as female and male."[148] Ferrand, a scholar of touch, would have liked Sheriff's and Young's perspective, and indeed she seems completely "bien dans sa peau," at ease with herself, her skin, her garment. She is relaxed, not rigidly posed. After all, she staged this sitting and appears in command of her artist, making confident eye contact, turning toward the beholder in a natural way, illuminated in her chosen activity. The book she is reading is an invention, as none of the available works on Newton in French, even blown up to folio size as this one is, looked at all similar.[149] This is perhaps symbolic, meant to imply that she has mastered them all, including Newton in the original. We see Ferrand interrupted in the semi-private space of her study, wearing a morning hat and indoor dress fashioned of fine satin, lace and ribbons yet informal.[150] This is not the large room where she holds her salon, rather her *cabinet* where she works, but she also receives here on occasion and seems to have invited us in. She is graciously suggesting to her viewers, if we follow the little finger of her left hand, that we too might have something to learn. Despite various interpretations of her expression—studious, absorbed, theatrical, imperious—all agree that she comes across looking very much alive and alert, as if cerebral vigor has energized her.[151] This life of the mind, this interiority, this pure intellectual power is hard for an artist to depict, but La Tour has done it.

Among the many commentators on the Salon of 1753, not one expressed astonishment that Mlle Ferrand was contemplating Newton. There were comments on La Tour's heavy-handed strokes, a seemingly new technique, but the influential critic Cochin retorted that the artist was a "scrupulous imitator of nature who puts so much truth in his portraits . . . [and] resemblances are such a perfect illusion that we think we are seeing the people he represents."[152] La Tour himself boasted, "I penetrate into the depths of my subjects without their knowing it, and capture them whole."[153] When Ferrand sat for him in 1752, he was trying to absorb some new advice.[154] Philosophically minded friends had urged him to be more of a *citoyen*, to do less portraiture of the vain wealthy—although they paid his bills—and more of "hommes illustres."[155] For the 1753 exhibit La Tour had certainly obliged, with a record-breaking eighteen pastels of famous scientists, artists, performers, engineers, and writers. One observer of the artist's remarkable collection of images that year commented: "It is easy to see . . . that he enjoys painting those who, like himself, knew how to achieve celebrity in the arts and sciences. Posterity which likes to research the lives and know the features of those who were important in their day, will find in the works of M de La Tour the most faithful portraits of those who do honor to the century in which we live."[156] Ferrand's membership within this group speaks volumes. She was indeed "important in [her] day" and wished to be remembered as a Newtonian from her earliest youth to her last breath.

Dear Elisabeth,

You seem to have lived and breathed math since girlhood, tutored first by a game theorist who would have taught you probability theory. Do you realize how fortunate you were to have had such an opportunity in your youth? How modern that was? Many young girls today are discouraged, by parents and teachers alike, from doing math and science because those are said to be difficult "masculine" fields. You were launched early, then known in 1733 as *the* person for good math talk with the visiting Bernoulli brothers, then busy all through your close friendship with Clairaut and your correspondence with Gabriel Cramer, and finally touted at your death as someone who really "knew her mathematics."

Another young girl not yet born when you died, Sophie Germain, shared your passion but did not have the encouragement of her family. On the contrary, they absolutely forbade her to do math. So she snuck Montucla's *History of Mathematics* under her blankets and read it by the light of a tallow candle. As she devoured the words, she was smitten by a subject so elegant, so powerfully beautiful, so completely enthralling for Archimedes that it rendered him oblivious to the enemy soldiers who burst in to kill him. It is a miracle that Sophie did not burn the house down, and amazing too that she grew up to be a great mathematician despite parental obstacles and despite, like Jeanne Barret, having to pretend to be a man to achieve her ends. Anyway, Sophie's is another story . . .

Now, about your portrait as a woman of science. How wonderful that you chose to be remembered this way, meditating on Newton. For me it is particularly exciting, thanks to your arrangement with the splendid Quentin de La Tour, to have an idea of what you looked like. That is why you grace the cover of this book. I do not have illuminating likenesses of the other five women in my story. There are no known portraits of field naturalist Jeanne or anatomist Marie-Marguerite. Gillaume Voiriot's painting of astronomer Reine drawing an eclipse map and surrounded by scientific instruments is lost, and so is botanist and illustrator Madeleine Françoise's self-portrait, which would surely have been revealing. Although there are two portraits of chemist Geneviève, in neither is she associated with her science. The first, by Charles-Antoine Coypel, depicts her at fourteen just before her marriage, bored with the frivolities of her station as the daughter of a wealthy tax farmer, gazing vacantly and holding a piece of music. The blue bow she wears shows up again in a much later portrait of her by Alexandre Roslin—well, she would only have been thirty, but transformed—where we see her wanly smiling, the unnaturally fire-red cheeks perhaps a reminder of the scars from smallpox that almost killed her, dressed expensively but severely, almost nun-like, closed in with nothing particular to say to us.

But YOU, there you are looking out at us, smart and valorous in the face of your illness, radiant as you reflect on the mathematical physics you have been reading, determined to leave your mark as a Newtonian. You should know that there were other portraits of women exhibited by La Tour in that same Salon of 1753, but they depicted singers and actresses. You stand out as the female thinker.

I believe it's fitting that you are buried in Saint Sulpice, the church with the huge gnomon, a scientific instrument, right in its midst, with a meridian line built for astronomical measurements of equinoxes in the early 1700s when Rome was finally seeing the light and considering the removal of Galileo's works from the Index of Prohibited Books. It might interest you to know that it was one of the few churches not desecrated or repurposed by the revolutionaries, who appreciated that it already had a rational and useful function. D'Alembert made clear that you were an atheist—how could you not be with close friends like Helvétius?—but given that your final resting place was a church, perhaps this science church is at least somewhat more acceptable to you. You can even converse there with Montesquieu, your neighbor in life on the rue Saint Dominique as well as in death now, who many called the Newton of politics, seeking regularity and laws in systems of government . . .

I marvel at the risks you took hiding the royal fugitive. Although it was your idea, you were assisted in this effort by your love, Antoinette-Louise-Gabrielle. What had drawn you two women to each other, how old were you when you decided to live together? How did you spend your time when not in cerebral exchanges with the members of your "bonne compagnie"?

Did you have other women friends for intellectual exchange? I think especially of Suzanne Marie de Vivans, Marquise de Jaucourt, only two years older than you and who lived from 1738 close by on the rue de Grenelle just parallel to your rue Saint Dominique. She was the sister-in-law of your salon member Louis de Jaucourt and like him contributed articles to the *Encyclopédie,* but hers, of course, were never identified as hers. Your colleague Condillac knew her well and bequeathed his beloved library to one of her sons. So my guess is that your social circles overlapped and that she would have been good company for you. But who knows, because friendships between women tend to go unrecorded, and much correspondence between them disappears.

Fourteen years older than Condillac and unafraid to tell him he was wrong, you were, as you said yourself, unfailingly frank. By the way, did you think your hypothetical statue could be a woman? It's a feminine noun anyway, so "la statue" doesn't betray anything. Much has been made of that possibility by modern scholars, but I'm asking you.

Most importantly, I would love to know if you ever met and talked science directly with astronomer Reine, perhaps through Clairaut who knew and worked with you both. I like to think so. But even if you didn't actually meet her, she very likely saw La Tour's portrait of you contemplating Newton at the Paris Salon of 1753. These state-run shows, held in the Louvre and referred to as spectacles, and sometimes as popular festivals, were open to all classes and widely, enthusiastically, even turbulently attended—for pleasure but also inspiration and edification. More contemplative beholders could go through at a slower pace, and art critics often planted themselves before particular paintings for long spells, judging technically, responding emotionally, then writing detailed reviews. What an impact your scholarly scientific image could have had on the bright clockmaker's wife! Reine was thirty years old when your portrait was displayed, and the vision of you might have empowered her, moving her soul as good art was and is supposed to do. Married five years earlier but childless, she had met the astronomer Lalande in 1751, by which time she was already beginning to do calculations. She could have stood before your portrait, and perhaps so did the aristocratic Geneviève d'Arconville. Thirty-three at the time, already deep into her science courses at the Jardin du Roi, Geneviève had given up the high life for serious study.

Yes. You could definitely have been a role model, the kind of inspiring person that female scientists today wish they had found early in their lives to urge them on. "You cannot be what you cannot see" is a mantra oft quoted by activist women. It is so important to have in the mind's eye someone who does new, different and great things, a visionary who challenges convention and tradition, and who thus rouses others to realize their dreams and stake their own claims.

Experienced as you were in lobbying for men at the Académie des Sciences, might you even have motivated Reine to set her sights on membership of her own in such an institution, which she unprecedentedly achieved?

TWO

Astronomer and "Learned Calculator"

Nicole Reine Lepaute (1723–1788)

Mme Lepaute was the only woman in France who acquired true knowledge of astronomy. . . . She had enough spirit to be imperious when it was useful; but she had enough judgment to yield in situations where resistance would have been dangerous.
—JOSEPH-JÉRÔME LALANDE, *Bibliographie Astronomique*

[Never] dishonor a title as beautiful as "académicienne."
—NICOLE REINE LEPAUTE, January 1762

"MADAME LE PAUTE," THE CHRONICLER BACHAUMONT RE-ported on 9 October 1780, "wife of the artist so renowned in clockmaking, is no less celebrated herself for her knowledge of astronomy and her usefulness to the Académie des Sciences in this area. She wrote just recently to Madame Necker to engage her help in obtaining from the Controller General of Finance a pension that she thought she merited for her services and for the singularity of a woman devoted to lofty research. Madame Necker responded very respectfully, but added that her husband insisted she never solicit from him any favor out of tenderness for her; that in addition, she had no doubt that by addressing him directly she would succeed. Madame le Paute therefore turned to him, received just as flattering a response in which he told her it was too late this year to alter the existing budget; but that he will take care of her next year, and that if anything changes before that in his work with the King, he will keep her in mind."[1]

Reine Lepaute was a confident woman who knew her worth.[2] She had been a confident girl too, retorting when her sister taunted "I am the fairest" with "But I am the smartest," and she had a history ever since of asserting herself.[3] Here, in 1780, she was asking for financial compensation that she believed was merited for her decades of scientific accomplishments. But already twenty years earlier she had expressed to her friend and collaborator, the astronomer Joseph-Jérôme Lalande, her frustration at being excluded from the Paris academy for which she had done much work. As such inequity troubled him, he had immediately begun campaigning for her at the provincial Académie des Sciences de Béziers, a town in Languedoc renowned among astronomers for its limpid, serene skies. Lalande had observed there several times, and Lepaute had done calculations based on observations of the 6 June 1761 transit of Venus as seen from Béziers. Later that same year she had become a member of the city's academy, making her the first woman of science elected to such a body in France. Her admission was based on a number of mathematical papers she submitted which, as is the case with so many writings produced by the women in my story, have not come down to us.[4]

Appreciative as Lepaute was of her new and unprecedented election in the fall of 1761 and grateful for Lalande's role in helping her secure it, she wished to ensure it not lead the academy to welcome just anybody and thus diminish the status of those already admitted. In particular, she needed confirmation that this was not a simple act of gallantry. It was critical, as she saw it, to maintain the same high standard for membership as had always been upheld. On 16 January 1762, a few months after getting her good news, she wrote a confidential letter to the Béziers academy's secretary, Jean Bouillet, concerning a prominent mathematician's anticipated manipulations:

> M. Clairaut has in his house a girl whom he keeps and whom he picked up three years ago at a dressmaker's shop, rue Saint Honoré. Her name is Gourlier. He painstakingly taught her to add numbers, but that is the sum total of her capacity. She knows nothing of the three-body problem and has never been able to understand it. Nonetheless this girl had the impudence to be jealous of the recompense you accorded me for the ten years that I've consecrated to astronomy. She is already announcing that a note has been written for her and that she counts on being admitted to our Académie on M. Clairaut's recommendation. Everything that I am telling you here, sir, regarding the status and ignorance of this girl, you can verify elsewhere, in order to guard yourself against this subject and not dishonor a title as beautiful as "académicienne." But, if you please, sir, never mention me in this connection.[5]

Having learned that Clairaut was promoting someone she saw as completely undeserving, Lepaute was determined to thwart these efforts. Writing in a clear and sure hand, she insisted that the Béziers academy inform itself about Gourlier's lack of knowledge, and about "what she was before she belonged to M. Clairaut," suggesting that they ask several men of science who would substantiate what she was telling them about the other woman's stupidity.[6]

That Lepaute was so exercised in 1762 over the pretensions of Alexis Clairaut's mistress—whose name was spelled numerous ways by others, Goullier, Goulli, most often Gouilly—must be understood, if not excused, in light of this paramour's damaging jealousy. Much appears to have changed since Clairaut's genuine and open veneration for Mlle Ferrand's mathematical prowess many years earlier. That Ferrand was thirteen years older than he, and Lepaute ten years younger, might have been part of the reason. Now, although he privately referred to Lepaute as the "savante calculatrice," the learned calculator, whose enthusiasm was boundless and whose "ardor for astronomy astonishing," publicly he had denied her any kudos for her work.[7] He had expunged her name from his writings and failed, back in 1759, to mention the key role she had played in accurately calculating the return date of Halley's Comet, a spectacular prediction for which Clairaut gained international glory but which he could never have made without Lepaute's help. His envious mistress had insisted that her lover never mention his female colleague, but because her role in this slight was widely known Lepaute had up until now chosen the high road. This second outrage, however, the furtive attempt of this other woman to gain academic membership, was more than she could tolerate, and she insisted finally on setting the record straight and standing up for herself. Clairaut, after denying her proper credit in 1759, should not now, three years later, be allowed to dilute her hard-won academic honor. Lest we think her catty—she herself did not want it spread about that she had registered this complaint—or prudishly snobby about the questionable background of Clairaut's chosen companion, we must look at Lepaute's indefatigable labors in her own rise to recognition. She was only claiming what she felt was her due for the decade of astronomy to which she had already devoted herself by 1762.

And amazingly, she succeeded in doing her science within a marriage, despite existing laws and customs that bound wives in feudal constraints and deprived them of individual personhood. We must realize how difficult it was for

Lepaute to stake out some turf for herself, because Clairaut was not alone in attempting to erase her from the picture and deprive her of well-earned acclaim. Her talented but difficult husband, Jean-André, was silent about her achievements throughout their forty-year marriage, rarely acknowledging the importance of her mathematical skills to his clockmaking career. His artisanal gifts were widely acknowledged, as was the fact that he had chosen an admirable domestic partner, but his wife's outstanding intellectual ability was never advertised. Both Clairaut and Lepaute's own spouse seemed bent on keeping her in the shadows. If not for Lalande, who called attention to her contributions in many of his own publications during her lifetime, revealed them in private letters to his friend the Genevan naturalist Charles Bonnet, and then detailed her science in long eulogies after her death, we would not know about the accomplishments of this smart, gutsy astronomer. Even Lalande, certainly more honorable in this regard than other men, was unable to escape entirely the gendered expectations of his time, stating in his glowing portrayal of Lepaute's scientific work, for example, that her most important contribution was nurturing an eager young nephew, whom she brought to Paris from the provinces to be trained in astronomy as a protégé of Lalande himself.

Mme Lepaute was born Nicole Reine Etable on 5 January 1723 in the Petit Luxembourg Palace (the part facing the rue de Vaugirard and connected to the Orangerie). The Etable family was from Haute Normandie and anchored in service to the court. The name de la Brière was assumed by one of Reine's brothers and is often attached to hers but, contrary to what some have written, she did not use it.[8] Several older siblings had been born in Versailles, where her father Jean and her uncle were footmen to the Duchesse de Berry, the wild and favorite daughter of the Regent Philippe II, Duc d'Orléans.[9] After this lady's death, Etable was moved with his family to another palace, the Luxembourg, where Reine and more children were born, and where he next became lackey to Elisabeth d'Orléans, queen dowager of Spain, a second notorious daughter of the regent, married very young to the sickly Spanish crown prince who then ruled for only seven months in 1724 before dying of smallpox. Unhappy and despised in Madrid, Elisabeth returned to Paris in 1725 and was given the Luxembourg for her home. Young Reine grew up glimpsing the unfortunate royal for whom her father worked, a broken, idle woman who faded into madness and oblivion and died forgotten in 1742, and she surely also saw the furnishings and magnificent artworks of the luxurious apartments. This east wing of the palace was opened in 1750 as the first public museum, exhibiting about fifty of the Luxembourg's great paintings, including twenty-four by Rubens.

All this was going on while Reine lived there, as a child, an adolescent, and then as a wife when she became Mme Lepaute. Family lore had it that she knew she was intelligent from a young age, perhaps because she repeatedly heard this said of her or because she had a genuine thirst for learning and enjoyed it. Her eulogist wrote that she was always studious, devouring books as soon as she could read, and poring over them all night long, but although her father may have secured for her brothers a good education it seems likely that she was self-taught.[10] She met the two most important people in her life at the Luxembourg, first the man she would marry, Jean-André Lepaute who came to install Paris's first horizontal clock, one of his commissioned masterpieces, on the front of the palace facing the magnificent gardens.[11] She was twenty-five, he twenty-eight when they wed on 27 August 1748, and the king soon named him one of the *horlogers du roi*, granting him an apartment in the Luxembourg in honor of his skills so the couple could continue to live there. Next came her meeting in 1751 with the young astronomer Lalande, age nineteen—she was nine years his senior—when he arrived to use the small observatory above the porch in the cupola of the palace, and that was the beginning of his lifelong involvement with the Lepautes and his scientific collaboration with her. The trio were to be intensely entangled for several decades. For much of that time they lived together, an unorthodox arrangement, a probably chaste but nonetheless handy ménage-à-trois that suited Reine perfectly and that she may very well have orchestrated herself.

The clock business was fiercely competitive and M. Lepaute, now assisted by his younger brother whom he had brought from Lorraine and trained, was involved in numerous ugly public feuds with his rivals throughout the 1750s which revealed him as prickly and insecure. (The botanist Philibert Commerson, a friend of Lalande's and like him a native of Bourg-en-Bresse, whom we will encounter in the next chapter and who, significantly, named many plants after his acquaintances, chose for Jean-André Lepaute the spiny, always rigid "Peautia Xerastate.")[12] But if Lalande had reservations about the man, he still persuaded him to start building astronomical clocks, which he then used and truly admired, so he supported him in several disputes with such rivals as Jodin, LeRoi fils, and a young man called Caron, the firebrand who would soon take the name by which he is much better known, Beaumarchais. M. Lepaute was a talented and extremely proud man, but these other clockmakers placed articles in the papers and circulated inflammatory *mémoires* accusing him of stealing their ideas, making inferior timepieces, not honoring his promises, and, to pre-empt his inevitable retorts to their attacks, "imposing on the public his black calumny."[13] Out of loyalty Lalande argued that Lepaute's clocks were of the highest quality, innovative, and completely original. Lepaute himself fought back against all the nasty imputations, stating, "If I stayed silent my advance-ment and my fortune would suffer. . . . My brother and I have debts now." He went on to claim that he was repeatedly harassed at his lodgings, where one rival "cursed me and demanded that I come out—there are witnesses. . . . This has been ruinous."[14] His litany of injuries exposed the genuine chagrin of a man whose reputation was at stake and who felt wronged on all counts, undermined financially, defamed unjustly.

One accuser, LeRoi, produced a notarized act signed by both Jean-André and his wife; LeRoi had absolutely insisted that Mme Lepaute be present at transactions because he believed she could control her husband and provide some steadying ballast. He clearly saw her as the one with class, influence, and far greater intelligence. Monsieur Lepaute, said his rival, was particularly stung to be treated as a "mere worker, a craftsman good with his hands but not able to think well and incapable of producing anything alone; [he claimed it] caused him even a considerable violation to be talked about as just a tradesman."[15] And Jean-André did have quite a chip on his shoulder; he in turn mocked the lofty pretensions of this accuser, calling him "a man of letters, superior to me, with flowery, delicate, polished style, the graces of light banter, fine pleasantries." Next came the dispute with Caron (Beaumarchais), the most damaging of all, because the Académie des Sciences not only ruled that Lepaute had indeed sto-len an escapement mechanism from the younger man but also promised to pro-tect Caron against future encroachments of this kind.[16] Lalande again defended Lepaute whose precision astronomical clocks he needed for his own work, faith-fully striving to save the couple's livelihood and dignity.

During this increasingly troubled time in the early 1750s Mme Lepaute stood by Jean-André, but it must have been frightening, embarrassing, and at the very least distracting, particularly when she was then trying to carve out time within the marriage to pursue her own scientific interests. Lalande reported, "She had too good a mind not to have intellectual curiosity and so she made observations, she calculated, and she wrote descriptions of her husband's work."[17] But her spouse did not see her as a learned person in her own right, saying of his wife, "It is known that she has integrity (*probité*) and that she is credible. In truth, she is not renowned in clockmaking, but she *is* for the fine conduct of her household."[18] Either she was managing to keep her astronomical computations to herself, or her husband saw her talking with Lalande but could not grasp the exceptional level of her thinking.

An only child, Lalande had almost become a Jesuit but grew instead into an outspoken atheist, co-author in his later years of a *Dictionnaire des Athées anciens et modernes*, writing in 1805, "The sky seemed to everyone to be proof of the existence of God. I believed it at nineteen; today I see there nothing but matter and movement."[19] He was ambitious, vain, opportunistic, but smart, energetic, and generous too. As a reluctant law student in Paris he happened to lodge in the Palais de Cluny where, on the tower's roof, the astronomer Joseph-Nicolas Delisle kept one of several small observatories.[20] Lalande simply fell in love with science and was soon studying mathematical physics with Pierre-Charles Lemonnier, King Louis XV's personal astronomer, who sent him to Berlin to measure the lunar parallax and that of Mars to help determine their distances from earth. Lalande saw science as a force of progress and truth; he later traveled to Italy where he beseeched Pope Clement XIII to take Copernicus and Galileo off the Index of Prohibited Books.[21] Physically unattractive, as even his charitable portraits betrayed, he was so described by a recent biographer: "His aubergine-shaped skull and shock of straggly hair trailing behind him like a comet's tail made him the favorite of portraitists and caricaturists. He claimed to stand five feet tall, but precise as he was at calculating the heights of stars he seems to have exaggerated his own altitude on earth." Many wondered if his atheism might have been his revenge against God who had made him so unsightly.[22]

Lalande was a voracious scholar and taskmaster who recognized the importance of cooperating and collaborating with all kinds of personalities in order to get good science done. He said of himself that he did not hold grudges, that he was "an oilcloth for insults and a sponge for praise." His philosophy of life evolved as he aged. "I am rich, but I have neither whim [*fantaisie*] nor needs. I have few servants, no horses. I am sober, my clothes are simple, and I go on foot. I rest where I happen to be. Money for me is useless." And of women he said simply: "I always sought to contribute to the instruction of women, but my

passion for them stayed sensible. Never did they disturb my fortune or my studies."[23] With Mme Lepaute he had a very intense friendship, certainly practical
and intellectual, but with elements of the fraternal, the platonic, the collegial,
some have even suggested the romantic.[24] When Lalande traveled to England,
Italy, and Holland, he wrote to her extensively, his only other correspondents
being his mother and his male colleagues. Reine Lepaute became and remained
one of the pillars of his life.

Although he admired her much more than he did her husband, Lalande defended Jean-André in his continuing feuds with competitors.[25] On 18 December
1754 the beleaguered clockmaker appealed to the Académie, stating that yet
another rival, a M. Mazurier, would be presenting a clock to them based on an
idea he first saw in, and therefore stole from, Lepaute's shop.[26] This nasty vying
for priority was constant, and being one of the *horlogers du roi* did not seem to
offer any protection. Something had to be done to save Lepaute's name and
enterprise because these controversies were taking a serious toll on his business.
First, the devoted Lalande lent the couple 3,600 livres on Christmas Eve 1754,
to assist them.[27] And then together, for damage control but also to stake an original, tangible, and visible claim against future infringements, the Lepautes and
Lalande decided to write a definitive book, addressing the professional competition in a new and unique way, allowing Jean-André to stand out as the singular clock expert in a category all his own. The result, an authoritative *Traité
d'horlogerie, contenant tout ce qui est nécessaire pour bien connaître et bien régler les
pendules et les montres,* was to serve several purposes. It would demonstrate
Lepaute's prowess and establish him as the foremost clock specialist. Also, the
book might benefit him financially, through an advance from the eventual publisher and from sales, and he could use it for self-defense, including in it reprints
of the many articles Lalande had written in his behalf and referring readers to
other pamphlets in his favor. In the *Traité* he would criticize Caron once more,
for creating frivolous ornamental things that cheapened the clockmaking business and gave it a bad name. And he would again denounce LeRoi, Jodin, and
Mazurier.[28]

The *Traité d'horlogerie* appeared at the end of 1755. It was very well received
on its own merits and also thanks to Lalande's public relations skills. By now
this driven young astronomer had been elected a member of the Académie des
Sciences, a fact broadcast repeatedly in the volume, which was a joint effort
in every sense. Lalande wrote a good third of the book, but he selflessly wanted

it to be announced with two authors, "M. et Mme. Lepaute," and he requested the writer La Beaumelle, his friend, to advertise it this way to journalists of his acquaintance. But in the end only Jean-André's name appeared on the title page.[29] This was a great injustice, Lalande believed. Leading up to the book's publication he kept trying to have Mme Lepaute's role made clear. He had written in the July *Mercure de France* an homage to her contribution, then in the December *Journal des sçavans* a short mention of the *Traité* praising the tables of oscillations she did that filled pages of the appendix. After the volume appeared, he was still paying tribute to her. In the *Mercure* of March 1756 there was a long article giving each of the authors appropriate credit, including a rave of Mme Lepaute's "application to the mathematical sciences already known to several savants" and her pendulum table. The review particularly admired the style of the whole work, which, as Lalande would later reveal to Bonnet, was almost entirely rewritten before publication by Mme Lepaute, who knew how to make her husband's material more accessible to a broader audience.[30] She, being someone who naturally liked to take charge though inconspicuously at this early stage, also engineered the marriage of her brother-in-law to the daughter of the *Traité*'s publisher Chardon in order to "establish" her husband's younger sibling and business partner securely in his career.[31]

The *Traité* discussed the delicate precision, theory, and reasoning required for clockmaking, arguing that the correct reckoning of time is "indispensable" and among "the real necessities of life."[32] The authors believed that in their field there were too many mere workers and too few veritable artists with knowledge of physics and astronomy, like themselves. The book included the contributions and accomplishments of some other respected clockmakers, but also named those who were inferior and even crooked. There were lists of famous public clocks made by Lepaute and his brother for Les Ternes, l'Hôtel des Fermes, la Verrerie Royale, and for many other buildings, and there were detailed engravings of internal clock and watch parts. Mention was made of Lepaute timepieces presented to the king at Marly on 23 May 1753.[33] Technical chapters on the internal mechanisms and movements abounded, along with much disdain for clocks with "bizarre novelties," such as suns rising and setting, decorative images of people, or anything with the name of Caron who falsely claimed originality. The authors of the *Traité* presumed that the "wise and impartial public" would recognize M Lepaute's new escapements to be "without contradiction the most perfect of all."[34] Lalande's letters defending Lepaute's clocks in the *Mercure* of

August 1754 and July 1755 were excerpted. And Mme Lepaute's computation of the number of oscillations per unit time of pendulums of various lengths was praised, her work summarized at the end of the book. "This table calculated with very great care, using logarithms, presents in intervals of 100, from 1 up to 18,000 vibrations per hour, the length of the pendulum, so that when we have a clock movement whose last wheel must make a certain number of turns per hour, we find what length pendulum we must apply" (figure 5).[35]

With the 1755 *Traité*, then, Mme Lepaute emerged somewhat from the background, for while she did not appear on the front page as a co-author the way Lalande had hoped, he saw to it that she was at least acknowledged loud and clear for the contribution of her table and the mathematical skill it required. The book did seem to improve her husband's fortunes as well, and his status as a serious, appreciated clockmaker grew. In addition to many new private orders, the Lepaute brothers eventually made more public clocks—for the École Militaire, Hôtel de Ville, Hôtel des Invalides, and the chateaux of Choisy, La Muette, Bellevue, Saint Hubert (near Rambouillet), for the Palais Royal and the Jardin du Roi.[36] In 1757 the king granted to Jean-André quarters in the Galeries du Louvre. There was possibly even a change in his attitude, as a grand-nephew portrayed a pleasant, relaxed side of the man, describing him as fun-loving, generous, enjoying art and the society of artists.[37] These words may have been some tactful whitewashing by a descendant, but in any case a period of family peace now set in.

The relative calm in the Lepaute household finally allowed Mme Lepaute to turn more freely to her own interests. Up to this point she had been the dutiful wife of an artisan, albeit one who fancied himself an *artiste* as Paola Bertucci has identified tradesmen with *esprit*.[38] She had thus far functioned very much in the craft tradition of many women in this early modern age, assisting their husbands and claiming little credit for themselves. Even the impressive pendulum chart she produced and got thanks for was a contribution to his field of clockmaking, not a sign of work in an independent area. Now, however, she was about to embark on something very different, doing astronomical research of her own. Other female astronomers of her century, for example, quite a few in Prussia like Maria Winkelmann Kirch, toiled for the most part with and for their spouses and advanced in their area only because they were married to men recognized in the field.[39] For the Kirch daughters, too, it continued as a family business.

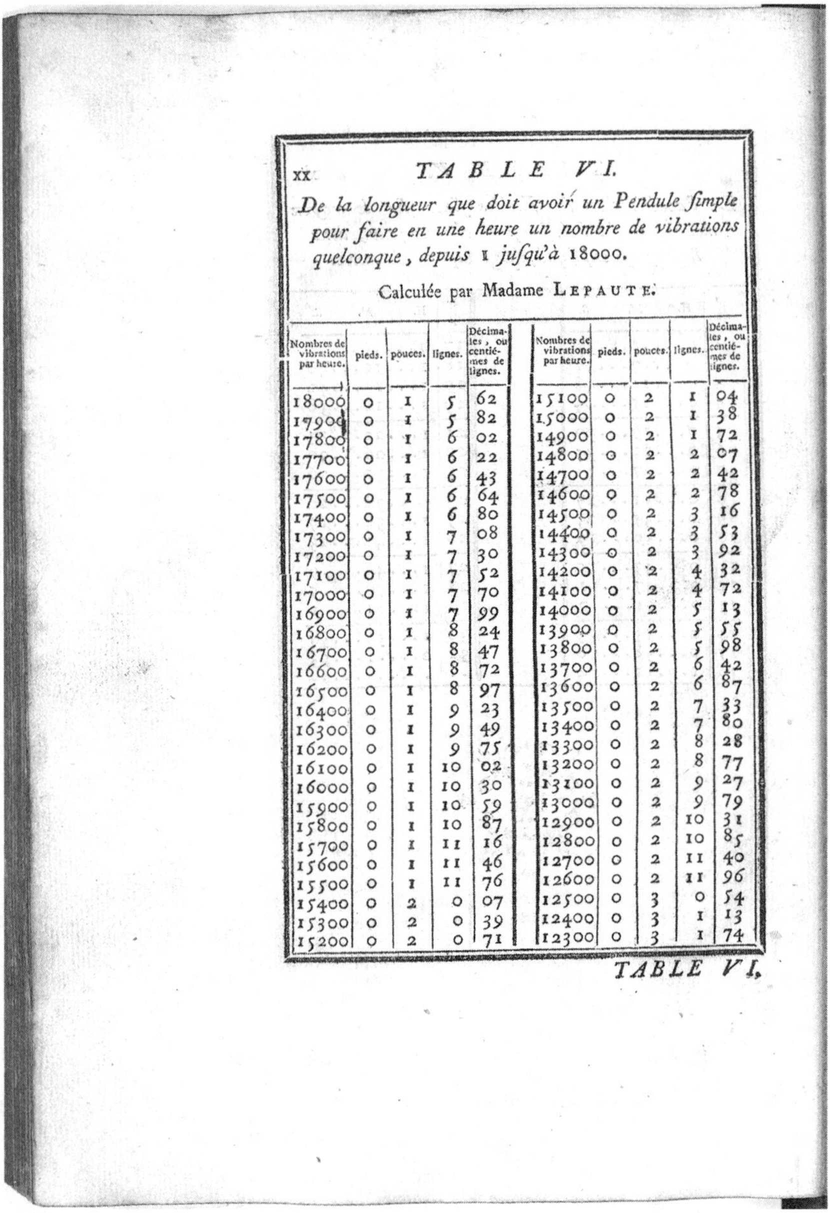

TABLE VI.

De la longueur que doit avoir un Pendule simple pour faire en une heure un nombre de vibrations quelconque, depuis 1 jusqu'à 18000.

Calculée par Madame LEPAUTE.

Nombres de vibrations par heure.	pieds.	pouces.	lignes.	Décimales, ou centièmes de lignes.	Nombres de vibrations par heure.	pieds.	pouces.	lignes.	Décimales, ou centièmes de lignes.
18000	0	1	5	62	15100	0	2	1	04
17900	0	1	5	82	15000	0	2	1	38
17800	0	1	6	02	14900	0	2	1	72
17700	0	1	6	22	14800	0	2	2	07
17600	0	1	6	43	14700	0	2	2	42
17500	0	1	6	64	14600	0	2	2	78
17400	0	1	6	80	14500	0	2	3	16
17300	0	1	7	08	14400	0	2	3	53
17200	0	1	7	30	14300	0	2	3	92
17100	0	1	7	52	14200	0	2	4	32
17000	0	1	7	70	14100	0	2	4	72
16900	0	1	7	99	14000	0	2	5	13
16800	0	1	8	24	13900	0	2	5	55
16700	0	1	8	47	13800	0	2	5	98
16600	0	1	8	72	13700	0	2	6	42
16500	0	1	8	97	13600	0	2	6	87
16400	0	1	9	23	13500	0	2	7	33
16300	0	1	9	49	13400	0	2	7	80
16200	0	1	9	75	13300	0	2	8	28
16100	0	1	10	02	13200	0	2	8	77
16000	0	1	10	30	13100	0	2	9	27
15900	0	1	10	59	13000	0	2	9	79
15800	0	1	10	87	12900	0	2	10	31
15700	0	1	11	16	12800	0	2	10	85
15600	0	1	11	46	12700	0	2	11	40
15500	0	1	11	76	12600	0	2	11	96
15400	0	2	0	07	12500	0	3	0	54
15300	0	2	0	39	12400	0	3	1	13
15200	0	2	0	71	12300	0	3	1	74

TABLE VI.

Fig. 5. Mme Lepaute's table of pendulum lengths from the 1755 *Traité d'horlogerie*. Only her husband's name appears as the author of the book, but Lalande revealed that she wrote much of it. [Bibliothèque nationale de France]

But, suddenly, Mme Lepaute had the chance to spread her wings and do computations that had nothing to do with her husband or with clocks. And just in the nick of time, for she and Lalande now began calculating in their race with Halley's Comet, which was speeding toward the sun, soon to be visible to the naked eye. But how soon exactly? The newspapers wrote of the anticipated event, teasing that one day it would abruptly appear in the sky, and encouraging all the "curieux" to keep their eyes peeled for the surprise. Lalande himself penned many of the articles stirring up excitement.[40]

Applying Newton's law of universal gravitation to comets, Edmund Halley had been the first to say that despite their extremely elliptical orbits these unusual heavenly bodies would be regular if infrequent visitors, and he predicted that the one seen in 1682 would return in 1758. Many had forgotten about this forecast, but Lalande and Lepaute remembered and realized that there was a problem with Halley's date for the homecoming of his comet—it now bore his name—because the perturbations on its orbit caused by Jupiter and Saturn, the two farthest and biggest of the then-known planets, would delay its appearance, although they had no idea by how much. They needed to involve Clairaut, for in connection with his earlier study of the moon he had done original work on the three-body problem throughout the 1740s, work which could now be brilliantly applied to figuring out the comet's path and speed. He had contributed a five-page article to the Lepautes' *Traité d'horlogerie* on oscillations in spherical lenses and knew the authors well.

Clairaut of course also remembered Halley's predictions, but it is not clear that he would have mobilized to undertake this task. It was likely, as he soberly put it, "that a body which travels into regions so remote and is invisible for such long periods, might be subject to totally unknown forces, such as the action of other comets, or even of some planet too far distant from the sun ever to be perceived."[41] Clairaut had demonstrated expertise with sophisticated calculations a decade earlier, but times had changed. There were many references to his having to be roused from an intellectual slumber to participate with Lalande and Lepaute, for though prodigiously gifted he spent most of his time these days enjoying life rather than pushing himself and no longer believed in working hard. He was already receiving a handsome pension of 10,000 livres from the king for having accompanied Maupertuis on his Lapland voyage in the 1730s to determine the earth's shape and verify Newton's theory that it must be flattened at the poles. The famous mathematician was sitting pretty. But he was needed for this job, which if successfully done would bring him unrivaled glory, and it was on that basis that Lalande approached him. Initially unmotivated, Clairaut eventually succumbed to the younger man's entreaty.

Starting in 1757, then, Lepaute, Lalande, and the newly committed Clairaut huddled together, working around the clock for half a year, energized by the thrill of using Newton to foresee the astronomical future and of stimulating people around the world with their chance to behold this dramatic phenomenon in the sky. Their calculations showed that the comet would come much later than Halley's prediction, but only as the computations progressed could they see the great length of the delay. Lalande later reminisced about the experience:

> Mme Lepaute helped us so much that we would not have dared to undertake this enormous work without her, where we had to calculate for every successive degree, and for 150 years, roughly two revolutions, the distances and forces of each of the two planets in relation to the comet. I gave her credit on this score in my *Théorie des comètes*, p. 110. Clairaut also cited her in his book on the comet where he profited from her immense work. But he suppressed this article to please a woman jealous of Mme Lepaute's merit and who had pretentions without the least bit of knowledge. She managed to get this injustice committed by a smart but weak savant whom she had subjugated. We know it is not rare to see ordinary women deprecate those with knowledge, accusing them of pedantry and contesting their talent to take revenge on their superiority; [those who are learned] are so few that the others have almost succeeded in getting them to hide what they know. Clairaut wrote to me "Mme Lepaute's ardor is astonishing." In another letter he called her "la savante calculatrice." It is hard to understand the courage demanded by this enterprise, if one did not know that for six months we calculated straight from morning to night, often during meals and that as a result of this nonstop work I contracted a malady that changed my constitution for the rest of my life. But it was important that the result be given before the comet's arrival, so that nobody could doubt the accord between the observation and the calculations that served as a foundation for the prediction. That is what actually happened. The comet was delayed more than 600 days by the action of Jupiter and

Saturn, and this slow-down was announced at the public opening of the Académie des Sciences in November 1758.[42]

And indeed, Clairaut stood up at the Académie on 14 November to make the confident prediction that the comet would reach perihelion, the position closest to the sun, on 13 April 1759 plus or minus a month. D'Alembert had earlier given up on such "frightful" reckoning, explaining on 20 November to the Italian mathematician and astronomer Paolo Frisi after hearing Clairaut's pronouncement, that "the dislike I have for the great calculations kept me from searching the disturbances that Jupiter and Saturn ought to or may cause on this Comet; but I have no doubt that they influence it."[43] Then as spring approached everyone waited and watched the sky, on the streets, at the royal court, in the salons, at the Académie itself. The anticipation made for quite a stir. And the comet did pass closest to the sun on 13 March, just within the margin of error, a "beautiful confirmation" of Newtonian theory, although it was much more dazzling with its bright tail shining after perihelion when it was headed back out and away. Some, who wanted to see "the Philosophers plunged into uncertainty and trouble," hoped the comet would not come and that way they could prove that the Newtonians and Newton himself were wrong.[44] D'Alembert jealously criticized the prediction that allowed for a month of leeway, but Clairaut retorted that because d'Alembert had not wanted to tackle the mathematical complexities of a comet that appeared only every seventy-five to seventy-six years, he should "leave in peace those who had the courage to undertake them." Clairaut thought that the most important thing was to have finally silenced the anti-Newtonians.[45] In fact, he decided it was the propitious moment to publish Mme Du Châtelet's complete translation of and brilliant commentary on Newton's *Principia,* a manuscript that had languished in his care since her death ten years earlier, realizing that readers would now be more interested in the great man's science than even before. And Lalande rejoiced at the Académie: "The universe sees this year the most satisfying phenomenon that astronomy has ever offered us, an event unique to this day, which changes our doubts into certainties and our hypotheses into demonstrations."[46]

Lepaute, one of the rare women on the continent to know advanced mathematics, realized that her role in this colossal Newtonian enterprise and the successful prediction had been indispensable, her work "immense" and "courageous" to use Lalande's terms, the game-changer that made possible their sprint to beat the comet. He at least had immediately acknowledged her essential contribution, admitting in print in 1759 that because the gravitational force of Jupiter meant all of Halley's tables had to be recalculated, "the immensity of details would have terrified me if Madame Lepaute, experienced since long ago with great success in astronomical calculations, had not shared all our work."[47] Seeing this published acknowledgment by Lalande of her pivotal role, Lepaute felt secure enough at that moment to ignore the usurpation of due credit by Clairaut and his mistress. Maintaining her dignity and rising above the insult, she even socialized with the couple, joining them and Lalande in hosting the Swedish astronomer Bengt Ferner when he visited Paris in April 1761, taking him to an evening gathering at the home of the learned apothecary Antoine Baumé on the rue Saint Denis.[48] That she escorted distinguished foreigners in this way—Ruggero Boscovich, founder of the Milan observatory, was another—also demonstrates Lepaute's autonomy, the scholarly life she led, independently of her husband, in scientific circles.

The comet prediction was an extraordinary French accomplishment, and it boosted international enthusiasm to advance the progress of science. France was at this moment fighting and losing what came to be called the Seven Years' War. It had started in 1756 and was already disastrous for the empire; by the time the Treaty of Paris was signed in 1763 the country had lost most of its overseas colonies—Quebec, Montreal, part of Louisiana, Guadeloupe, bases in West Africa, and Pondicherry in India. What better way to lift morale than to get credit around the globe for having predicted a spectacular cosmic event? The hope was that science would allow France to recapture some luster despite ignominious military and diplomatic defeats. As Lalande extoled the triumph of astronomy and the glory of the human mind, hairdos, boats, streets—in Paris a rue de la Comète appeared—plays, poems, and fashions celebrated the achievement in a wave of *cometomania*.[49]

Lepaute and Lalande next turned their attention to another rare astronomical phenomenon, the upcoming transit of Venus across the sun expected 6 June 1761, the first of two paired happenings eight years apart which would make possible nothing less than the measurement of the distance from the earth to the sun, and with that knowledge the distance to the other planets in the solar system. Many astronomers had abandoned the outdated Observatoire royale and watched both transits from other vantage points.[50] Lalande himself watched from the Luxembourg observatory where he had first met Lepaute ten years earlier.[51]

Recording this 1761 transit was an unprecedented crusade, an international mobilization of astronomers all with the same goal in mind. Over one hundred observers, from England, Germany, Denmark, the American colonies, Holland, Italy, Portugal, Russia, and Sweden traveled to more than fifty carefully chosen locations, Pondicherry, Siberia, Vienna, and the Isle Rodrigue off Madagascar among them. While each star-gazer was eager to be singled out for a discovery—the inevitable thirst to be first—they did coordinate their efforts and share their results.

Pooling information was the whole point of this venture, an innovative and modern idea. As it turned out, because nothing like this had ever been tried before, because the war was ongoing, because of various dramatic mishaps, and finally because the weather was bad for viewing in most places, the results of the 1761 transit were not sufficiently accurate or organized, the measurements not numerous or rigorous enough. The turbulence of Venus's own atmosphere made visibility fuzzy even where the local weather was clear. And although the "caress" lasted five hours and thirty-seven minutes, estimates of the distance from earth to sun varied widely and wildly between seventy-seven million and ninety-six million miles. This first transit, therefore, was something of a rehearsal, a trial run for the second one, expected on 3–4 June 1769, when everyone would be better prepared. Lalande, a pioneer of this kind of cooperation, was tapped to orchestrate the next transit and make for the Académie a mappemonde showing where all observers should be strategically stationed in both hemispheres.[52]

In the meantime, much could be learned even from the imperfect results of 1761. Lepaute calculated all observations of the passage that were made by sky-watchers in Béziers, work which led to her election as a member of the academy of sciences in that city. Lalande was friends with Jean-Jacques Dortous de Mairan, eighty-three years old and one of the three founders of the Académie de Béziers, an institution crucial for the cultivation of science throughout the south of France. He sent Lepaute's "learned calculations" and some impressive papers she had written on the transit—these have since been lost—to Jean Bouillet, the academy's secretary. She was soon admitted as a member of the academy, thanked them, and then received the following letter on 22 November 1761 from one of the officers, La Rouviere-d'Eyssautier:

> The modesty that reigns in the thank you note that you addressed to us
> gives new luster to your merit and confirms more and more the high

regard we have for it. The Académie de Béziers, by associating you to its work, has found a way to ally the graces with the muses, it knew how to procure by one worthy choice the satisfaction of having in one of its members the most amiable qualities united to the abstract sciences of geometry and astronomical calculation. The *beau sexe*, madame, that sex always made to please, and of which you are, by your talents, the ornament and the glory, has often produced students of Apollo but rarely disciples of Euclid. The Scudéry, the Deshoulières, the Chéron, and several other modern Sapphos only blazed a path to academic positions across the flowered fields of the beaux-arts, and to arrive at the same goal you alone wanted to follow, from a most tender age, the spiny path of mathematics. What an advantage for us, I dare say, and what a stimulus for several of our colleagues who, full of the same zeal that animates you, devote their attention and their vigils to the progress of astronomy! Surprised to recognize in you, Madame, a master rather than a disciple, those from whom you ask enlightenment will someday need help from your erudition, and will place you in their school next to the Agnesis and the Duchâtelets [*sic*]. As for me, I see this epoch as one of the most flattering of my life, because in my capacity as director, interpreter of the feelings of this company, I find at the same time the occasion to make you understand my own, which will always be inseparable from the respectful consideration with which I have the honor to be, Madame, your very humble and very obedient servant.[53]

This missive, at once patronizing and appreciative, probably both irked and pleased Lepaute, filled as it was with fawning flourishes while at the same time making clear the uniqueness of her scientific accomplishments. Lepaute wanted to guarantee, however, that the academy remain focused on the excellence of her election, and not cheapen it by admitting someone with ambitions but lacking the bona fide qualifications. Two months later, she complained about Clairaut's Mlle Gouilly trying to gain admittance into the same academy, as if her own admission might have somehow blazed an easy trail for others less worthy. Lalande agreed that "the girl" was undeserving, saying she had "pretensions without the remotest kind of knowledge." Gouilly never did become an academician, but the fierce envy of this woman whom Clairaut referred to as his "pupil," "little companion," and "logarithmière" had already succeeded in depriving Lepaute of comet acclaim in 1759.[54] Unlike Ferrand, who was willing and even glad to inspire Condillac from behind the scene, and Mme d'Arconville who, as we will see, published prolifically but always behind the veil of anonymity, Lepaute expected credit for her work and did not see herself as an "invisible technician."[55] Astronomy meant the world to her, she was good at it, and she wanted others to know. The man who had robbed her of due recognition should not now succeed in lobbying for someone else.

And Mme Lepaute was not alone in disapproving of Clairaut. There was by the early 1760s a general consensus that he had become, scientifically, a shadow of his former self and that his mistress was leading him around by the nose. The astronomer Jean-Baptiste Chappe d'Auteroche said he had too many women, too much wine, and did too little work. Cassini said Clairaut was very smart and had been a great mathematician but was now terribly debauched. Le Gentil said that he had been a genius at twenty-five—this during the period he was involved with Mlle Ferrand and Gabriel Cramer—but had done nothing since.[56] When Clairaut died suddenly of a stroke in 1765, Diderot described his colorful friend as almost the equal of Euler, Bernoulli, and d'Alembert but less "penetrating," a gourmand despite digestive problems and an "inflammable" heart, who loved the female sex madly and shared his abundant money generously with his friends, lavishing his fortune chiefly on his most recent mistress, "a little *gouvernante* extremely pretty who was in charge of his household, to whom he taught enough geometry to help him in his computations, and who is left at his death in widowhood. A sudden and violent malady took him in just four days, he had time for no arrangements in favor of this companion of his labors and his pleasures; her fate preoccupies and interests all the *gens de lettres* at this moment."[57] Nor did Mlle Gouilly disappear from view, for she was a true femme fatale, said to be so beautiful that the king himself would have taken off his hat to her had they crossed paths. She was wooed by many suitors, several of whom literally blew their brains out (*brûler la cervelle*) when she cruelly spurned them, and she would go on to have quite a career as mistress to a number of academicians and savants.[58] She and the man she eventually married, one Leblanc, were sponsored by the geometer and astronomer Dionis du Séjour, who left them a house and money in his testament.[59] It is unlikely that Gouilly was quite the nitwit Lepaute wanted Béziers to imagine, although at the time Lalande did confirm her dim opinion of Clairaut's woman.

By now another project was in the works. Since 1759 Lepaute had been working with Lalande on the *Connaissance des temps,* an almanac published annually by the Académie des Sciences and useful for astronomers, surveyors, ship captains, and navigators. The two would do this job together until 1774. Once again Lalande claimed that he would never have agreed to become editor of this periodical, which involved tremendous ongoing computations, had it not been for Lepaute's help. She did, he said, the work of several people, relieving him of a painful responsibility.[60] This almanac had always reliably provided the position of the planets, their passage at the meridian, their rise and set, but now under its new team of directors it was transformed to be more appealing to general readers. Lepaute contributed "Observations," and on 10 March 1759 Lalande told Bonnet that she was doing most of the work for the forthcoming number. Then on 16 July he elaborated, "I have the pleasure of living with a woman of mind who did the work for a year almost alone, and with whose help I am delivered of the weight of this burden." When Bonnet asked the identity of this "new Du Châtelet" Lalande replied: "The muse who willingly does for me the *Connaissance des temps,* as for the one being done now I play only the smallest part, is Mme Lepaute."[61] So she produced by herself the volume dated 1760, and it is refreshingly different. Even though many of the articles were written by others she seems to have chosen, gathered, and assembled the team. The work was, as Lalande admitted privately, essentially all hers.

When Lalande said he was "living" with Mme Lepaute he meant exactly that, as he cohabited with the couple most of the time. When they were installed in the Luxembourg he was living on and off at the nearby Royal Observatory, an easy five-to-ten-minute walk.[62] Jean-André gave his address as the Luxembourg in the 1755 *Traité,* but shortly after that they moved to Place de la Croix Rouge (now rue de Sèvres) where they had more room, and Lalande joined them there from 1758 to 1764. Next the three of them moved together to the rue Saint Honoré where the Lepaute brothers also had their first clock boutique, near the Croix du Trahoir. They stayed there from 1765 to 1772, when they all moved again to the Palais Royal.[63] Not until 1777 did Lalande leave the Lepautes to assume his post as professor at the Collège Royal, where quarters were provided for him. By then Jean-André had retired from clockmaking and begun his descent into delirium.

Lepaute and Lalande changed the *Connaissance* fundamentally. Both before and after their years in charge it was a quite specialized and indigestible compilation of charts, tables, and navigational information for each day of the coming year, regarding twilight, sunset, moon rise, the positions of constellations, when different planets would pass by the meridian—without a doubt useful for astronomers and sailors but of little or no importance to anyone else.[64] The pair of new editors took it as a challenge to make this almanac an enjoyable, edifying, and sought-after yearly volume, with the requisite rigorous information, of course, but also full of articles and observations of more general interest. There is a world of difference already in the very first volume they did, that of 1760, which included advice on how to safely watch an eclipse of the sun and a discussion of the historical contributions of various thinkers to the field of astronomy, for example, a detailed story of the satellites of Jupiter starting with Galileo, who first saw them through a telescope that he fashioned. Lepaute and Lalande involved the reader and made astronomy feel like a living science, dynamic, full of controversy, suspense, and joy. There was, not surprisingly, a victorious article on the calculations for the much-heralded return of Halley's Comet in 1759, a "prediction which has just been realized as exactly as the authors hoped it would."[65] There was now a novel section for amateur astronomers, "observations to be made each month," listing celestial phenomena that could easily be seen by any reader who might wish to study the night sky. The 1763 volume applauded the clocks of Lepaute as an indispensable help to astronomers and reviewed what was learned from the cooperative effort watching the transit of Venus in 1761. The huge number of people who participated in this project demonstrated "the progress that the taste for science makes in the public every day."[66]

In 1763, as the Seven Years' War drew to a close, Lalande went to England for several months; he had not been able to travel to France's declared enemy during the long conflict. He of course paid his respects at Newton's mausoleum in Westminster Abbey, visited Greenwich, saw astronomers, clockmakers, and various instrument and magnet makers. The Royal Society was giving John Harrison a hard time for his work determining longitude at sea, telling him he needed to make three more watches which would then need to be tested on a voyage to Jamaica. Lalande lamented this and believed that Harrison was not sufficiently appreciated.[67] He visited John Wilkes in prison and conversed with him about his North Briton attacks on the king, saw a glass harmonica, visited Portuguese and German synagogues in London, looked through the royal microscope, met the king himself and gave him the *Connaissance des temps* volume for that year,

1763, which contained Mme Lepaute's table of parallactic angles for eclipses. And he wrote of all these adventures to her, to his mother, and to a few men of science and clockmakers. He did not write to her husband.[68]

Soon after this trip Lalande was taken seriously ill, feeling fragile enough to write his testament in 1764, and then in 1766 adding a codicil with six new items, the first two concerning his mother and the third concerning Mme Lepaute. He left to her all of his Paris furniture, his watch, his books of literature and those on the history of astronomy, and his *arrérages*, debts owed to him by the Académie des Sciences, as thanks for the work she contributed to their present joint project, the *Connaissance*. "I beg Mme Lepaute to believe that the love of science and the public good alone could prevent me from giving her a more extensive sign of my just and respectful gratitude," this an explanation for his leaving his money, scientific instruments, and the rest of his library to the Académie where they would be of more widespread, general use. Once again we see that Lepaute continued to be central in his life.[69]

And she worked autonomously too, independently of Lalande. She was now busy calculating the eclipse of the sun predicted for 1764, which she began studying two years in advance. The plan was to make a map of Europe with the moving shadow of the eclipse on it. It would be not a total eclipse that darkens the sky, turning daylight to night, but instead an annular eclipse that would leave a peripheral crown, a halo of light, a ring of fire, and it promised to be very beautiful. Like comets, eclipses had been experienced as calamitous portents in earlier times, unpredictable signs from the mysterious beyond that had, could, and would again cause plague, famine, even signal the end of the world. But such superstitions were mostly things of the past; eclipses were demystified in the climate of the Enlightenment and treated now as understandable natural phenomena. Halley can be said to have "invented" the eclipse map by superimposing on a normal map a shaded band to mark the path of the eclipse, with the moon's shadow (umbra) indicated by a darker ellipse. Lepaute wished to provide a predictive, not a retrospective eclipse map, and following Halley's model she created an image of beauty and usefulness, producing a rather technical explanatory booklet to go along with it. The French were known to be great scientific cartographers in this period, so the bar was high. First she did calculations for the whole of Europe and published a chart showing the track of the solar eclipse for every quarter of an hour. It depicted the path of centrality as a band of dark gray. Then there were thin black lines showing the decreasing amount that the sun would be obscured by the moon from various vantage points. The map was printed in brown sepia, the eclipse information in black.[70] Lepaute proudly presented the image, along with her eight-page pamphlet, to King Louis XV on 12 August 1762.[71] Whether she had other meetings with the monarch we do not know. Her husband, as one of His Majesty's clockmakers, had certainly presented clocks at court but this was *her* work, *her* royal audience. It must have been thrilling (figure 6).

Lepaute's *Explication de la carte* was very detailed and not really for popular consumption. First published in 1762, it was entirely her work. A longer, more detailed version came out in 1764 shortly before the eclipse itself, when the newspapers were rousing readers about the upcoming spectacle, and this second edition shows the hand of Lalande, who naturally added plugs for some recent writings of his own. Both versions were densely packed with numbers and measurements but also contained passages that had a broader explanatory reach, and the later one, quoted here, included mention of an additional chart by Lepaute made specifically for viewing the event from Paris: "Total eclipses of the sun are very rare phenomena, but annular eclipses are even more so." A total eclipse had been seen from the capital in 1724, and there was no record of an annular one visible from there. "We call annular those where the moon passes directly in front of the disc of the sun, but has a diameter too small to hide it entirely." So there was to be a crown of light around a dark disc, like a thin ring, a "great new spectacle and a precious new observation." Earlier annular eclipses, like one in Scotland in 1748, led to some incorrect conclusions about the diameter of the moon. "But now we have the discovery of heliometers more accurate than any instrument known in 1748 to guide us. Now we can verify to the precision of one second, and without leaving France, the true diameter of the moon during an eclipse." To prepare astronomers and the curious for observing this big eclipse, "we traced on a map of Europe the route of the shadow and of the penumbra of the moon, based on an exact computation done by the method of astronomical projections, employing all the elements given to us by the state of perfection of this science including the effect that the flattening of the earth produces in the parallaxes of the moon." That the earth was not a perfect sphere had been previously hypothesized by Newton and proven by the two expeditions of the 1730s, to Lapland and to Peru.[72]

"The astronomical tables," Lepaute went on, "teach us that on the first of April 1764 at 10 hours 32 minutes and 7 seconds of the morning, the moon will

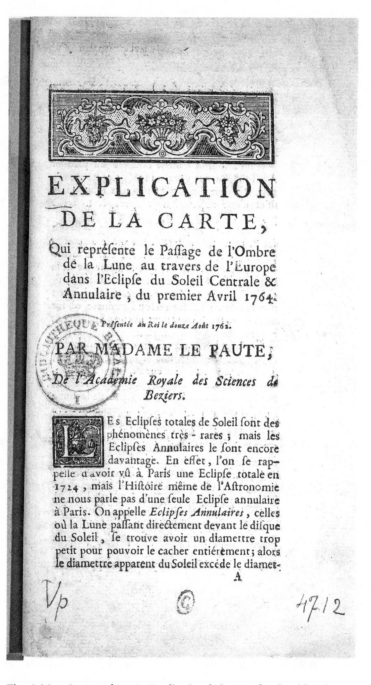

EXPLICATION
DE LA CARTE,

Qui repréſenté lé Paſſage de l'Ombre
dé la Lune au travers de l'Europé
dans l'Eclipſe du Soleil Centrale &
Annulaire, du premier Avril 1764.

PAR MADAME LE PAUTE,

*De l'Académie Royale des Sciences de
Beziers.*

LES Eclipſes totales de Soleil ſont des
phénomènes très - rares ; mais les
Eclipſes Annulaires le ſont encore
davantage. En effet, l'on ſe rap-
pelle d'avoir vû à Paris une Eclipſe totale en
1724 , mais l'Hiſtoire même de l'Aſtronomie
ne nous parle pas d'une ſeule Eclipſe annulaire
à Paris. On appelle *Eclipſes Annulaires*, celles
où la Lune paſſant directement devant le diſque
du Soleil, ſe trouve avoir un diametre trop
petit pour pouvoir le cacher entiérement; alors
le diamettre apparent du Soleil excéde le diamet-

A

be in conjunction with the sun," and she provided many technical measurements. "It is on these bases that we did the calculations necessary to trace the apparent route of the Eclipse on the map of Europe. The principal line passes at the tip of Spain a little above Cap Saint Vincent, crosses part of Portugal and Spain to Saint Ander. This same line next passes over Normandy, from Nantes to the surroundings of Rouen; after crossing the Sea of Germany, it passes into Norway and across Swedish Lapland. Then it leaves the continent around Cap Wardhus, the northern-most part of the Mer Glaciale. All the countries situated on this line will see the central and annular Eclipse." Lepaute then explained the two other lines running roughly parallel to that one, "one northwest and the other southeast. . . . All areas under this first line will see the lower border of the sun razed [razé] by that of the moon at the middle of the Eclipse—Brago, Lago, Quimper, Cherbourg, London, Cambridge. The [second] line . . . passes at Palos in Andalusia, celebrated city from which Christopher Columbus embarked for the discovery of the new world; to Bordeaux, Poitiers, Blois, Paris, Brussels, Groningen and Tornea in Lapland. . . . Between these two lines are all the countries of Europe where the moon will appear entirely over the sun. The closer a country is to the middle line, the more it will see the center of the moon coinciding with that of the sun, at the moment of the middle of the Eclipse, and the luminous ring approaching exact roundness that we have shown in a particular figure of the central Eclipse to the right of the image." There followed discussions of how the eclipse would look in Tunis, Naples, Esclavonie (part of Croatia), Poland, and how "it gets lost in the deserts of Moscow."[73]

Turning next to her other chart, for Paris only, Lepaute wrote that in addition to the eclipse there would be viewing opportunities not normally possible in daytime: "Because of the ring we will not be in complete darkness but will still be able to see Venus and Jupiter to the left of the sun and Mercury to the right, maybe even some stars. That is why I published a figure of the state of the sky as it will be in the middle of the eclipse, with the detail of the phases for Paris, 12 principal phases from quarter hour to quarter hour. In it we also see the sun surrounded by the principal stars, the head of Andromache, the head of Aries the ram and the tail of Pegasus above the sun, the tail of the Whale under it. These are the stars most noticeable in this part of the sky." The little book ended abruptly, businesslike, no small talk, a strange mix of information for the layperson and details for the savant, stylistically inconsistent but revealing of how much Lepaute knew about our place in the universe and what she thought the reading public was capable of understanding. She appears to have had quite a high estimation of that public.

Her special eclipse chart for Paris was approved for printing and distribution by the police chief Sartine on 22 March 1764, about a week before the event itself. This was her opportunity to become more widely known and flaunt her academic credentials. Titled "Figures of 12 principal phases of the grand Eclipse of the Sun to be observed 1 April 1764, calculated for Paris by Madame Lepaute of the Royal Academy of Sciences of Béziers," this easy-to-read diagram circulated in large numbers, Lalande reporting that several thousands were going around.[74] A single sheet about ten by seventeen inches, hot off the press, it depicted in black and white the moon moving over the sun every quarter of an hour starting at 9:11 a.m., the black shape coming from the right and appearing larger and more round as it moved left until 10:39 when it was almost central, not quite covering the larger white circle of the sun behind it, and then passing away up toward the left of the solar orb.[75] Two insets on the lower left and right explained, as had her booklet, what planets and constellations might be visible. The piece was published and sold "Chez Lattré rue Saint Jacques près la fontaine Saint Severin."

At the bottom center of the sheet, between these two insets, Lepaute's other eclipse map was advertised. "One can find at the same publisher a map on which we have traced the shadow that the moon will cast on Europe, calculated by the same woman [par la même Dame]" (figure 7).[76]

By including this notice on her chart for Paris, Lepaute was reminding her public in no uncertain terms that she was a woman. But beyond that, her other chart, the one for all of Europe, was deliberately made without the participation of any men at all. Lepaute had proudly created and advertised a project that was entirely the work of women, as the decorative cartouche revealed: herself, Mme Lattré, a copperplate engraver, and Elisabeth Claire Tardieu from a map-publishing family who did the ornamentation.[77]

When the Europe chart was first announced it had not gone unnoticed that it was an all-female enterprise. The *Journal de Trévoux* of June 1762 devoted five pages to advertising and explaining this uniquely practical eclipse map made by three women, pointing out that it was "new proof that the knowledge of science and the fine arts is not an exclusive privilege reserved for only one gender; women also will distinguish themselves in this noble career [*cette noble car-rière*], when they will desire to busy themselves usefully or, rather, when we see the abuse of giving them merely a frivolous education."[78]

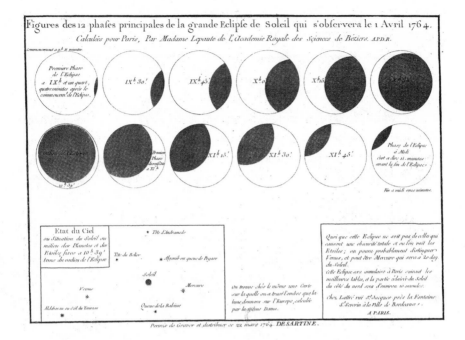

Fig. 7. Mme Lepaute's image of the 1764 eclipse as it was to be seen from Paris. [Bibliothèque nationale de France]

But in 1764 when the eclipse was imminent and Lepaute's map in great demand, others were less enamored of her success and sought to compete with her venture. Lalande related an incident in which a rival map and globe maker named Charles Desnos created a copy of Lepaute's chart claiming it to be by a Polish woman, one Dame Kraziowna, and which "seemed to have been meant to throw a varnish of ridicule on Mme Lepaute."[79] The news even spread to America where the *New York Mercury* reported: "Letters from Paris say that M. Desnos has been condemned by an arret of *Parlement* as a Plagiary. He had assumed the name of Krasiowna, calling himself a native of Poland, and had attributed to himself the description which Mademoiselle LePaute, a French lady, had given of the place in which the eclipse of the 1st April next will be visible."[80] Desnos was duly convicted of plagiarism and forced to give up the copper plate,[81] but not before advertising his copy for half Lepaute's price, seemingly bent on undermining her.[82] In addition, the *Gazette de France* of 19 March mistakenly announced that the upcoming eclipse would be total, which no astronomer believed or had ever said, but because much of the reading public relied on such periodicals they expected complete darkness and therefore decided Lepaute's chart must be wrong. There were also some who criticized her calculations, but others vociferously defended the "*académicienne de Béziers*."[83] While Lepaute must have found such disputes unpleasant, she became better known as a result of the publicity and was quickly vindicated.

Yet what a disappointment lay ahead. The eclipse did occur as predicted on 1 April 1764, but the miserable rainy weather made it impossible to see from Paris, the much-anticipated phenomenon "hidden from the eyes of astronomers." Quite a few other cities, however, enjoyed the heavenly spectacle exactly as Lepaute had advertised.[84]

Meanwhile the joint work with Lalande on the yearly *Connaissances* continued. The 1764 issue contained a history of eclipses since Kepler and naturally advertised Mme Lepaute's solar eclipse map. "It is the result of a very long succession of calculations done by the method of projections, to determine by longitudes and latitudes all the parts of France where the eclipse will be seen in different aspects."[85] The volume for the next year, 1765, praised the Englishman Harrison's work on determining longitude, his special clock that stayed accurate and stable even in a tossing ship, alerting the public to the "beautiful results of this experiment and the present state of the question that has enjoyed such celebrity."[86] That year Lalande made a trip to Italy during which he was very pleased to meet, in Naples, the famed female Newtonian Mlle Ardinghelli, an informal foreign correspondent of the Paris Académie whom he knew by reputation and came to admire still more after their encounter. Like Sauveur Morand, who encouraged female anatomists (see chapter 4), Lalande genuinely appreciated the women who elucidated the physical workings of nature.[87]

By the 1766 volume of the *Connaissance* the editors were already preparing readers for the second transit of Venus now only three years away, with suggestions on where to go to watch it. Lalande explained that the 1761 transit had been useful but not seen well due to weather and war, which had made certain key locations off limits. This next one "has for us a completely different importance . . . the last that the present generation of humanity will have the chance to see," as there would not be another transit until 1847. The upcoming transit was quite simply "the most important phenomenon that astronomy awaits at present."[88] The *Journal des sçavans* raved about this 1766 volume of the *Connaissance* with its tables of movements and eclipses of the satellites of Jupiter and its new theory of Mercury, a planet little known because it was hard to observe. But

Lalande had "battled all the obstacles; harsh seasons, inconvenient hours; in-constant weather" and "finally made more than 20 observations of Mercury in chosen positions" so that now the dimensions of its orbit were known. There were also reviews of good books about astronomy. Mme Lepaute's eclipse map, still advertised for sale even though that event was long past, now had value as a document of enduring historical interest.[89]

The 1769 passage of Venus, although its recording too was fraught with difficulties, yielded more accurate observations and so had a considerably better outcome than the earlier transit. Lalande took charge of organizing the international cooperation. Once again the government had to be convinced to fund voyages by a number of observers to predetermined points around the world, but this time it took less persuading because the project, talked about in high places ever since the first transit, had gained significant momentum. The uniqueness of these two planetary developments occurring so close together, and the unsatisfactory results of the first, cast the transit of 1769 as a second chance and one not to be interfered with by any unpatriotic or unenlightened equivocating.

Even before the 1761 transit Lalande was framing these events, and the extreme good fortune of his day in being able to witness them, in a dramatic exhortation to take full advantage of a treasured opportunity. "The chance that this celebrated phenomenon gives is one of those precious moments whose advantage, if we let it escape us, will never be compensated for, not by the efforts of genius, not by the most constant work, not by the magnificence of the greatest kings; a moment that the past century envied us, and which will be in the future, if I dare say it, a blight on the memory of those who neglected it."[90] This second transit in 1769, using more precise clocks—some made by Jean-André Lepaute—and more coordinated observations, narrowed down the distance between earth and sun to between 92 and 96 million miles, close to what is understood today to be 92,960,000 miles.

Mme Lepaute seems not to have been involved in calculating this second transit as she had so actively for the first, because at just this time she was engaged in a very different kind of activity, having invited to Paris a boy, a nephew of Jean-André's living in Lorraine, to study astronomy with her and Lalande, and thus becoming a sort of surrogate mother to him. The point here is not that this was a chore in any way, for Lalande emphasized her generosity and strong sense of family. Childless herself, this was a hugely important move in light of the domestic pressures she was under and given the social expectations of her day. She had produced no progeny to whom the knowledge, skills, and techniques of the family could be transmitted, and had she not brought young Joseph Lepaute to Paris in 1768 and given him the opportunity to inherit this artisanal and intellectual capital it could have been lost. She had, about twenty years earlier, been cast in something of a maternal role when her husband had invited his own younger brother, Jean-Baptiste, to help him in his clockmaking business, and since then she had helped raise and teach astronomy to one of this brother's sons, a very talented and precocious fellow whose parents, in the end, preferred that he enter the more lucrative legal profession. Lalande lamented that they were seduced by money and so did not let the boy "profit from his happy education and take a path much loftier for the mind and for glory."[91]

The arrival in February 1768 of the new sixteen-year-old nephew from Lorraine, who took the name Lepaute d'Agelet, then just Dagelet, absorbed Mme Lepaute and prevented her from working on the 1769 transit herself. But Lalande, aware more than anyone else of her intellectual strengths and willing at numerous junctures to list and laud them, would later say in praise: "Among the services [Mme Lepaute] rendered to our science we must cite principally the great care she took in 1768, to bring from Montmedi a nephew of her husband's then 15 [sic] years old, to devote himself uniquely to astronomy. This Lepaute d'Agelet, who was accepted by the Académie des Sciences and went to Australia in 1773, and then around the world with Lapérouse, proves in a very important way that Mme Lepaute was useful to astronomy."[92]

Well. We must ponder the full import of this comment, disappointing as it is on the face of it, and even below the surface of it. But it is a window into the eighteenth century, revealing a fundamental presumption that this was what a married woman was to do, that family must come before all else. We saw no such expectation in the case of the single Mlle Ferrand, nor will we in the stories of Mlle Basseporte or Mlle Biheron that follow. They were not wives. It is noteworthy too that while Mme Lepaute was never one to downplay her science, on the contrary standing up for herself as we saw to the Académie de Béziers and to the finance minister when requesting a pension commensurate with her contributions, she still always signed her letters "femme Lepaute."[93] This was primarily how she was supposed to see her role, making it all the more extraordinary that she prevailed in doing science as well. Lalande stressed the equilibrium she achieved between her work and her duties as a wife. "Her calculations never prevented her from taking care of the household affairs, the account ledgers were always next to her astronomical tables. She had taste and elegance, without it interfering in her studies. Foreigners who visited her because of her competence [*mérite*] added to the reputation of M. Lepaute, and so were helpful to them both."[94] Even in Lalande's eyes, her life needed to exhibit an exemplary, respectable balance.[95]

Thus young Dagelet was definitely one of her projects, and she and Lalande became his intellectual parents. Johann III Bernoulli, whose father had admired Mlle Ferrand decades earlier, reported on his trip to France in 1768–69 seeing the teenager already at work. Commenting that the Luxembourg observatory in the dome on the middle of a terrace above the entrance was dangerous, as that part of the once-beautiful palace was crumbling to ruins, Bernoulli was glad that Lalande had gotten a better setup at Collège Mazarin (Quatre Nations), today the Institut de France. There he had a room with a mobile rotating roof that was equipped with an astronomical pendule by M. Lepaute. An additional room farther away had more instruments and another Lepaute clock. Now, by the time Bernoulli's book about his travels was published in 1771, Lalande had moved again, to the Palais Royal where he had a small observatory right in his own home.[96] Some things seemed to be going well for Jean-André too, Bernoulli reporting that he was making numerous astronomical clocks and getting orders all the time from near and far. The Comte de Mercy, Austrian ambassador from Empress Maria Theresa, was having him make two clocks for Vienna, and it seemed that people were impressed by his inventive spirit.[97] Not everyone was a fan, however. At the same time another customer complained bitterly, calling his expensive Lepaute timepiece "so grossly neglected in the execution, so unequally finished, that it was not difficult for the meanest novice in the trade to point out the cause of the irregularity."[98]

Whether the family clockmaking business was going well or poorly, Lalande and young Dagelet carried on their work together. The talented disciple observed the 1769 transit of Venus, a vicarious participation for Mme Lepaute, and through continued diligence was selected for an expedition to the South Seas in 1773.[99] Soon after his return he became a professor at the École Militaire, and inspired by his aunt's earlier map he made one modeled closely on hers predicting the 24 June 1778 solar eclipse, which was "total over parts of Africa and America but partial across France and England."[100] Elected a member of the Académie in 1785, he was invited that same year to join the coastline-mapping expedition headed by a renowned naval officer, the comte de Lapérouse. He had not loved the earlier sea voyage and would have been glad to stay on dry land, but Lalande and Lepaute encouraged him to go. Lapérouse so appreciated Dagelet's enormous help on the journey that he named a newly discovered island after him in May 1787.[101] Tragically, the entire expedition perished off the island of Vanikoro in 1788. There is a story that Dagelet took an assistant with him on the trip, a favorite student from the École Militaire named Roux Darbaud, and that he chose him over another student, a certain Napoleon Bonaparte, who was also eager to go and wild about science but not as gifted in astronomy. Had Dagelet selected this other young man for the doomed voyage, history would have been profoundly altered![102] A full year after the ships were expected to reappear, fear turned to certainty that they had been lost forever, and Lalande would refer to Dagelet, so full of zeal and courage, as the most recent martyr to astronomy.[103] Mercifully Mme Lepaute died without knowing that her beloved nephew had met such a fate, but Lalande would always feel deeply guilty for having pushed him to sail with Lapérouse. All this, however, was down the road.

Not one to stay idle, Lepaute had assisted with a second edition of the *Traité d'horlogerie* that appeared in 1767. She frequented and influenced men of science, for example, the naturalist Philibert Commerson, who had recently arrived in Paris and whom she encouraged to join Bougainville's round-the-world voyage which sailed the following year (see chapter 3). She socialized at dinners with Buffon, Malesherbes, Turgot, and the Neopolitan economist and humorist Galiani, and of course with other astronomers. She continued working on the *Connaissance des temps* until 1774 when she switched to a similar kind of effort for the *Éphémérides des mouvemens célestes pour le méridien de Paris,* also under the auspices of the Académie des Sciences.

But along the way she had to navigate a new crisis caused by her closest colleague, which might well have made her wonder why she associated with male partners at all. In the 1750s she had put up with her husband's continual feuds, and now, two decades later, it was Lalande's turn to embarrass her.

The public had for some time been following scientific progress with great interest. Such involvement and enthusiasm was wonderful, except when it got out of hand as it did now. Lalande was careless in speaking to some colleagues about the remote possibility that a comet might hit the earth around 20 May 1773, word leaked out, and the rumor spread like wildfire causing inevitable hysteria. The police had to request that he do something to mollify the public.

The chronicler Bachaumont reported that women were miscarrying in their panic, that it was no laughing matter, that the newspapers should immediately publish some reassurance.

On 7 May 1773 Lalande attempted in a letter to the *Gazette de France* to dispel the distress he had caused, declaring that there would be no impact, but two days later Bachaumont reported that people were still profoundly frightened and that devout believers were praying furiously to avert the disaster. The chronicler went so far as to call Lalande a "base and vile" publicity seeker who wrote gratuitous nonsense and based everything on vague hypotheses.[104] Indeed, the mischievous Lalande might secretly have enjoyed the power he had to both scare and sooth. He quickly penned *Réflexions sur les comètes qui peuvent approcher de la terre,* meant to be delivered as a paper to the Académie but now expedited, shortened, and published separately for the general public. The censor of this pamphlet, the mathematician Montucla, wrote in his wordy 8 May approbation that he saw nothing in this work that would "justify the terrors [among the people]. It seemed to me, au contraire, useful for calming them by showing that the dreaded event, while in the order of the possible, is in that order of the possible to which all reasonable beings pay no attention, given its unlikelihood following the laws of probability."[105] Would this really have calmed anyone's nerves?

In his brochure, Lalande explained that he had told a few friends about calculations he was doing and regretted that misleading distortions leaked out.

> Soon it was being said that I had announced a comet that in a year, a month . . . in eight days would cause the end of the world. This popular noise was terrifying. I offered reassurances briefly in the *Gazette* but this was not enough to exonerate me from all the absurd things imputed to me almost everywhere in Paris and even in the provinces. Similarly a small derangement I discovered on Saturn's movements in 1769 had led to public gossip that Saturn had vanished and was lost. It was even printed in the newspapers. This year's rumor is scarier and the multitude of letters I have received and questions on this subject made me realize that it was indispensable to publish without delay this part of my scientific paper. There is nothing to fear, because the number of combinations necessary [for this collision to occur] is immense as is the number of chances that distance them from occurring.[106]

This pamphlet, ostensibly designed to comfort, nonetheless depicted some frightful impact scenarios—total incineration by proximity to the sun, or vast tidal swells that would submerge the earth to the highest mountains, making it likely that only those on the sea in boats might survive, or the knocking of the earth so far from the sun that it would need to join other "systems in other worlds." But the overall argument was that a real hit would be "infinitely rare and difficult," the probability "almost null." Lalande pointed out that someone on the earth dies every second according to mortality tables, but people do not worry that it will be themselves.[107] Voltaire wrote a satirical *Lettre sur la préten-due comète* on 17 May mocking the panic caused by past inaccurate predictions that the earth could be reduced to powder, quoting from Moliere's *Femmes Savantes* some of the nonsense the women were told and believed, and musing in feigned comfort that if the comet that is supposed to hit 20 May 1773 is the same size as earth, we will do as much damage to it as it will do to us.[108] There were many other caricatures and songs, but the alarm of the folks in the street was very real.

The Académie was mortified by Lalande's sensationalism, and so was Mme Lepaute, for it seemed he was really getting a kick out of the whole affair, and she became guilty by association. On 15 May 1773 in the *Procès verbal de l'Académie des sciences* there appeared a disclaimer that Lalande's *Mémoire* was not published with the Académie's approval. Another fierce critic was the astronomer Cassini de Thury (III), who wrote his own leaflet criticizing Lalande for hiring a stable of calculators and directly attacking Mme Lepaute. He accused Lalande of heading a "veritable manufacture of negligent clerks and ignorant workers that he pays to furnish [material] to twenty gazeteers and journalists with whom he has made prior engagements. Second-in-command directing Lalande's factory is an *académicienne* from I don't know which academy and useful to Lalande's glory who by this method will soon find herself ready to give birth every month to a small book [*un in-douze*] on astronomy and a large [*un in-quarto*] one of calculations." Lalande fired back, and the war of letters and circulars continued for months, Condorcet and Turgot adding their voices disapproving of Lalande's behavior. Finally an intervention by Macquer, the famed chemist, brought about mutual forgiveness and peace in November.[109] But Mme Lepaute, buffeted around and explicitly insulted in this fracas, must have felt, not for the first time, that the men in her life were more hindrance than help.

And, indeed, the last two volumes of the *Connaissance* that she worked on with Lalande represented damage control. They continued to provide the expected information but now also to dispel rumors about heavenly catastrophes. In that of 1775, their last and longest of all at 379 pages, Lalande boasted of their extreme importance in supplying solid science to counter rampant superstition. In a clear reference to the panic over earth's possible collision he sought to deflect the blame from himself, deplored the gullibility of most people, and argued that the misunderstanding was the fault of laymen tittle-tattling. He had simply explored the possibility of various celestial events and had not "foreseen or imagined to what degree the poorly informed public can be affected by the words of the ignorant."[110] All the more reason for an essential almanac like the *Connaissance des temps* to provide reliable information and accurate facts.

Around this time Lepaute's husband's health began to decline. Although Jean-André may have soldiered on in his shop for a few more years, most Lepaute clocks thereafter seem to have been made by his younger brother Jean-Baptiste, and the quarters in the Galeries du Louvre were officially turned over to him in the autumn of 1776 with all the privileges and advantages of the other artists working there.[111] But Mme Lepaute did not retire from Paris quite that soon. When she and Lalande undertook the editorship of the *Éphémérides des mouvemens célestes* in 1774, it meant a long-term commitment, for this periodical was published only every ten years, with predictions for the next ten on tides, positions of the sun, moon, and planets, enabling astronomers to prepare way in advance for observations.[112] Lepaute did both volumes 7 and 8, which covered the period 1775–93. The preface for volume 7, dealing with the years 1775 to 1784, made clear the contributions of many distinguished savants who calculated for the sun and moon (Guerin fils), the four satellites of Jupiter (Wargentin), Mercury and Jupiter (Jeurat), Venus and Mars with the conjunctions of the moon and the planets to the stars (l'Emery), and then stated that "the calculations for Saturn were done by Madame Le Paute who, for many years now, applies herself with success to astronomical computations."

Volume 8, covering from 1784 to 1793, contained predictions that would outlive her. This publication appeared in 1784 which suggests that Lepaute was still busy in the early 1780s despite her deteriorating vision and her husband's worsening dementia. Another woman, Mme Du Pierry, new on the scene, also contributed a bit to this volume. She had by this time become Lalande's much younger protégée and lover, but in his preface Lalande explained how busy he was with other things, noting as so often before that Lepaute alone did much of the work, this time the calculations for the sun, moon, and all the planets.[113] Since 1776 Lalande had been devoting time and energy to founding the masonic lodge Les Neuf Soeurs, fortifying his friendship with Benjamin Franklin there and famously inducting Voltaire to membership just seven weeks before his death. Lalande had earlier become professor at the Collège Royal and was increasingly absorbed in his courses and business there as well.[114] The busy man was therefore more grateful than ever to Lepaute for taking over.

Her responsibilities were far greater than those of other women doing astronomical calculations. Fortunée Briquet's article on Lepaute in her 1804 dictionary of famous women mentions that similar ephemerides for Bologna were computed by the Manfredi sisters and those for Berlin by the three Kirch sisters,[115] and Jeanne Peiffer has studied the way mathematical work became divided and such computations gendered in this period.[116] But in Lepaute's case this was just a small part of her activity, which also included sophisticated charts, eclipse maps and a booklet explaining them, scientific papers for the Béziers academy, records of observations by others, tables of parallactic angles, and numerous other kinds of written and editorial contributions.

What did her female contemporaries think of her? Lalande lamented that "ordinary" women deprecate those who have superior knowledge, accusing them of pedantry because they are jealous.[117] Beyond Gouilly's attempt to wipe Lepaute out of history, however, we have no record of the ways other women viewed her scientific work, but her much better known predecessor Emilie Du Châtelet was resented terribly, lambasted and reviled even by women not at all "ordinary" who had every reason to be proud of their own abilities and fame in other arenas. Mme de Graffigny whom we met in the last chapter, best-selling author of the quite feminist *Lettres d'une Péruvienne* and once an admirer of Du Châtelet's, later felt humiliated by her. Appalled by what she perceived as intellectual one-upmanship, she gloated unapologetically when Du Châtelet died: "The news of the Monster's death is great. I can't hide that it delights me. I feel the true happiness of no longer having a declared enemy in the world."[118] Mme Du Deffand, one of the most established *salonnières* of her day, criticized Du Châtelet for showing off with her science and gave a brutal description of her as tall, withered, and narrow-chested, with enormous feet, small head, pointed nose, flat mouth, teeth far apart and decaying, and a yelping laugh. She figured out "what made her prefer to study the abstract sciences rather than the more general and pleasant branches of knowledge. She thought she would gain a greater reputation by this peculiarity, and a more decided superiority over all other women." Without talent, memory, taste, or imagination Du Deffand went on, Du Châtelet made herself into a geometer to appear above other women although she really did not understand mathematics any better than Molière's buffoon Sganarelle understood Latin.[119] The Marquise de Crequi, a woman of letters and Du Châtelet's first cousin, whom we met in the last chapter as the recipient of d'Alembert's comments about Ferrand's atheism, described Du Châtelet's skin like a nutmeg grater and went on cruelly about "la docte Emilie," "la divine Emilie": "a colossus in all her limbs, a marvel of strength and a prodigy of clumsiness." She was always impossibly pedantic and must have bribed Voltaire and Clairaut to treat her like a savant. Hearing references to Du Châtelet's "sublime genius" or "profound knowledge" caused Crequi to burst out laughing.[120] It would seem, unfortunately, that women of science in particular were resented by their female contemporaries for trespassing on terrain more properly left the preserve of men.

Only the gracious writer Mme Du Bocage seemed to like Du Châtelet, and vice versa, as we see in a letter from Voltaire.[121] According to Lalande, Du Bocage was in possession of the famous unattributed portrait of Mme Du Châtelet—a painting that, he said, closely resembled a seemingly lost but very similar one by Voiriot of Mme Lepaute calculating the eclipse—which suggests that Du Bocage did not feel threatened by other learned women. We know that Clairaut's dinner parties with Lalande, Lacaille, Pingré, Buffon, Malesherbes, Turgot, Galiani, and Watelet also included both Lepaute and Du Bocage.[122] Did the two become friends? Mme Du Bocage was a feminist of sorts, author of a play called *Les Amazones*, performed in 1749 to mixed reviews—male critics including Condillac did not like her penetration of the stage or the depiction of an upside-down world where women with physical strength, discipline, and virtue made rules that men had to follow—and later another work again about female power, *La Colombiade*, in 1756. Thirteen years older than Lepaute, Du Bocage may have had the confidence to appreciate and support her as she had Du Châtelet, but no record of this relationship remains except for their mutual attendance at some social events. We simply do not know if Lepaute frequented other women eager to converse about ideas, or if so, how they regarded her scientific activity.

She probably stopped working in the early 1780s, shortly after requesting a monetary reward for her scientific contributions in the form of a pension from the Finance Minister Necker. Polite promises notwithstanding, it is unclear whether any pension was ever granted. In 1782 Louis-Sébastien Mercier mentioned among other active "auteurs nés à Paris" "Mad. le Paute, auteur de divers mémoires d'astronomie."[123] Lalande was by then completely smitten with Mme Du Pierry, an intelligent widow he had recently met, referred to lovingly as his "astronomette,"[124] and to whom he soon dedicated his *Astronomie des dames*.[125] First published in 1785 and with many subsequent editions, the book enjoyed great popularity. It is unlikely that Lepaute would have wanted the dedication of a work that seemed to suggest in its very title the need to simplify or at least alter astronomy for a female readership, even if Lalande had not meant it to be insulting. In any case about that time, having completed what needed to be done for volume 8 of the *Éphémérides*, Lepaute retired and moved with her failing husband to relaxing Saint Cloud. On a bluff along the Seine just west of Paris, the fashionable town was known for its healthy air, magnificent parks, cascading water fountains, stunning views, and for its easy accessibility by pleasure boat.

Did Lepaute stay in touch? Had she, for example, actively celebrated the discovery of Uranus in 1781, the most distant planet in the solar system as then known and, invisible to the naked eye, the first one found by telescope? Only William Herschel's relentless following of it through instruments made with lenses he ground himself, in which he was partnered by his talented sister Caroline, finally yielded the fact of its periodicity. Completely thrilled by this discovery, Lalande delivered a rapturous *Mémoire sur la planetète de Herschel* to the Académie 22 December 1781, and he would surely have shared his excitement with Lepaute. But her concerns were elsewhere by this time, first because she was slowly going blind as a result of the eye strain from decades of calculations[126] and also because Jean-André had completely lost his mind. Lalande described movingly her devotion to the ailing man: "For seven years Mme Lepaute showed the heroism of virtue in the care she took of a sick, delusional husband separated from society. She had the courage to shut herself in with him in the house where he needed to be placed in the early stages of his insanity; she left Paris and retired to Saint Cloud with her invalid, to get him better air, and to be less distracted from the attention she wanted to give him without break and without other help, and to which she sacrificed her time, her pleasures, and even her health with an assiduousness and courage that has few examples."[127]

After some years the Lepautes eventually moved back to Paris, perhaps because Marie Antoinette purchased the Saint Cloud chateau in 1785 and in 1787 began disruptive renovations that shattered the calm of the place. Lalande, delighted to see his old friends back and attentive to the end, bought them tickets in 1788, "to distract them" from their infirmities, for the Duc d'Orléan's new Cirque du Palais Royal, a huge half-subterranean spectacle space of food, entertainments, boutiques, and gaming that ran the length of the park and was the talk of the capital.[128] But Mme Lepaute was struck down by a "putrid fever" on 6 December of that year as Lalande lamented, "taken from her family, friends, and from the sciences. The husband was oblivious to this, and did not long survive her, dying 11 April 1789." Both expired in Paris, she on the rue du Dauphin (today rue Saint Roch),[129] predeceasing him by four months, and he in his brother's residence nearby.[130]

Lalande wrote two eulogies of Mme Lepaute which, as was his wont, simultaneously praised and discredited her. The first was at the time of her death in the *Journal de Paris*.[131] Her scientific interest and skill, he wrote, should be an inspiration to other women. Describing her studious inclinations from earliest childhood, he reiterated his conviction that neither the calculations for the return of Halley's Comet nor those for the *Connaissance des temps* would have been undertaken without her courage and rescue. Enumerating her many other astronomical and mathematical accomplishments, he said that Lepaute's principal contribution was bringing Dagelet to Paris and tutoring him. In December 1788, when Lalande wrote this version of her eulogy, no one had yet concluded that Lapérouse's expedition with her nephew aboard was lost at sea. He explained that his tribute to Lepaute was both to comfort himself after this dear friend's death and to encourage other women to follow her example.[132]

Fifteen years later he wrote another homage to her, which resembled the 1788 obituary but was three times longer and embellished with much personal detail. In his 1803 *Bibliographie astronomique avec l'histoire abrégé de l'Astronomie* he asked his readers' indulgence for his need to mourn Lepaute again. He began with a baldly egocentric comment quite dismissive of her but then improved somewhat. "Her loss was less great for astronomy than for me, but I will be pardoned for this little *hors d'oeuvre* due to my sensitivity. It will be a consolation for me and an object of emulation for a sex which we should include more in our work. Mme Lepaute deserves to be cited among the small number of women of mind who give an example to their sex with their striving and taste for the abstract sciences. . . . She was the only woman in France who had acquired veritable knowledge of astronomy."[133] Trying to revive her memory, he wrote of her physical appearance, her personality, and what she had meant to him. "When we speak of a woman we allow ourselves to speak of her looks, and to ask, Was she pretty? . . . Without being remarkable of face, she had most of the charms of her sex, an elegant figure, an adorable foot, such a beautiful hand that Voiriot, painter to the King, having made her portrait, asked permission to copy [the hand] to keep a model of the best in Nature. He has used it since in his tableaux. The portrait of Mme Lepaute was placed in my cabinet next to a rare portrait of Copernicus. . . . [She] is represented tracing the image of the eclipse of 1764 which she had just calculated, with a globe next to her. The portrait resembles a bit that of the Marquise Du Châtelet which is chez Mme Du Bocage in Paris."[134]

The artist of this lost portrait, Guillaume Voiriot, seemed to have been in the Lepaute/Lalande circle. He painted Lepaute's husband holding his *Traité d'horlogerie* in 1761, his clockmaker brother Jean-Baptiste holding a piece of paper the same year, and his brother's wife sitting at a table near a porcelain coffee pot and stirring the contents of a cup with a spoon. Oddly, there is no portrait of Reine Lepaute to complete this formal series. Perhaps, instead, she asked Voiriot to do the very different painting of her that Lalande described, surrounded by iconic scientific instruments and tracing her eclipse map. The fate of that painting is unknown, but Lepaute, like Ferrand, clearly wished to be depicted doing science.

The only image said to be of Mme Lepaute today is probably not of her, for it shows nothing but a bust and is not stylistically similar to the ones of the other family members. Nor does it show her hands, of which the artist was, according to Lalande, so enamored. The portrait that once hung on the wall of Lalande's study does indeed sound similar to the famous unattributed one of Du Châtelet holding a compass that is now in the Chateau de Breteuil; one of Voiriot's descendants believes that he was Du Châtelet's artist also.[135] Much later, around 1780, Voiriot did a portrait of Lalande in a blue-green costume with lace collar and cuffs. He sits and faces frontally resting on an astronomical engraving and holding a pair of compasses in one hand and a ruler in the other. Instruments, including an armillary sphere and an astrolabe, are on the table, and a telescope can be seen behind him. It is a flattering portrait of the ugly man, nothing like most others which depict him as grotesque. Voiriot must have been a kind, good friend of this group.[136]

Lalande included in his 1803 homage two gushy poems to Lepaute, the first of which appeared in the *Mercure* in 1776 and was written by La Louptière, a former editor of the *Journal des Dames*.[137] The other was by Lalande himself to the effect that she surpassed Hipparchus and Ptolemy by being both the "sine of the Graces and the tangent of our hearts." He mentioned that his compatriot Commerson, the naturalist on Bougainville's voyage, named the beautiful orb-like flower we know today as the hydrangea for Mme Lepaute, calling it *Peautia Celestina*.[138] Then Lalande revealed his deepest feelings for her, implying that he was much more able to appreciate her than was her own spouse.

> Her husband had for her that consideration made of respect, but which a rare merit inspires in those who know how to feel it. She was, however, full of thoughtfulness for him; she served him with eagerness and in small details that another would have found beneath her given the loftiness of her character and her mind. Her company was useful and dear to me. She kept me away from dangerous liaisons; she introduced me to the charms of a comfortable life with pleasant and learned people; she tolerated my faults and helped to diminish them. She had a strong enough personality to be imperious at times, when that could be useful, but she had the sense to yield in situations where resistance would have been dangerous. Finally, she was so dear to me that the day I attended her funeral was the saddest I have ever spent since the one when I learned of my father's death, the most respectable and tender of all fathers.[139]

Lalande had earlier said that the loss of his father in 1755 had almost made him die of sorrow.[140] Finally, he tried to gain some perspective: "This interesting woman is often present in my thoughts, always dear to my heart; the moments that I spent near her and in the heart of her family are my favorites to recollect, and their memory, mixed with bitterness and pain, spreads some sweetness on the last years of my life, as her friendship was the joy of my youth. Her portrait, which I have always before my eyes, is my consolation, when I think that a philosopher should not complain about the laws of necessity and the losses that are an inevitable consequence of the laws of nature."[141]

Here again we see Lalande's tendency to compliment and belittle Lepaute all at once. Paying tribute to her brilliance, he then felt the need to give a physical description of her, speaking of her "charms" and using adjectives like "elegant," "adorable," "beautiful" for various body parts. With Bonnet he had referred to her as his "muse." He admired her diligent performance of house-hold duties and her devotion to a husband he considered unworthy and ungrate-ful. Most of all he saw her assuming a proper maternal role toward Dagelet and in that way fulfilling her domestic responsibility. A proud descendant of Mme d'Arconville, the chemist of my last chapter, assured me when we met in Paris that his ancestor, despite the scientific work she did, was always "très correcte dans sa famille." The unmarried women in my book were never subjected to such criteria for approval. They did not have to be attractive or socially savvy, keep house a particular way, raise a younger generation to whom they could pass on a legacy of values and skills. This was the burden of the wife, and even in the eyes of an admirer like Lalande that role overshadowed everything else. If her spousal duties were not performed with balance and grace, nothing else mattered.[142]

That Lepaute did not allow marriage to smother her is a testament to her remarkably robust sense of self. It was a characteristic she shared with the other women in my story.

Her final resting place was almost surely the Église Saint Roch, for she died on the same street as that church.[143] Her bones likely lie in the ossuary under the altar, along with Diderot, d'Holbach, and the Abbé de l'Epée who invented sign language, so she is in bracing company. The facade of this baroque church would get a Lepaute clock in 1835, made by someone in the next generation of the family business. Recently restored, this clock has a strikingly bright blue enamel face and gold hands. Another Lepaute timepiece, modernized and less elaborate than the original, continues to grace the front of the Luxembourg Palace where Jean-André installed it when our story began. Although capital letters rather than his ornate signature indicate that it is a more recent replace-ment, it still boasts the words LEPAUTE, H.ger DU ROI and ticks away as a steady, visible reminder of the girl born there who became an astronomer.

Dear Reine,

What was it like, married to a man you so surpassed intellectually and who, while immensely talented in his trade, brought down upon himself such opprobrium from his rivals? You were simply out of his league. I marvel that you stayed true to him, devoting your last years to caring for him in his senility, losing your own eyesight, health, and life in the bargain. Was it out of guilt? Very clever bringing your savant into the household. And Lalande was always true to you in his fashion, giving you challenging problems to stimulate your fine mind, putting your portrait on his study wall next to Copernicus, and telling your story.

Lalande's constant presence is probably why you were not smothered in your marriage, why your ambition was able to survive wedlock. Women in science today speak of great difficulties combining their research with family obligations, the divided loyalties, the conflicting demands, the sometimes impossible juggling and balancing it requires. They also say it is essential to choose the right partner. Not clear that you did that, given the extent to which your husband downplayed or misunderstood your brilliance. But at least you were able to set up the radical threesome without his objection, perhaps because Lalande very early on persuaded him to start building astronomical clocks for which he had the skills and found a whole new, eager, and lucrative market.

You stuck up for yourself, did not lose focus, and overcame systemic obstacles by getting elected to a scientific academy. Extraordinary.

Bad male actors in your life: Clairaut—although he said that your ardor for math astonished him—tried because besotted by another woman to deprive you of proper recognition. Then Lalande, erudite and cultured, who sincerely admired you, nonetheless liked to divert attention to himself. He was more than a little eccentric; he ate insects, spread spiders on bread claiming their delicate taste resembled strawberries and artichokes, and swallowed caterpillars whole all in the name of philosophy, reputedly getting you to overcome your revulsion and do it too, to please him. Did you? Imagine, the chemist Macquer wanted Lalande to marry his daughter. Lalande tells us that he refused out of friendship to the family for he was sure they could and would find better prospects. Funny thing for him to say, quite self-aware. Then your own husband, inventive, innovative, but in the habit of claiming credit for work done by others—young Beaumarchais, great seducer of public opinion, caught him at it and publicly shamed him—and he did the same thing to <u>you</u>. And even your portraitist Voiriot, who painted you doing science, instead became obsessed with the beauty of your hand and received your permission to use it as his model in other works. So much for his appreciation of your intellect!

But if it hadn't been for Lalande we might never have known of your accomplishments. Someone else may eventually have come forward, maybe the Swiss naturalist Charles Bonnet to whom Lalande confided his dependence on your indispensable calculations. Books and movies today—*Hidden Figures, Code Girls*—are bringing to light the work of brilliant but unsung women like you. The truth will out.

Well, the botanist Commerson knew you in Paris and clearly revered you, and later while on Bougainville's voyage he named for you the flower we now call the hydrangea. It reminded him of the celestial orbs you studied with such success. Did you ever find out that he had done this? Most of these namings were buried in his manuscripts, and he died before getting home. He referred to you numerous times as one of his best friends. So I wonder how you treated his so-called housekeeper whom we will next meet, the smart young Jeanne Barret. You, seventeen years older than she and more worldly-wise, could have encouraged her during your visits with Lalande to Commerson's rue des Boulangers apartment.

When Jeanne arrived in Paris in the fall of 1764 the April eclipse, for which you had made two beautiful charts, was long past and weather had conspired to render the phenomenon invisible, a huge letdown for you. Then the king's favorite, Mme de Pompadour, had died later that month sucking all the attention away from everything else. What was your frame of mind and what work were you doing since then? Without any effort you would have detected that Jeanne was much more than Commerson's domestic. If you saw them when they first came to Paris that September, during the third trimester of her pregnancy, their intimacy would have been obvious. How much did the couple trust you? How much did Jeanne trust you? I like to think you were not judgmental. Did you have a chance to notice how much she already knew about science? Did you urge her to allow her reach to exceed her grasp, to aspire much higher than her station? Did you give her advice?

I wonder how affected you were by Lalande's atheism? He said, "I lived among the most famous atheists, Buffon, Diderot, Holbach, d'Alembert, Condorcet, Helvétius; they were persuaded that it is necessary to be an imbecile to believe in God." And how much do you think Jeanne was influenced by Commerson's indifference to religion as he railed against the pieties of his priest brother-in-law?

Here are some things that will surprise and please you. First, I recently took a walk on rue Nicole Reine Lepaute in Paris. Dijon has also named a street after you. But I realize that this one in Paris is not far from where your parents once owned a home, in Petit Gentily, now the Maison Blanche neighborhood. You would of course remember that on 6 July 1751 it was sold and the proceeds divided up between you and your siblings. Must have come in handy given that your husband's business was having difficulties just then. Anyway, this vicinity, adjacent to the modern Bibliothèque nationale, has both streets and buildings dedicated to scientists; in fact your rue Nicole Reine Lepaute is perpendicular to rue Albert Einstein. Nearby, other important women are being honored: the midwife Louise Bourgeois has a street, Sophie Germain has a math institute, and there is a café named for the feminist Olympe de Gouges who believed in equality of the sexes and wrote a *Declaration of the Rights of Women*. Streets all around Paris are named for so many of the men in my story— Condillac, Clairaut, and Lalande, of course, and others we will meet in later chapters: Jussieu, Thouin, Daubenton, Buffon, Linnaeus, Rouelle, Bougainville, Morand, Franklin, Diderot, I could go on and on. But you are the only woman in this book to have a street named for you in Paris.

You were also "invited" to Judy Chicago's feminist installation *The Dinner Party*, a gathering and celebration of great women throughout history. It must be hard for you to picture such a thing, but it represents an imaginary, grand coming together of accomplished, creative women from all ages and all spheres of inquiry. They can feast and talk forever. You are there in association with the German-English astronomer Caroline Herschel with whom you may have corresponded toward the end of your life, although there would have been a language barrier to surmount, but in any case, of course, such letters have not survived.

And—prepare yourself—going beyond earth, as befits an astronomer, a small but clearly distinct heart-shaped lunar crater is named after you, roughly ten miles in diameter, with walls almost a mile high, in the southwest margin of the Palus Epidemiarum on the moon's near side. Finally, much farther away in space, the outer main-belt Asteroid 7720 (4559 P-L), orbiting between Mars and Jupiter with a periodicity of about six years made an appearance—through a telescope of course—not long ago, observed from earth on 25 April 2016. It too bears your name. These are things I believe you would want to know.

THREE

Botany in the Field and in the Garden

Jeanne Barret (1740–1807) and
Madeleine Françoise Basseporte (1701–1780)

This extraordinary woman.
—1785 award to Barret from **KING LOUIS XVI** through the
minister of the Royal Navy for her contributions to science

Nature gives plants their existence, but Basseporte preserves it for them.
—J.-J. ROUSSEAU

JEANNE BARRET AND MADELEINE FRANÇOISE BASSEPORTE
were intrepid women who deliberately positioned themselves and thrived in
their scientific work surrounded by some of the most celebrated naturalists
of their day. Bernard de Jussieu, generally revered as the "Newton of botany,"
presided over botanical demonstrations in the Jardin du Roi for half a century
in pursuit of the underlying laws of his science. George Louis Leclerc, Compte
de Buffon, became the director of the Jardin in 1739. Louis Antoine de Bougainville, whose circumnavigation of the globe took place in the 1760s, was not only
a sea captain but a lover of science who had earlier, after studying mathematics
with Clairaut, published a treatise on Newton's integral calculus. The naturalist
he chose to accompany him on his journey, Philibert Commerson, a close friend
and compatriot of Lalande's from Bourg-en-Bresse, was intimately tied to the
Jardin and sent to Bernard de Jussieu, whom he called his master and "the father of modern naturalists,"[1] letters and shipments of flora—countless seeds,
cuttings, dried specimens, and whole live plants—thus assisting him in the formulation of his famous "natural system" of classification, an alternative to the

artificial Linnean one.[2] These distinguished men, Jussieu, Buffon, Bougainville, and Commerson, taught and learned from their collaborations with Barret and Basseporte, both of whom contributed significantly, although in very different ways, to revealing and documenting the orderliness behind the world's floral profusion.

Forceful, tenacious women who fashioned unprecedented plotlines for their lives, they escaped circumscribed gender roles and used their resulting freedom to investigate nature from the 1730s through the 1770s, one inhabiting flower worlds, the other discovering and exploring new ones.[3] Barret, disguised as Commerson's valet, accompanied him on Bougainville's voyage, hunting, gathering, drying, and curating exotic plants from distant lands. She became the first woman to sail around the world and was responsible for the discovery of countless new species. Basseporte, a protégé of both Bernard de Jussieu and Buffon, devoted her professional life to dissecting, exposing, illustrating, and thus immortalizing plants in the king's gardens of Paris, of the Trianon, and of various other royal chateaux. She was the first and only woman to be granted the salaried post of *dessinateur du roi* at the Jardin, and she held it for nearly half a century. Barret captured vegetation wild, Basseporte tamed it into imperishable scientific images. As Rousseau put it, flowers are ephemeral but her botanical illustrations gave them everlasting vigor. "Nature gives plants their existence," he wrote in admiration, "but Mlle Basseporte preserves it for them."[4] Lest it be objected that neither woman *really* did science, one *just* an assistant botanizer and the other *just* an artist, the men closest to them and best aware of their skills and knowledge considered them veritable botanists, expert enough to hold positions of great responsibility, and formed partnerships with them based on mutual need and respect. For the women were at least as much assisted as assistants. In a very real sense, it was Barret who used Commerson to escape servitude in the countryside and realize a new life of discovery for herself. Basseporte got the male staff of the Jardin to accept her as one of them.

The two women had in common their devotion to botany and the boldness to seize unique opportunities, but they were otherwise quite different. Barret was a Burgundian peasant girl, scrappy and bold (although the naval surgeon Vivès would describe her as diminutive). A risk-taker, she made the daring and literally transformative decision to cross-dress as a man and, hiding in plain sight, travel to far-off places where no European woman and hardly any European men had ever been before. She was, like Ferrand and Lepaute, a nonbeliever, ir-reverent enough at one point to open a billiards saloon and serve spirits during Sunday mass, a menace to public order for which she was reprimanded and fined. Basseporte was the daughter of a wholesale wine merchant, born and baptized in the very center of Paris on the Isle Saint Louis, a *petite bourgeoise* who went no farther from the capital than the royal gardens of nearby Versailles, Compiègne, Fontainebleau, and Bellevue. She won and secured for almost fifty years a coveted job, one never held by a woman before or since. She was, like Biheron and Thiroux d'Arconville whom we will meet in the next two chapters, almost surely Jansenist, and so devout, but also rebellious. Barret came on the scientific scene in the 1760s and appears to have left botany about fifteen years later; Basseporte was already working by 1730 and stopped only in 1780 when she died, a prime example of creative aging. Barret's story is brief and spectac-ular, Basseporte's marked by steadiness and longevity.

Jeanne Barret—this is how she signed her name, although it is spelled Baret and sometimes even Baré by others who wrote about her—was born 27 July 1740 in La Comelle near Autun. She was a provincial domestic who literally remade herself to have a fuller, richer life. Although we don't know how or when she met Commerson, recently compiled genealogical records give clues as to why she may have been eager to leave home. Her mother had died when she was just twenty months old, after which her father remarried three times, once when Jeanne was two, again when she was five, and yet again when she was ten, this last marriage yielding her three half-siblings: a short-lived boy born when she was eleven, a girl when she was fourteen, and another girl when she was fifteen, this last child arriving three months after the death of their fa-ther in 1755. So Jeanne had lost both of her parents when she was still young and had dealt with three stepmothers in rapid succession, the first two dying soon after joining the family and the last no doubt favoring her own children. There may have been a dispute over inheritance—Barret's later tearful confession to Bougainville about having been orphaned, then destitute after losing a trial could have had more than a grain of truth to it. But she had two older full siblings as well, with whom she felt closer. Her sister, also named Jeanne, was three years her senior and married a miller the same year their first stepsister was born, and her brother Pierre, six years older than she, married at the time their second stepsister was born. Both of these true siblings, then, left home to start their own families elsewhere, while our Jeanne remained alone with the young brood of her father's fourth and now widowed wife. A second marriage for her sister Jeanne took place in 1759 in Thil-sur-Arroux, about ten kilometers south of La Comelle. It is very likely that our Jeanne escaped at that time to go live with this sister and help with her niece and nephew.[5] This also might explain how she met and got hired as a governess for his son by Commerson, who lived in Toulon-sur-Arroux, about another few kilometers down the same river, and whose medical practice extended through the region.

The versatile Philibert Commerson had earlier done some fine research for Linnaeus on Mediterranean fish, and his reputation as a naturalist had obviously spread far beyond his native Bourgogne. His medical degree was from Montpellier, but he was also known as a passionate and indefatigable botanist. In 1762, suddenly a widower with an infant son, Commerson hired Barret as "gouvernante" for his child and housekeeper for himself. At some point the two became lovers and Barret declared her pregnancy in 1764, as required by the local authorities.[6] Although she discretely refused to name the father, it seemed prudent for the couple to avoid scandal by leaving the region before the birth, and Lalande, who had been serving for years as a correspondent and link between Commerson and Bernard de Jussieu, prevailed upon his friend to come to Paris and join the great men of science there.[7] Two-year-old Archambault, whom Barret had been hired to care for, was left in the countryside to be raised by his reluctant and sanctimonious uncle François Beau, a priest and the brother of Commerson's dead wife. Beau disapproved mightily of the new liaison, and the naturalist reprimanded him from Paris for insinuating nasty things about his relationship with Barret, for "the attacks you have claimed on my reputation by supposing a second marriage as false as it is ridiculous. . . . [T]his kind of Inquisition [you] charitably carried out from a hundred leagues away on my private morals." Then Commerson concluded his accusatory letter dramatically: "Hatred drives you to injustice."[8] Soon after they arrived in the capital in late 1764 Barret gave birth to a boy who was sent off to a wet-nurse in Dreux and perished there.[9] Commerson spent the next two Parisian years frequenting men of science and other luminaries, impressing them sufficiently to be selected for Bougainville's expedition. He was in particular involved at the Jardin du Roi, which was the reason that he chose lodgings nearby on the short, elbow-shaped rue des Boulangers, a street appreciated by bakers for the purity of its well's water.

How did Barret, bereft (perhaps conveniently) of her illegitimate newborn, console and occupy herself in the capital? Her official employment there as Commerson's housekeeper began on 6 September 1764,[10] two years before the voyage with Bougainville was even conceived of. During that time much energy must have gone toward enhancing Commerson's social connections and his scientific career. Would they have stayed in Paris indefinitely had the voyage not been proposed to him in October 1766? Although we cannot know Commerson's thinking, he was a driven, ambitious man, and returning to the provinces to resume his paternal duties and practice as a country doctor must have seemed an unglamorous alternative. Without childcare to keep her busy, Barret figured out ways to further her partner's botanical reputation and to improve herself.

Being in Paris was already an adventure for her, and sharing a home with Commerson, she surely met the people in his circle. Some of the contacts and friendships he formed can be surmised by looking at the names he later bestowed on newly found plants during his travels. From their very arrival he frequented his countryman Lalande and through him Mme Lepaute, whom Commerson considered among his best friends, she one of the people who later encouraged him to accept Bougainville's offer.[11] He would name plants for both of them, the strikingly orb-shaped hydrangea for Lepaute, calling it *Peautia Celestina* in recognition of her mastery of the mechanics of the heavenly bodies.[12] (For her husband he named a prickly, rigid shrub.) He dedicated plants to some personnel from the Jardin du Roi where he went daily—the head of the Cabinet d'histoire naturelle L.-J.-M. Daubenton, the eccentric but hugely popular chemistry professor Guillaume-François Rouelle, and the head gardener André Thouin.[13] Others for whom he would designate new species included his old medical school chum Cleriade Vachier, Diderot, Turgot, d'Alembert, Mme de Pompadour's brother the Marquis de Marigny, the chemist Macquer, the polymath La Condamine, and the astronomer Véron.[14] These, then, were much more than casual acquaintances, and Barret must have known them too. Diderot later wrote admiringly of her in his *Supplément au Voyage de Bougainville*, both Turgot and Daubenton actively sought to salvage Commerson's missing collections, quite possibly at her urging when she returned to France after the naturalist's death, and Vachier helped her claim her due inheritance. Lalande may even have been in on the masquerade plot, as he would refer to it later in his obituary of Commerson, insisting that while the naturalist had been faithful to his wife until her death he had indeed taken a disguised woman with him on Bougainville's voyage, but that "his taste for pleasure never interfered with his work," and that he could not have chosen a better person, discrete and lacking the short-comings of her sex, to accompany and assist him.[15]

Commerson probably received visits from these men of science and philo-sophes in the third-floor Paris apartment he shared with Barret, modest and not fit for elaborate entertaining but fine for the meeting of minds. Here she would have served guests in her capacity as housekeeper while overhearing and ab-sorbing the heady conversations, maybe even participating in them for she was already quite knowledgeable.[16] About Commerson having instructed her in bot-any there can be absolutely no doubt, for this was a foundation, a prerequisite for his most meaningful relationships. While courting his wife he had confided to a close friend that despite his fiancée's fine figure and inherited wealth, what he loved most about her was that he had been able to instill in her his botanical enthusiasm.[17] He professed to successfully "inoculate" all those he taught and thus to make them true botanists in their own right. "So it is that I infect every-one who approaches me with my own *botanomania*," he told his friend Louis Gerard.[18] That he did the same with his new partner Barret is certain.

But beyond this training, she must also have profited from the teaching available at the nearby Jardin. Women were welcome to attend the public botany lectures in the garden presided over by Bernard de Jussieu, whom Commerson considered the greatest botanical sage and his "venerable master."[19] Barret could greatly benefit from these acclaimed courses, and perhaps she and Commerson went on some of the famous *herborisations*, Jussieu's botanizing excursions held in and around Paris every Wednesday, leaving at four in the morning and often including an overnight stay. Thouin reported that they took a double-barrel rifle or shotgun and a pistol, a change of clothes, many pairs of shoes, a hunting knife, and a pruning knife to blaze trails or cut specimens. Other requirements were white iron boxes for seeds and cartons for drying samples to be later inserted in the *herbiers*, a writing case to jot down impressions, thermometers, barometers, compasses, and flasks of volatile alkali for snake bites. And of course books.[20]

And Barret would have met Basseporte, an omnipresent fixture in the Jar-din who had been a member of its official staff since 1736, listed right alongside all the male personnel in the yearly *Almanach Royal*.[21] She may even have re-ceived art instruction from Basseporte, who was known throughout her life to take on needy female pupils and apprentices. The botanical artist was often asked to coach individuals before their travels, as in the case of the young man about to set out on a survey of the flora in Guadeloupe to whom she gave a crash course in plant illustrations, enabling him later to sketch and paint what he saw.[22] Obviously Barret could not have divulged to Basseporte, or to anyone for that matter, her transvestite scheme to accompany her lover on the voyage. And we do not know how long that plan was in the works. But honing her drawing skills with Basseporte would have been a felicitous opportunity.[23]

Commerson, unquestionably brilliant, was something of a madman—ardent, impetuous, obsessive, extreme in all things, but aware of his own "immoderate zeal that makes me ill," his "fury to see everything." Lalande's *éloge* of his good friend explained that such raging passion for work, such compulsive studying, sleep deprivation, and indifference to food, inevitably led to debilitating weakness, total exhaustion, and collapse at times, to frequent leg ulcers that almost resulted in amputation, and to other serious ailments self-diagnosed as pleurisy, gout, rheumatism, dysentery, and nephritis. Commerson even predicted to Lalande, "I'll die from overwork."[24] He was nothing if not theatrical, speaking in many letters to several friends of profound, immobilizing despair. Yet in his voracious curiosity and eagerness to find and collect everything new that he could get his hands on he climbed rugged mountains, braved all weather, refused or dismissed guides to penetrate treacherous terrain, and was caught swimming in a crater of an active volcano after heavy rainfall had turned it into a reservoir. Spurts of superhuman energy punctuated by periods of debility and prostration seem to have been the pattern, and the devoted Barret had signed on to put up with these frenzied excesses.

We must remind ourselves that just about every hardship Commerson would describe on the voyage in his customary sensational style—and he spared no gruesome details—was shared by Barret. But we have no record of her complaints, instead just glowing reports from astonished crewmembers of how hardworking and constant the young valet was through all adversity. She had known her partner for years before they sailed, and she was no stranger to his quirks. As the departure of Bougainville's voyage drew near, she must have decided that being apart for an indeterminate period while Commerson circumnavigated the globe was unacceptable. He could not picture gathering worldwide flora during fleeting landfalls without her energetic assistance, and she, unable to countenance the idea of life back home as a servant for someone else, hitched her fortunes to this turbulent man. He told his brother-in-law Beau that once the voyage was proposed to him, refusing would have meant retreating to a crushingly mediocre and obscure existence.[25] Barret clearly agreed and could not imagine missing out on the trip. As she would later confess to Bougainville after her unmasking, what opportunities were there, really, for the average house-keeper or maid? She knew she had a better chance at everything if she were a man. Besides, she was eager for adventure, explaining to the bemused captain that the prospect of a round-the-world journey "piqued her curiosity."[26] Because women, by a royal ordinance of 1689 renewed in 1765, were strictly forbidden to travel on the royal fleet, together she and Commerson cooked up the scheme to pass her off as a male domestic accompanying him.[27]

It was considered prudent for voyagers to leave a will in case they did not return, and Commerson wrote his testament on 9 November 1766. But before it assumed its final form on December 14–15, the eve of his departure from Paris, he added a wily P.S. stipulating that Barret would remain in their rue des Bou-langers home in his absence, getting her customary wages, in order to receive and organize his specimens as he sent them back.[28] He finessed this to sound like she was staying behind, and he entrusted her with the huge task of spending a year supervising and helping to curate his herbaria and collections should he die on the expedition.

> I leave to Jeanne Barret, also known as de Bonnefoi, my housekeeper, the sum of 600 livres in a single lump sum, in addition to the wages I owe to her since 6 September 1764 at a salary of 100 livres a year. I declare in addition that all the bed and table linen, all the women's clothing that might be in my apartment belong to her as well as all the other furniture, such as beds, chairs, tables, commodes, with the exception only of the herbaria, books, and my personal effects given to my brother as specified above. I desire that this furniture be given to her without difficulty after my death, and that she be able to stay for an-other year in the apartment and that the rent continue to be provided for her, so that there be enough time to put in order the natural history specimens to be sent as explained above to the king's collection.[29]

By 1766, then, when Commerson charged Barret with this heavy responsibility—he stipulated in his testament that she was to be assisted by other botanists, ei-ther Michel Adanson or Louis Gerard—she had clearly acquired an impressive botanical education beyond what he had taught her. Bernard de Jussieu's bot-any courses at the Jardin would have provided just that.

In mid-December the couple left Paris, Jeanne Barret possibly already in disguise practicing for the role she would need to play as "Jean Baré," and headed southwest to the port of Rochefort where the storeship *L'Etoile* was waiting.

By the time Barret departed Paris to embark on the high seas, Basseporte had already been working in the Jardin for several decades. How exactly had she steered herself to the impressive and permanent rank she achieved there? Why and when did she set her sights on and decide to occupy this vibrant, male-dominated science center instead of remaining the portraitist that she had initially been? She had managed to make herself indispensable and bring new prestige to her job in the garden, her reputation spreading far beyond it, her networking skills apparent from the beginning, her understanding of scientific sociability such that "with her the situation of *peintre du cabinet* [took] on a social importance that it never had previously."[30] At her death she would be the subject of a lengthy obituary in the yearly *Nécrologe des hommes célèbres*, the only female among the fifteen people eulogized in the 1781 volume, and her entry, written by a man of science, one of the longest.[31] Who was this singular woman?

Baptized 28 April 1701 in the church of Saint-Louis-en-l'Isle on an island in the center of Paris, she had early shown artistic talent, coming to the attention of Robert de Sery (also spelled Seri), the Cardinal de Rohan's painter, with whom she studied. He had access to the prelate's palace and there, it was said, young Basseporte sat month after month copying the great masters even in the dead of winter. She and Sery then set up an art school for girls at the new home to which she moved in the block of houses now numbered 71–85 rue Vielle du Temple, and she turned her talent to teaching.[32] From the start she was business-minded and enterprising. She made money as an independent artist as well, and she may have taken lessons from the famed Venetian pastelist Rosalba Carriera, who visited Paris during Basseporte's youth and clearly influenced her move into portrait painting to further support herself and her widowed mother. So similar were their styles that a picture of a young woman done by Basseporte in the 1720s was acquired by the Rijksmuseum in the belief that it was by the famed Italian.[33] Both women disliked pandering to vain patrons who insisted their likenesses be more flattering, would not reliably pay for the finished work, and grew fickle in recommending them to others.[34] The indignity of depending on clients for her reputation, the capriciousness of her sitters' whims, and the need to continually drum up potential business led Basseporte to abandon such work for flower painting, where the subjects were not only naturally beautiful but also uncomplaining, and thus far preferable.

She taught decorative floral art throughout her life to impoverished young girls, providing them means to a livelihood, and Mme de Genlis, future governess of the royal children, reported that Basseporte remained expert at it herself even in old age.[35] Abbé Gabriel Charles de l'Attaignant, whom the chronicler Bachaumont called "the great chansonnier," wrote a poem about the perfection of Basseporte's flower painting. It told the story of a bee, then a butterfly, and then a breeze, all attracted to a gorgeous rose. But quickly in succession all three, after alighting, flew off in a state of confusion. Upon investigating, the poet discovered the "imposture," as this was no rose but a painting by Basseporte, "happy rival of nature," that had fooled them all.[36] She had occasional private commissions, tulips for the Prince de Conti, other flowers for Buffon which somehow ended up in the library of the revolutionary Mirabeau's father, these images for individuals "precious for their rarity" today, as most of Basseporte's preserved works are in the king's official collections.[37] In these personal, informal paintings she was clearly having fun, following no particular rules, several different kinds of plants on a single sheet with butterflies, beetles, moths, spiders, flies, and ladybugs. Some flowers had chewed leaves and chunks missing, others burst exuberantly through their gold borders.[38]

But always restless to challenge herself, Basseporte next moved to the much more demanding and intellectually rewarding work of botanical illustration, a "genre thorny with difficulties" as her eulogist said, placing her skills in the service of science. Through her closeness with Sery, who had actually moved in with Basseporte and her mother and lived with them through his final illness, the young artist met the Jansenist Abbé Pluche, who was already at work on his natural history book, *Le Spectacle de la Nature,* for which she did illustrations in the first volume of 1732 and the second of 1735. This became a wildly popular best seller, the only publication of its kind before Buffon launched his many-volume *Histoire Naturelle* in 1749. Pluche's work, enhanced by Basseporte's detailed illustrations, did much to spread the taste for science among a broad reading public.

Her training in this new field had begun even earlier, however, when she reached her legal age of majority in 1726 and started lessons in the Jardin with the resident botanical artist Claude Aubriet. Bernard de Jussieu was already there, his good friend the agronomist Duhamel de Monceau a constant visitor, and both were intimately allied with the Comte de Maurepas, minister of the royal household who in that capacity oversaw the king's garden. The three men backed young Basseporte from her arrival, watching as the aging Aubriet received more and more excellent but unacknowledged assistance from his talented pupil, who was in fact producing numerous images that the old man appropriated, signed, and got paid for. On 30 April 1735, after she had been working for nearly a decade and due in no small measure to the influence of Maurepas at court, Basseporte received the royally granted right to succeed Aubriet *en survivance* as the king's botanical illustrator. This was extraordinary, and although she did not get paid or officially take over until 19 July 1741, and Aubriet did not die until 1742, she had the guarantee that this job was hers for the duration and the satisfaction of seeing it publicly proclaimed in the *Almanach Royal* every year starting in 1736.[39] Duhamel also tapped her to do many illustrations in 1737–38 for a treatise he was preparing on fruit trees. C. F. de Cisternay Dufay, head of the garden from 1732 to 1739, was a devotee of Basseporte's too, and he praised her in yet another, different circle, one that included Buffon who would soon follow Dufay as the Jardin's director in 1739. So she had created good will and a kind of broad security for herself at a quite young age, unprecedented for a female in that scientific setting.

Bernard de Jussieu was a perfectionist who published almost nothing, but he was an outstanding teacher, and we get a glimpse into his mode of instruction in an elementary lesson he sent to a new disciple. To this correspondent in Cayenne in a series of letters starting in 1736 he wrote: "Exactness . . . becomes more and more necessary. . . . We should not adhere solely to the form of the petals, and the part which, in the flower, changes into fruit; it is essential to particularize the figure of the calyx, its composition, the different figures of the petals, the part they occupy, their number, their division, the number of the stamens, whether they stand alone and distinct or whether, united in several bodies or a single one, they spring from the sides of a calyx or petal." He explained the three parts of the pistil—the lower which is the ovary, the middle which is the style, the upper and last which is the stigma—and went on to caution that these parts might be multiple. "Their figure, situation and proportion vary, and all this requires details . . . and seeds have their appropriate forms." There are also "bodies that serve to secrete, in the interior of the flowers, a juice of honeyed liquid modern botanists call *nectarium*." Some plants, he went on, are hermaphrodites, others simply either male or female. "You will not only find in the [botanical] occupation a source of pleasure, but will be enabled by your researches to correct, reform and authenticate . . . information."[40] The great master taught Basseporte patiently when she was first learning, probably in just this way.

The Jardin was an exciting place. Great fanfare surrounded the planting in 1734 of a cedar of Lebanon—rumor had it that this special gift from arborists in England was small enough to fit and be transported in Bernard de Jussieu's hat—which then grew dramatically to everyone's surprise and delight. In 1738 Linnaeus came from Sweden to Paris and spent most of his days in the Jardin. He was so smitten with Basseporte that he mentioned her in almost every letter he wrote to Bernard de Jussieu over the following decades, greeting her along with the garden's famous naturalists and also Clairaut, whom he had earlier met during the latter's Newtonian mission to Lapland. That Basseporte was gracious and charming is mentioned by many of her contemporaries, who also commented on her beauty. Her comeliness could have been a liability for one so bent on her profession, but she insisted that her work be taken seriously and forcefully discouraged the distraction of suitors.[41] In fact, her most meaningful relationship appears to have been with another woman, the anatomist Mlle Biheron.

With Linnaeus, however, whose stature as the Sage of the North was so grand, she played along with the flirtation. Always in elegant Latin, he wrote to Jussieu, "Send my compliments to my very dear Mlle Basseporte, the ornament of her sex; I speak of her in my dreams and if by chance I become a widower, she will be my wife whether she wishes to or not." This was on 12 April 1749, more than ten years after his Paris visit. A bit later, in an undated letter he added: "Transmit also I beg you my homages to my very sweet and suave fiancée, Mlle Basseporte; she has not yet gratified me with the child promised long ago, neither in her mind nor from her hand. If she could paint for me a little B. Jussieu in-quarto that I could insert among the [pictures of] botanists on the walls of my museum, she would thus give me a son who would be very agreeable to me."[42] Basseporte obliged, and with Jussieu's permission and willingness to sit for it the portrait of him was made.[43] But Linnaeus also called on her botanical expertise, at least once that we know of, asking Bernard de Jussieu to forward to Basseporte a request for a dried and depicted specimen of *Lonicera*, which he suspected might be a species of *Linnaea*. Lonicera is the honeysuckle, an arching shrub, or mostly a twining climber vine—and Linnaea is actually in the same family, a relation that Linnaeus was trying to work out. Linnaea, named for himself, was his favorite, a twin-flower which he had found in Lapland.[44]

Gaining confidence and perfecting her networking skills, Basseporte flourished even more after Buffon's arrival as director in 1739. When Aubriet died in December 1742 she mustered the courage to officially protest the fact that artworks done by her but signed by him, and for which she had never received any compensation, were going to be considered as part of his estate. She did not divulge the reasons for her action, stating that she would in due time, but then she withdrew her complaint the very next day.[45] Her objections would have been entirely justified, especially because she was of very modest means and Aubriet a rich man with a second apartment in Passy, but the fact that she desisted was not an act of cowardice. It was widely known in the Jardin that the deceased had taken unfair advantage of her talent for years, so overnight Basseporte chose dignity and forbearance instead of an ugly confrontation with Aubriet's heirs. Not wanting to appear grasping or peevish to admiring colleagues she kept her grievance to herself, which endeared her still more to the higher-ups in the garden.

Basseporte knew a lot of people and understood how to make felicitous connections, encouraging and assisting worthy contemporaries in need so that they too could become "useful citizens," as her eulogist put it. That same year, 1742, vowing to "cast off the chains that bind the wings of this genius," she was instrumental in helping the gifted but impoverished Rouelle, then laboring away in a pharmacy shop, to obtain a permanent post in the Jardin, where he became a famous and beloved lecturer on chemistry. Through her wealthy friends she secured equipment for him, crucibles and furnaces, and in this way "creat[ed] another great man for France." A second male protégé of hers with a gift for sculpture, P. H. Larchevêque, received with her backing a commission in Stockholm and became a favorite monument maker of the Swedish king. She even helped Bougainville's older brother Jean-Pierre secure the proper education to become a classicist.[46] The budding geographer Edmé Mentelle got a boost from her to begin his serious training despite crippling shyness.[47] And Marie-Marguerite Biheron, an art student of hers with more interest in the human body than in plants, became with Basseporte's energetic sponsorship a renowned anatomist and anatomical wax modeler (see chapter 4). She might also have inspired another strong woman, the wife of Buffon's close collaborator Daubenton, author of a novel about a self-reliant female Robinson Crusoe.[48]

Unfailingly diplomatic in her devotion to botany and to the Jardin, Basse-porte maintained friendships with both the Bernard de Jussieu and the Buffon camps, although the two men were often at odds and many perceived them as polar opposites, the one retiring, modest, and austere, the other a savvy show-man. Malesherbes, future director of the book trade and friend of the philo-sophes, studied with Bernard de Jussieu from 1746 to 1749 and would become an accomplished amateur botanist.[49] When the first volumes of Buffon's magis-terial *Histoire Naturelle* began to appear in 1749, Malesherbes was highly critical of it, although the manuscript he wrote detailing its shortcomings was not pub-lished until after his death. This very polemical *Apologie des naturalistes* was full of praise for all the smart, hard-working botanists whose ideas Buffon neither understood nor acknowledged, and in particular Bernard de Jussieu, "truly *the* naturalist of the nation." Buffon by contrast, wrote Malesherbes, knew no bot-any and never went into the garden to observe and study a plant. He was in-stead just a "cold compiler incapable of reasoning on nature which he only knows [from books]." Lazy and speculative, Buffon was a man of the salon and the private study, his multi-volume *Histoire Naturelle* a "confused mass of hypoth-eses, digressions, errors, and trivial worn-out beliefs."[50] Basseporte, who knew Malesherbes well and was aware of this critical manuscript—which incidentally came to light only because it was safely preserved by Mme d'Arconville (see chapter 5)—nonetheless stayed loyal to both Jussieu and Buffon. His hugely popular *Histoire Naturelle,* after all, was making science accessible to a wide read-ership, and she even provided a dazzling anatomical illustration for one of the first volumes to appear.[51]

Mme de Pompadour greatly appreciated Basseporte's scientific and artistic taste, summoning her frequently to the chateaux of Fontainebleau or Bellevue. For King Louis XV himself she was always on call, responding to requests from his ministers whom she knew well enough to lodge with them on her trips. Maurepas, for example, asked that she rush to Versailles in 1744 to paint an ex-otic gift sent to the monarch, and d'Argenson urged her to hurry to Compiègne in 1750 to capture a fresh fruit just received from the Indies before it perished. It must have pleased her to be so needed, although the calls were sometimes urgent: "I count on your not delaying an instant and coming to S.M.'s order and if you do not leave this very evening you will leave first thing in the morning to arrive here before His Majesty gets up. I will put you up at my house." The king, Basseporte was told, was always grateful for her diligence.[52] In 1759, Louis invited Bernard de Jussieu to create a garden at the Trianon of Versailles that would reflect the botanist's new "natural" method of classification that he had been developing. This alternative to the Linnaean groupings, which were based only on sexual characteristics, had been in the works for some time.[53] Thomas-François Dalibard, friend of Basseporte's and author of a book on botany, whom we will meet again, was among others growing impatient for Jussieu to finish this work that he had undertaken so long ago but was always attempting to perfect.[54]

Excited about his new plantings, the king took an interest in Jussieu and Basseporte during their many trips to Versailles. It was reported that he encouraged them to dispense with formalities when they addressed him, as the rearrangements took quite a few years.[55] When the regular Trianon gardener Antoine Richard questioned Basseporte's ability to continue her work as she aged, Bernard de Jussieu rushed to her defense. In a letter dated 20 December 1763, when she was in her early sixties, he wrote that she had not slowed down at all, that she had just done an exquisite rendering of a special plant for him, one raised from a seed that his brother Joseph de Jussieu sent from Peru. Jussieu suggested Richard contact the Marquis de Marigny, Pompadour's younger brother and director general of the king's buildings and grounds, if he needed more convincing. Nobody could rival or replace Basseporte, concluded Jussieu, and he thanked Richard for reassuring the king that she was not helped by students in any serious capacity, that she merely gave them lessons but that they could never possibly do her botanical work for her.[56] After Pompadour's death in 1764, Basseporte made fewer trips outside of Paris, although she was occasionally asked to paint rare plants at the royal chateaux, such as Duchesne's Chilean strawberry, which had given fruit only twice since 1716 but in July 1765 suddenly produced four berries, one of which was almost ripe. Bernard de Jussieu asked Basseporte to immortalize this unusual occurrence in Compiègne where it would be presented to the king, and to "paint these strawberries on their stalk which has never been done."[57]

That same year Antoine Laurent de Jussieu, Bernard's seventeen-year-old nephew, was called from Lyon to Paris to study with his uncle, and he immediately got to know Basseporte, whose doings he faithfully reported on to his mother in regular, detailed letters for the next fifteen years. The transplanted teenager was homesick, confiding that much as he appreciated the privilege of studying with his famous relative he found the atmosphere in the Jussieu house on the rue des Bernardins overly severe, even oppressive, and was glad for Basseporte's warmth and hospitality during his frequent visits to her home in the Jardin. He reported that his uncle visited Rouelle on the rue Jacob as many as four times a week, that Malesherbes and Duhamel were steady guests but that there were very few others, and that with the exception of their bossy, faithful housekeeper Jussieu did not tolerate women around him.[58] Basseporte, however, was a colleague, and had obviously learned early on to respect Bernard's privacy and impress upon him that she was as serious as any man about her and their joint botanical studies.

Many academicians relied on her as well.[59] In those days before photography, only drawings could capture the plant being studied and ensure its eternal visual existence. Her finished images, under the guidance of Duhamel de Monceau and other members of the Académie des Sciences who employed her, but especially her work with Bernard de Jussieu, had to represent the whole class of plant being studied, its essence, its invariable traits. She depicted not individuals with their anomalies, variations, and idiosyncrasies but composites that revealed the universal characteristics, the constants, the typical regularities for each plant. Only long, patient, and keen observations of particular specimens could lead to this general knowledge of the pure reality of each species, this ability to see past surface differences to the underlying "truth." It has been called "four-eyed sight," for it was a combination of the naturalist's and the artist's vision. Basseporte grew attuned to Jussieu's intellect, and in the process became a botanist in her own right through drawing. So closely did they collaborate that "the social and cognitive aspects of the relationship between the naturalist and the artist blurred."[60] Recognizing her straddling of these two worlds, art and science, Adanson's 1763 *Famille des Plantes* referred to Basseporte as "Peintresse Botaniste du Roi."[61] Her colleagues protected her in that position for five decades. They respected her mind and her hands, the work she did with "mindful hands."[62]

Before meeting the plant she was to study and depict, Basseporte did research about it. Was it edible? Useful for healing? Poisonous? What season was best for capturing its full glory? Some would perish immediately when transplanted, others would disintegrate, some bloomed only at night, some were so short-lived as to seem to be dying before fully open. Extreme heat, cold, rain, bugs, dirt, wind, and too much sunlight could make some plants close or wilt prematurely. Where did each plant originate, even if it had been naturalized to the Jardin? Did it normally grow in grasslands? Marshes? Forests? On sea cliffs?

In prairies? At alpine heights? Looking, watching, seeing, focusing, paying rapt attention over a long period of time were all required to detect the tiniest relevant detail. Basseporte's penetrating gaze led to knowledge, her sight led to insight. But she also handled plants to study their innards, to find and analyze what was hidden. The toolkit in her satchel included magnifying glasses, microscopes, and fine dissecting knives similar to the instruments used by jewelers.

Generally Basseporte would make sketches in the field, taking notes for herself about color and anything else that the quick rendering might not show, and then complete the work later indoors—but only after repeated visits to the plant in the ground. Sometimes the easel would come outside, in a wheelbarrow with paints, chalks, and brushes. Her work was challenging, for while the main goal was scientific accuracy, the representations had to be aesthetically pleasing as well, so that the recorded image would have sympathy, understanding, and soul.[63] Basseporte skillfully imparted to her illustrated plants, as a later enthusiast noted, some of the life that circulates in them with the sap.[64] Just as Enlightenment writers sought to amuse *and* instruct, botanical pictures were to be beautiful as well as educational and technically impeccable. Why else would the king have wanted a collection of them for his *Cabinet?*

Basseporte's depictions were realistic but also unreal, revealing in a single image things not normally seen together in order to communicate maximum information. She played with space by showing both the top and bottom of a leaf, or displaying the inside of a flower usually shielded from view, or greatly magnifying some tiny detail, or bringing foreground and background simultaneously into focus. She played with time by showing different parts of the plant's life cycle all at once: seeds, buds, flowers, wilts, fruits, roots, cross-sections of light stems, and mature branches. Using a wash with muted overtones, she made the leaves appear three-dimensional with little strokes of white gouache and nuanced patches of different hues, detailing their ramifications and nerves very freely and putting them into relief with accents and different shades of green (figure 8).[65] Her images were intricate, some bold, others subtle and seemingly seen through a veil.[66] She departed from Aubriet by using artificial light sources, glints, and reflections, in a style that was uniquely hers and that some regard as the most original among the paintings on vellum, a kind of parchment made from calfskin.[67] During her long tenure she did countless undated pencil sketches "for M Jussieu," and 313 finished *vélins* for the royal collection, 295 of which are flowers and plants.[68] They depicted vegetation from Peru and Chile, India and

Fig. 8. Madeleine Françoise Basseporte's painting, "Magnolia," on vellum, for the king's collection of botanical works, *Les Vélins du Roi*. [Muséum national d'Histoire naturelle, Paris, France. ©RMN-Grand Palais/Art Resource, N.Y.]

China, the Himalayas, Corsica, the Alps, Mexico, the Antilles, Barberie, and Virginia.[69] Her images enlightened, introducing rare exotic vegetation that would not have been seen otherwise in France.

But botanical illustration was always, primarily, art in the service of science. As A. L. de Jussieu explained, "[Basseporte] drew all the new objects that Bernard de Jussieu directed her to and traced under his supervision the details presented by the distinctive character of each, . . . details always necessary to the naturalist and which for him add immense value to the work of the artist."[70] Condorcet's *éloge* of Bernard de Jussieu said he shared his knowledge so generously and taught so well that "he instructed gardeners and succeeded in making them true botanists."[71] The same was so of his draftsperson, Mlle Basseporte. In general, there was a leveling of ranks within the Jardin, as made clear by a *Mémoire* of Thouin's describing the democratic atmosphere that prevailed. It was a botanical school whose staff members "make no distinctions among themselves except those arising from merit and seniority. Since their work has no other purpose than the progress of science and public utility, each of them tries only to deserve in his own area the esteem of scientists and the affectionate regard of his fellow citizens."[72]

By the mid-1760s Basseporte knew all about Bougainville's planned trip, as the Jardin was abuzz with anticipation of all the wonders it would receive from Commerson. He was in almost daily touch with Bernard de Jussieu throughout his stay in Paris and even forged a friendship with the young nephew, Antoine Laurent, to whom he sent greetings in his letters during the voyage. When some shipments of specimens from the expedition's first landfall in South America got lost en route to France, including a magnificent plant he and Barret found in Brazil and named the *Bougainvillea* in honor of their captain, Commerson was so distraught that he almost refused to send any more, but then realized that despite such risks he could not frustrate Jussieu's hopes.[73] Had those plants and seeds arrived in a timely way, Basseporte would immediately have been called upon to inscribe them into the permanent record.

Commerson and Barret, after leaving Paris in mid-December 1766, traveled for several days and nights through Estampes, Orléans, Blois, Amboise, Tours, Poitiers, Niort—where they were almost crushed by a drunk postilion—La Rochelle, and finally arrived before Christmas in Rochefort, a busy port city on the Charente estuary.[74] They encountered delays before they could actually embark, and it is not clear just how they spent their month and a half waiting. Barret could have polished her male role still more in this navy stronghold where there were plenty of tough models to imitate. A penal colony supplied free laborers for sail and rope makers, shipbuilders, the gunpowder works, and cannon foundry, and of course there were always sailors everywhere. Commerson no doubt redoubled efforts to procure the latest equipment for his botanizing.

Then, as he finally boarded the storeship (*flûte*) *Etoile* on 1 February 1767—the frigate *Boudeuse* carrying Bougainville had cast off somewhat earlier from Nantes—he reported with feigned surprise to M. Bernard, a friend from his hometown of Bourg, that "they have given me a valet, paid and fed by the king."[75] There were 214 people on the *Boudeuse*, a three-masted sailing ship with square sails, and 116 on the *Etoile*, where the couple, now master and man-servant, were accommodated with all their books, gear, instruments, and collections. The stupefied captain of the storeship, seeing all the paraphernalia Commerson brought along—including 232 books in his traveling library—generously relinquished his own quarters, enabling the naturalist and "Jean Baré," his "valet," to cohabit in this relatively ample space.

After three months crossing the Atlantic, they reached the east coast of South America, where they were thrilled to go ashore and stretch their legs on solid land. Commerson's log revealed that he had been violently seasick a good part of the trip, which was surely unpleasant for the sturdier Barret who had to tend to him. They explored Rio de Janeiro, where they found and dedicated the *Bougainvillea*, Montevideo, and Buenos Aires. Commerson, so overjoyed to finally be on land and starting his botanical researches, bragged in a letter of 7 September 1767 from Buenos Aires, "You know my passion for seeing: in the midst of all the hostilities, despite being formally forbidden to go outside of the city . . . I dared twenty times to go out with my domestic in a dugout canoe directed by two blacks."[76] Barret, his "domestic," went everywhere with him, clearly indispensable for his work. Even François Vivès, the ship surgeon who resented Commerson for his superior medical knowledge and his airs, was awed by the devotion of his assistant, following him constantly up and down high mountains at Montevideo, carrying weapons, provisions, collecting bottles, specimen jars, cumbersome notebooks, and special papers for drying plants.[77]

Bougainville decided that braving the treacherous Strait of Magellan would be safer than sailing around Cape Horn and waited for the Austral "summer," but the weather was punishing as they entered the strait on 5 December 1767.

There were Patagonians to the north (Chile), Fuegans to the south (Argentina), and the ships traveled at a snail's pace, requiring fifty-two days to get through those three hundred nautical miles. This slowness, however, afforded felicitous opportunities for Commerson and his manservant to explore for plants. Barret was dauntless, displaying fortitude under the worst conditions. Later, once her secret was out, many marveled in retrospect at her extraordinary industry, especially during this passage. Dumbfounded by her strength, Bougainville used it as an excuse for his blindness to her true sex. "But how recognize a woman in this indefatigable Baré, already an expert botanist, following her master in all his *herborisations* in the midst of snow and on the glacial mounts of the Strait of Magellan, and carrying even on perilous excursions provisions, weapons, and herbals with a courage and force that earned her the nickname of the naturalist's beast of burden."[78] Prince Charles Henri de Nassau-Siegen, Maurepas's nephew, an energetic twenty-one-year-old nobleman on the expedition who befriended and often botanized with the couple, admired Barret enormously, recollecting in his log her vigor and resourcefulness in the face of extreme danger.[79] Vivès echoed in astonishment: "One had to have seen her at the Strait of Magellan, exhausted from the rigors of the cold, in the water for shells, or in the woods in the foamy mud and snow for entire days seeking plants. . . . One can say in praise that it is impossible to conceive the work that she did."[80] In another version of his journal, Vivès emphasized that she "braved the elements carrying things that would normally take eight or ten hands."[81]

At last, on 26 January of the new year 1768 the ships exited the straits, and now the wide Pacific beckoned. On 21 March they reached the Tuomoto Archipelago, on 2 April they could see Tahiti. Two days later, before they had even set anchor, the Tahitian prince Aotourou clamored on board to greet them, and on 6 April the French went ashore to claim the island for their king, only to discover that the Englishman Samuel Wallis had gotten there first. The next day Barret ventured onto the shore and, instantly surrounded by the natives, was immediately unmasked.

Beardless, high voiced, small in stature, refusing to urinate in the presence of others or to jump naked with the crew into the equatorial waters when the boats had "crossed the line" the previous year, Commerson's "valet" had aroused some suspicion among the sailors, but was so zealous and hardworking that for the most part they left the peculiar individual alone. To dispel rumors and discourage advances, as Vivès later explained it, "our little, fake man" told the other sailors that she was a eunuch, then doubled down to work more strenuously than anyone else. She was a hardy traveler, in contrast to Commerson who was at most times either seasick or nursing some fever or festering wound. But here on Tahiti, on 7 April, the truth of Barret's sexual identity was definitively established. Aotourou may have been the earliest to figure it out.[82]

Accounts differ as to what happened after Barret was rescued by one of Bougainville's officers and escorted back onboard. Oddly, the captain did not address the matter for almost two months, finally getting around to it and writing in his log on 29 May of a long confession he got from a tearful Barret, an artful mix of truth and lies, that she was an orphan from Bourgogne who, reduced to penury after losing a lawsuit, came to Paris to work cross-dressed as a manservant for a Genevan and then, because she was eager to sail around the world, tricked Commerson at Rochefort. She thus absolved her master of complicity and impressed Bougainville whose entry showed that he "admired her resolve, especially because she always conducted herself with the most scrupulous prudence."[83]

Once Barret was exposed, Commerson abruptly lost the helper upon whom he had relied for years. Arrangements were made to ensure her welfare on the ship, although Bougainville admitted that her modesty was sometimes compromised. She could of course no longer bunk with Commerson, but then where *could* she have safely stayed? Vivès reported that she befriended Aotourou and that the two of them happily did each other's toilette while the boats were still anchored. When the ships left Tahiti on 15 April Aotourou volunteered to return with them all to France, but he was soon transferred to the *Boudeuse* under Bougainville's watchful eye. Barret remained in men's clothing—there would not have been any female dress on board—but Vivès explained that she was much relieved not having to bind her breasts any longer, a procedure that had resulted in all manner of skin lesions and bruises to her ribs.[84] Being confined to the boat was no doubt boring, but in fact there were not many landfalls during the next part of the trip. On 2 May they saw Samoa (Petites Cyclades). On 5 June they neared the Great Barrier Reef of Australia, then headed north, designating the New Hebrides the Louisiades for King Louis XV, and Bougainville naming an island in the Solomons for himself. On 2 September they anchored at the island of Bouro (today Buru) in the Moluccas, where Commerson went ashore to explore alone but named a gorgeously colored bird for Barret, the "perruche à Baret ou de Bouro."[85]

Threading through islands too numerous to count, the boats arrived on 28 September at Batavia (Jakarta, Indonesia), where they stayed for over two weeks getting necessary medical attention and replenishing supplies and morale. On 16 October they sailed on to face the desolate Indian Ocean, Bougainville now heading for the French colony on the island called Isle de France (today Mauritius), off the east coast of Madagascar. Commerson, always flamboyant, described the hardships of the crossing in letters, but for this stretch the winds were favorable and they made good time, arriving at Port Louis on 8 November.[86]

Caught in an untenable situation with a woman on board, a grave breach of naval regulations, Bougainville needed to divest himself of the couple, who would surely have caused embarrassment on his return to France had they stayed with him. Impressed with Barret's behavior and intelligence, he had written in his log that "the court, I believe, will pardon her infraction of the law,"[87] but despite this sanguine prediction and his own lenient attitude he decided to leave the pair off on Isle de France. The local administrator, or intendant, of this French colony, a botanist interested in the cultivation of spices and aptly named Pierre Poivre, was keen on the idea. A close friend of Bernard de Jussieu, Malesherbes, Turgot, and Mentelle, he had met Commerson before leaving Paris to take on this colonial assignment, and he now invited him to join in studying the vegetation of the Mascarenes. Poivre's enthusiasm made unloading the couple tidy. Whether Barret's identity was discussed we do not know, but Bougainville soon reported tersely in his log that he had disembarked the naturalist and his valet, "fille en homme."[88]

Commerson's letters show that he missed his son and was therefore concerned about the delay in returning home, but he was seduced by the prospect of being able to explore this part of the world after so many frustrating months at sea. Poivre genuinely welcomed the help for his lively research program, which included impressive studies of climate, remarkably prescient conservationist efforts against deforestation, and the care of the unique Jardin de Pamplemousse at Poivre's residence, Mon Plaisir, the world's only tropical botanical garden, so rich in exotic fragrant plants that it could be smelled before it was seen. Growing there, among other things, were acacias, balsams, pomegranates, waterlilies, lilacs, hydrangeas, lychee, ginger, palms of all kinds, camellias, mangos, bamboos, cinnamon, Bengal roses, giant geraniums, calla lilies, gladioli, ebonies, bloodwood trees, and some nutmeg and cloves purloined from the Moluccas to break the Dutch monopoly, although these ultimately acclimated better on the nearby Isle Bourbon (today Réunion).[89] Commerson planted countless seeds and seedlings he had brought from South America, Tahiti, and other South Sea locations that thrilled Poivre and thrived in this park "beyond all our hopes." The intendant was also pleased to have intellectual companionship, which was in short supply in the rough town of Port Louis.[90]

It took Bougainville many weeks to get his ships repaired, stocked, and fortified for their return to France, but on 12 December the *Boudeuse* set sail from Isle de France, and on 1 January 1769 the *Etoile* followed. No longer urged by Commerson to stop repeatedly for specimens, the expedition made its way relatively swiftly back to France, the captain's frigate arriving at Saint-Malo 16 March, the storeship following some weeks later and eventually docking on 24 April at Rochefort.

How did Barret feel watching the vessel that had been her home for almost two years sail away? She may have been sorry, even frightened, to lose Bougainville's support, as he was indulgent and kind; perhaps she regretted the end of the odd relationship with Aotourou who first sniffed her out and then, according to Vivès, became her friend onboard. But she had trekked through snow and ice, torrential rains, howling winds, dense forests, peat so spongy she sank into it, rugged ravines, and she had endured hurricanes, seasickness and its stench, the menace of fevers, scurvy, mouth and lip ulcers, raw throats, the fear of piracy and shipwreck. Her food had been biscuits, lard, salt beef and salt cod, cheese, beans with oil and vinegar, some wine. When bad weather killed the livestock they had aboard and supplies ran dry between landfalls, they ate rats and even leather from the boat fittings.[91] Drinkable water was made from the sea with Poissonier's 1762 distilling invention the *cucurbite*, but only in small amounts. Commerson's histrionic letters had described the hardships he suffered, his life-threatening exploits in the name of science, collecting in hostile territories, near starvation, putrefying wounds, perilous terrains, labors "greater than Hercules," and his heroism at overcoming them. Of course Barret was right there along with him for most of this, and for her there had been the additional discomfort of her disguise, which covered her with sores from perspiring in the tapes and wrappings, and always, always the fear of being discovered. So standing now on dry land as the sails of the *Etoile* disappeared over the horizon she must have felt some relief, while believing she needed to stay by Commerson's side if there was any chance of one day getting back home.

The big question is what Barret did on Isle de France between the time she arrived there and Commerson's death five years later.[92] Did Bougainville's "fille en homme" description mean that she remained in men's clothing and continued to pass herself off as Commerson's male assistant? If so, how long did that last? How much did Poivre know? He gave Commerson lodgings at the Intendance in Port Louis. Did Barret stay in male garb and join the group of domestics there? Do the same, but as a female servant? Poivre at forty-eight had recently married the seventeen-year-old Françoise Robin, so he may have been sympathetic to Commerson's unorthodox liaison. But whereas on the *Etoile* Barret had been scrutinized, with many onboard required to keep ship's logbooks to record events and several among them fascinated by the peculiar valet, on Isle de France nobody was obliged to take note of her doings, and so the paper trail vanishes. Whether, and if so in what capacity, Barret was still working with Commerson from November 1768 until he died in March 1773 it is not possible to establish. But it is also difficult to envision that they completely abandoned each other.

In any case Barret's role would have been considerably diminished now. On the voyage in 1767 and 1768 she had been Commerson's sole assistant, hiking with him through thick and thin, collecting, organizing, drying, and labeling specimens for his herbaria, probably even sketching in many cases. Many of the images sent back were signed by him but some were not and may have been executed by her.[93] She had also nursed him when he was ill and given him hopeful companionship. Once on Isle de France, however, Commerson had Poivre's circle for conversation, and he had the assistance of the *noirs du roi*, slaves assigned to him to do his bidding, in addition to two slaves he found for himself—the tireless Joseph from Mozambique on Africa's easternmost coast, who knew everything about the island and whom Commerson even envisioned as a future companion for his son Archambault, and Farlat from Madagascar.[94] Poivre also provided two draftsmen for the naturalist, his cousin and godson, the twenty-year-old Pierre Sonnerat and nineteen-year-old P. P. S. Jossigny, so Barret's artistic services would no longer have been needed. Commerson apparently did not pay much attention to his newly assigned artists until 1770 when illness slowed him down, collecting got harder, and he invited others, especially Poivre's good friend Cossigny de Parma, to send him specimens to keep his "bureau botanique" occupied.[95] He made no mention of the presence of additional helpers; "les nez delicats" he brought up at one point, who objected to some of the smellier research, must have been Mme Poivre and her friends, because Barret was made of sterner stuff.[96]

And because he had gotten used to sallying forth alone after Barret's unmasking and confinement, he continued to do so, his sorties into the wilderness as reckless as ever. The writer Bernardin de Saint Pierre, future director of the Jardin du Roi who had made a trip to Isle de France to study the local vegetation, reported that Commerson climbed one day on the Corps de Garde, a parasol-shaped mountain with a sharp escarpment, where there is hardly any way to get up or down. With characteristic bravado he sent his guide back so he could concentrate on botanizing there, assuring the experienced native that he would be able to manage by himself. As darkness fell he got hopelessly lost, unable to find any path for descent. He had to eat some pea plant he found, edible though sparse, but the next morning he had no better luck getting down from the plateau and would have died up there had the guide not gotten concerned and gone back up to fetch him.[97]

Was Barret with Commerson when he left in a rush on 11 October 1770 for Madagascar where a small, unstable French outpost was about to be dismantled? It was to be a short trip. He landed at Fort Dauphin and was pleased with his accommodations and the welcome he received from officials there. He sent an almost Darwinian rhapsody about Madagascar, recognized today as a biodiversity hotspot, to his friend Lalande: "What an admirable country Madagascar is! It deserves not informal observers but whole academies. Madagascar, I hereby announce to naturalists, is their veritable promised land; it is there that nature appears to have retreated as if into a private sanctuary, and worked there on different models than any she has used elsewhere; one comes across the most strange, the most marvelous forms at every step."[98] Commerson's *herbier* included at one point 495 plants collected there, one with heart-shaped leaves that he named for his deceased wife. He went next to the Isle Bourbon, where he was unexpectedly held up for almost a year because of an infected wound, reading Horace for lessons on fortitude and firmness to help himself out of despondency.[99] Despite weakness and infirmity, however, he recovered enough for an ambitious multi-week trip to the island's active crater in October and November 1771, telling Lalande that while he already heard the bowels of the earth rumble, he would try to escape incineration and not get himself killed in nature's ovens and pyrotechnical labs. A description of this expedition names the men in the party, including some who climbed the Piton de la Fournaise, a restless basaltic shield volcano near the moonlike landscape of the Plaine des Sables. Barret could have been there if she had assumed male garb again and taken another pseudonym, climbing around on the unstable crust crumbling beneath her feet, breathing sulfurous emanations.[100] Commerson also botanized on Isle Bourbon, naming plants for Lalande and for his dear friend Vachier, whom he had known and trusted since medical school in Montpellier and who was now left in charge of all his affairs back in Paris, and for Poivre, Cossigny, Nassau, and other new colleagues.

It is, in fact, rather unlikely that Barret went with him to Madagascar and Isle Bourbon, at least initially, because his excursion was to be so quick, no more than a month. Tempting as it is to picture her along on all those adventures, she was probably instead assigned responsibilities in his absence. Perhaps to watch over and protect the existing collection which Commerson described to Bernard de Jussieu as already "immense"?[101] To receive shipments of additional items he sent back from the neighboring islands? To dry and organize recently gathered plants, a crucial skill that others on Commerson's new artistic staff might not have had? With his extreme ways and irascible temperament he had made enemies and become profoundly mistrustful, wary of his collections being tampered with and priority being denied him, so he may well have designated Barret to guard duty. But when his delay stretched into more than a year and news of his illness reached Poivre, Barret could at some point have taken a boat ride to work with and nurse him on Isle Bourbon. The trip there from Isle de France would have taken her just two or three days. Did Commerson send for her? Did she pop over more than once during his time there? Of this we have no record.

Had Barret split completely from Commerson she would have needed to become even tougher than she was, and more resourceful. Port Louis was a harsh place, full of bustle and threat, drunks, criminals, transients, whores. There were only three churches and all in bad shape, poor roads, hospitals with no medicines, much corruption among the locals, lots of "mauvais sujets," a severe housing crunch, insufficient food—too much coffee but not enough sustenance. An observer in late 1771 remarked: "It is not possible to imagine the chaos . . . that has been present in this colony for a long time, the divisions and the cabales . . . rival factions . . . no public tranquility." And then there was smallpox, ripping through the island since August 1771, causing numerous deaths each day in the city and still raging at the end of the year.[102] Aotourou, who had gone with Bougainville to Paris where he was displayed and studied as a "noble savage" by La Condamine among others, returned to Port Louis on his way back to Tahiti, only to contract smallpox and die, never seeing his home again.[103] Barret survived the epidemic but exactly how she spent her time and stayed safe we do not know.

Commerson returned to Isle de France in early 1772, narrowly missing the devastating hurricane of 29 February–1 March, sixteen hours of violent raging storm that smashed boats and killed countless people and animals. Once back he botanized with the admiring and loyal veterinary doctor François Eloy de Beauvais, a "grateful and patriotic friend" who wanted to make certain the brilliant naturalist got the proper credit for all of his discoveries.[104] Beauvais would soon be his sole supporter, for in October of that year Poivre left for France with the astronomer Alexis-Marie Rochon, who had just visited Isle de France for the second time, and all at once Commerson lost his other two allies. For him this was an unmitigated disaster. Too encumbered with his collections to accompany his friends when they sailed for home, his fortunes changed overnight. He and his numerous specimens, herbaria, and papers were summarily kicked out of the Intendance by Poivre's successor Jacques Maillart, and he was forced to hastily purchase a structure on the rue Ancienne des Pamplemousse to house everything. He still owned this sorry dwelling at his demise, when it was suggested that his estate sell it quickly before it was crushed completely by the next hurricane.[105] He felt increasingly sick—dysentery and rheumatic gout according to his own self-diagnosis—and extremely isolated. He wrote to Lalande that he was at death's door and that it was time to start writing the history of his scientific martyrdom.[106] His last known letter, of 27 October 1772, was a tragic plea for help and support sent to the professor of botany at the Jardin, Louis Guillaume Lemonnier. In it Commerson lamented that he was "like another Prometheus nailed to the rock" suffering repeated calamities, without funds, friends, "on the edge of his grave" and with the premonition that his collections, "watered in his blood and sweat" would, if separated from him, end in pillage and certain ruin. He was also deeply in debt, unable to function without his draftsmen, incapable of paying anyone to help him.[107]

Friendless, paranoid, and sure he was dying, Commerson made his way east, still botanizing when he was able. On 26 February 1773, he "had himself carried" from VilleBague in Pamplemousse in the middle of the island to Flacq, a town on the cooler, windward coast where he lodged at a plantation called La Retraite, chez one doctor Bézac. Commerson died just before midnight on 13 March, entirely destitute. He was buried the next day.[108] The inventory of his possessions showed their meagerness: in a small trunk mentioned by Bézac were found some old, worn-out clothes appearing to be from his Paris days and ill-suited for his occupation, only two pairs of shoes, a complete microscope in its case, a work on medicine, a few brochures and a leather natural history portfolio, candles, feather pens, and a packet of seeds. Cufflinks, an English silver watch, his dilapidated house in Port Louis where his collections were stored, and his two slaves rounded out the list of what he owned on Isle de France. As his brother-in-law Beau would plead to the French authorities, in an attempt to get a pension for Commerson's now-orphaned son, "This child is absolutely without fortune. His father, to support the different voyages that he made, sold almost everything that he possessed. . . . At his death his effects were sold at a vile price."[109] Commerson could not have been expecting help from Barret in this situation. We cannot even be sure that they were still in any kind of partnership.

But he was certainly thinking of her, and had been all along, revealing his affection and respect through the naming in her honor of a tall shrub with dark leaves and white flowers. He initially saw it on Isle de France and then found other species of it on Isle Bourbon. Calling this new family *Baretia*, invoking also her masquerade and her mysterious nickname Bonnefoi, as mentioned in his testament, Commerson had found a fascinating plant with ambiguous sexual characteristics. He wrote to Cossigny on 19 April 1770, captivated by its cryptic foliage, a fitting reflection of Barret's gender malleability. His letter included a detailed list of many specimens, but he singled out numbers 54 through 58, examples of a charming bush: "I am crazy about it, partly because of the singularity of its leaves, partly because it gives me a new genus whose character is unique: that is it has its stamens carried without filament at the edge of a little cup inside the flower fold that we call nectarium. I ask you to tell me, regarding the leaves, if the fullest do not occupy the summit of the tree and the most sinuous the lower branches. The fruit, sir, the fruit, don't forget it I beseech you. In the past I gave it the name of *bonafidia* for a reason."[110] He soon received another *bonafidia* from Cossigny, which gave him pleasure, one with ten stamens and five petals. "This way we have all the species and varieties drawn. We won't talk about it anymore until the fruits, that I will ask you for when they are ripe."[111]

Commerson found other samples of *Baretia* on all three of the islands, as we see in his herbaria.[112] His dedication to Barret is in Latin, a moving homage to her invaluable work and her right to recognition as a woman of science:

> This plant with deceptive finery and leafiness is dedicated to the valiant young woman who, assuming the clothes and temperament of a man, had the curiosity and audacity to travel the entire world, by land and by sea, accompanying us without our realizing anything. Many times she followed in the steps of the illustrious prince of Nassau, and in ours, crossing with agility the highest mountains of the Straits of Magellan and the deepest forests of the austral isles. Armed with a bow, like Diana, armed with intelligence and gravity, like Minerva, helpful and virtuous, inspired by some propitious god, she eluded the traps of beasts and men not without risking countless times her life and her honor. She will be the first woman to have circumnavigated the earthly globe covering more than 15 thousand leagues. We are indebted to her heroism for so many plants never gathered before, for so many herbaria constituted with care, for so many collections of insect and shells, that it would be an injustice on my part, and of any naturalist, not to render her the deepest homage by dedicating this flower to her.[113]

The morphological changeability of the leaves of this plant, adaptations for survival in a hot climate, struck Commerson as particularly suitable, given Barret's disguise. The leaves are lobed like oak leaves when young to facilitate transpiration, but the contours smooth out in maturity.[114] Commerson played with several names for the varieties of the plant that he discovered, finally settling on *Baretia oppositiva, heterophylla* and *Ababella Bonafidia*.[115]

There is near unanimity among tellers of Commerson's tale that Barret was with him until the end and closed his eyes. For example, Montessus, a descendant of the naturalist, wrote that his ever-attentive servant offered her warm hand, her consoling words, "exhorting him to be hopeful" in his abandonment.[116] We do not have hard evidence of this, although it is the sentimental favorite. For who but his faithful Barret, never paid since she fled her official Paris employment in December 1766, would have been at his deathbed? Once she boarded the *Etoile* she became entirely dependent on him, without any standing of her own, and Vachier sent her no salary during those years for she was no longer a housekeeper. She had certainly been devoted. Yet Bézac reported that Commerson was alone with only the local parish priest, which must have been small comfort to him as a nonbeliever. Perhaps this host thought it unnecessary or inappropriate to mention an attentive female servant who might have been nursing him?

Another plausible scenario, however, is that Barret was not there in Flacq, and was instead back west in Port Louis supervising his collections, which makes sense considering Commerson's fierce suspicion of the new intendant Maillart and others. She might have felt she could best help by taking care of his beloved specimens, more dear to him than life itself: "My plants, my beloved plants," he wrote, "have consoled me for everything; I found in them Nepenthes, curare, dulce"; only they unfailingly banished his grief, healed him, and provided sweetness.[117]

Even this storyline is problematic, for how involved with Commerson's material could Barret have been after his death, with no independent title to intervene? Yet trained in taxonomy, nomenclature, and the techniques of drying samples for herbaria, she was the most competent person for the job at a time when everyone else was ignoring the collections.[118] She had witnessed complete and passionate devotion to science—Commerson telling his friend Gerard, as if it were necessary, that "science is indeed essential to my being"[119]—and in the very first paragraph of his testament he had even bequeathed his body to a medical faculty for dissection, with an artificial anatomy to be made of it to perpetuate his usefulness, a wish impossible to fulfill under the circumstances of his death in a remote location. Barret knew of his other unfulfilled wishes too, for example, his quixotic dream of a universal academy of happy savants, which was just that, a hope beyond realization, but as he wrote, "When one builds in this way castles in Spain, it costs nothing to make them grand and beautiful."[120] He was a visionary, a man possessed, and this must have affected Barret in some profound, enduring way. How could she turn her back on thousands of notes, hundreds of herbals, countless cartons and catalogues that he and they had so painstakingly assembled over the years? Someone had to at least have kept track of where they were. It could well have been Barret, given the indifference, indeed the hostility, of most everyone else.

Maillart's visceral dislike for Commerson was revealed in a letter of 15 March 1773 two days after his death, reporting it to the minister in France and saying that although he would supervise the posthumous mailing of his collections if ordered to, the naturalist was a friendless monster: "He was known to be very debauched, and considered to be a mean man capable of the blackest ingratitude."[121] We can only imagine what this was meant to refer to, but clearly Commerson had offended and alienated those in positions of power. Meanwhile from Port Louis Barret sent a letter to Vachier on 17 March informing him that Commerson had died and expressing concern about the inheritance due her from the terms of his testament. Hers was an informal rather than official notification, as she may have been the only one who knew of Commerson's personal friends back home. Vachier did not receive her letter until August, almost half a year later, and answered her in September promising that he would honor the terms of Commerson's testament and that she did not have to send a *procureur* to represent her.[122] While waiting for this reply, which took months to reach her, Barret may have been part of the reason an inventory of thirty-four cases of Commerson's effects was made, a fraction of what they had together amassed, dated 9 November 1773. A doctor, one Saint Mihiel, was said to have boxed the material, and a shipment was sent to France that month on the *Victoire*. Jossigny was ordered by Maillart to accompany the cargo, but the young draughtsman had grown to dislike Commerson enormously, did this job grudgingly, and would not have been an advocate or even a careful steward of the priceless boxes.[123]

Had Barret been instrumental in getting this catalog done in early November, she may have felt that she could do no more and turned to her own need to make a living. By late December 1773 she had opened a billiards bar/cabaret, where she was fined 50 livres for selling spirits during mass on Sunday.[124] She soon took another big step, marrying on 17 May 1774 a French soldier named Jean Dubernat, a drum major stationed temporarily on Isle de France with his regiment.[125] As a military wife she would finally be able to sail home with her husband, and knowing Jossigny's animosity for Commerson, she may also have felt some urgency to get back to France to see that the precious collections were safely received.

Meanwhile, what of those collections? From Lorient, where they arrived in May 1774, the shipment traveled northeast to Le Havre, then Rouen, and finally to Paris in late summer where A. L. de Jussieu was supposed to order them. There is no mention in his letters of this material, suggesting that it may have been ignored. Buffon took some of it, found years later stashed away in his attic. Jossigny seems to have considered his escort duties done on 2 August 1774.[126] A letter from Beau reveals that he was in Paris that same day to stake his claim to Commerson's possessions for his ward Archambault. Commerson's rue des Boulangers residence was thoroughly examined between August and Novem-

ber 1774 and an elaborate inventory made of his specimens and other belongings, followed closely by Vachier and his notary, by Beau, and by Lalande who
had over the years lent Commerson considerable sums.[127] Basseporte's good
friends Dalibard and Jacques Barbeau Dubourg, both of whom had written
books on botany, were also involved looking through the scientific materials
the couple had left there when they departed eight years before.[128] A. L. de Jussieu was sometimes there as well. By November when the complicated cataloguing was finished the apartment was given up, its contents sold or distributed.

As for the thirty-four cases that had been sent to the Jardin from Isle de
France, there was some sparring between the botanists and the Cabinet du Roi
about the material, but then a deafening silence about this treasure. And there
things might have remained had Barret not returned, in 1775, to France.

While Barret was away on her journey and until she reappeared in Paris,
an absence of nine years, Basseporte's work in the Jardin continued unabated,
as demanding as ever although during that decade, as she aged from sixty-five
to seventy-five, her health fluctuated wildly; frequent asthma was a particular
problem according to A. L. de Jussieu's letters.[129] She soldiered on, however,
still enjoying and expanding her circle of acquaintances. Through her connection with the French electricity group spearheaded by Dalibard and Dubourg—
she knew these two men as botanists, but they were also physicists and, in fact,
Benjamin Franklin's French translators—Basseporte had met and impressed
Franklin himself and his traveling companion Sir John Pringle when they came
briefly to Paris in 1767 and 1769.[130] Franklin continued to send her greetings in
numerous letters to his translators, written from London and America. On at
least one occasion Basseporte entertained Franklin and Buffon at her home.[131]

Rousseau, whose passion for botany stretched back many years, was spending more time in the Jardin in the 1770s, enjoying Dubourg's 1767 *Le Botaniste
Français* with its homage to Bernard de Jussieu "who relates in his *moeurs* to
former centuries and anticipates by his wisdom future centuries to come."[132] As
early as 1763, Rousseau had told a friend that "often when I see around me the
abundance of rare plants, I become sad as I walk by them in my ignorance."[133]
The statesman Malesherbes, another fine amateur botanist of Basseporte's acquaintance, wrote to Rousseau in 1764 sharing his love of botany and noting
what a benign habit it was compared to hunting, which "lower[s] the standard
of kindness." Rousseau would talk for the rest of his life about the consolations

of botany, writing in 1775, "My dialogue with plants came only after humanity had refused me its companionship."[134] He wrote of being more "touched" by nature in the raw than by cultivated gardens and critiqued the appropriation and naturalization of foreign flora,[135] but despite this he had enormous appreciation for Bernard de Jussieu's efforts doing just that in the Jardin. Their respect was mutual, the older man eager for Rousseau's insights.[136] And Rousseau came to know Basseporte and appreciate how her illustrations gave plants eternal life, famously remarking that nature gave plants their existence but she immortalized them.

Whenever Basseporte fell sick, others hovered like vultures over her coveted position, including young Mme Vien, the former Marie-Thérèse Reboul, who had earlier been one of her students. On 31 May 1769 the minister Marigny in charge of the Jardin heard that Basseporte was dangerously ill with little hope of recovery and wrote a letter about replacing her, but Buffon would not hear of naming a successor and put an immediate stop to the discussion.[137] He had several times witnessed Basseporte's remarkable recuperative powers. Nothing would happen without his permission, and he was not granting it. Saint Florentin, the minister of the king's household, felt likewise and reassured Buffon, "I know, Monsieur, the services of Mlle Basseporte and I think as you do that it would be an injustice to give her a successor unless it would be with her consent. I will take no commitment on this position except in concert with you and only on the presentation that you will make of a subject capable of taking her place, about which you can be a better judge than anyone else."[138] It is extraordinary that she had all of these powerful men cowed.

Basseporte's devotees in the Jardin were numerous and steadfast, and not only those in her generation. She was also loved and esteemed by several budding botanists forty years her junior, among them Bernard's nephew A. L. de Jussieu; the Jardin's gardener André Thouin, whose sister she was tutoring; and botanist Jean-Louis-Maurice Laurent, a visitor who spent three years there from 1768 through 1771 and who would soon become the head of the naval botanical garden in Brest. When the botany professor Lemonnier found it hard to meet his obligations to the Jardin because he was also physician to the king, A. L. de Jussieu was chosen as his substitute, a huge boost to the young man's career. Basseporte wrote immediately to his mother on 21 June 1770, telling her what a wonderful job he did in his new capacity, that it was a testament to his fine upbringing by her in Lyon, that he thought and behaved at twenty like a person twice his age, all in a gushing, unpunctuated letter overflowing with delight and pride:

> Your son was approved by the king, the minister, M de Buffon and all the public applauded this choice, although it is only temporary, actually it is the way to make known his merit indeed he gave his first speech Tuesday 19th and his first lesson Wednesday 20th of this month and was applauded by a large assembly including the dear uncle who made clear how satisfied he was. He is right, [the young man] is well worthy of him and of you. He combines a real merit, lots of modesty and would not allow that we make the fuss and fireworks that the gardeners wanted, he was obeyed but the joy was universal, I swear to you that it was a great celebration at the Jardin the amphitheater resounded many times with the clapping of hands, applause well deserved I assure you that I felt too much joy to not share it with you, these are the fruits of your excellent education and of the great example he received. . . . He will give you much well-deserved satisfaction I congratulate you with all my heart."[139]

Basseporte had her fingers on the garden's pulse at all times.

What was the reaction of the Jardin staff in 1771 when they read Bougainville's best-selling *Voyage autour du Monde,* which included the story of the still-absent Jeanne Barret's masquerade as Commerson's valet? Basseporte would now suddenly have understood why the young woman mysteriously disappeared from the garden exactly when Commerson left Paris to sail with the expedition. Bougainville's rendition of Barret's deception, that it was born of eagerness to see the wide world, was not critical but admiring, which made others see it that way too. The captain wrote of her tearful confession to him that she was indeed a girl, how she had fooled her master by presenting herself to him in men's clothes at the moment of embarkation, how desire for better opportunities had made her decide to hide her sex, and how the prospect of a sea voyage around the world had "piqued her curiosity. She will be the first and I owe her the fairness to say that she always conducted herself on board with the most scrupulous good sense."[140] Soon after, Diderot wrote what was to have been a review of Bougainville's book for Grimm's *Correspondance Litteraire,* although it was not published until much later and then greatly expanded. This *Supplément au Voyage de Bougainville* saluted Barret's cleverness, bravery and strength of mind.

"Commerson's domestic . . . was a woman disguised. Undetected by the crew the whole length of the trip, the Tahitians guessed her sex at first glance." He marveled that she had chosen to go around the world, braving the rigors of such a voyage. "She always showed discretion and courage. . . . These frail mechanisms [*frêles machines*] sometimes have within them very strong wills!"[141]

Meanwhile Louis XV, once referred to as "le bien aimé," was losing his grip, and the years between 1771 and 1774 were politically fraught. The king encouraged his new chancellor Maupeou to effect a coup, dismissing the traditional *parlements* that were threatening to hamstring royal wishes and replacing these magistrates with yes-men, causing Malesherbes to write passionate remonstrances against such illegitimate, despotic measures. Basseporte and Biheron were enamored of Malesherbes's daring and Biheron sent a manuscript copy of his writings to Franklin. The entire Jussieu household was against Maupeou as Antoine Laurent's letters attest.[142] But there was botanical work to be done no matter the political situation. Now that the Trianon was completely modified according to Bernard de Jussieu's natural method, the Paris garden had to be replanted along the same lines, and everyone, including Basseporte, threw themselves into that enormous labor starting in 1773. The next year, when the king died suddenly of smallpox, his new young successor Louis XVI called Maurepas back—he had been disgraced in 1749 for an epigram that offended Pompadour—and also recalled and appointed Malesherbes and Turgot to powerful positions in the ministry. So the friends of the Jussieu family were back in power, and shortly thereafter Basseporte received an additional 400 livres pension beyond her usual wages.[143]

She and her Jardin colleagues got another reminder of Barret's part in Commerson's work when in early 1774 his friend Vachier published his will, *Testament singulier de M. Commerson, docteur en médicine, médecin botaniste et naturaliste du Roi, fait le 14 et 15e décembre 1766,* in an attention-grabbing sixteen-page pamphlet. This was the first they heard of Commerson's death, which must have surprised and saddened them. Now the large role he entrusted to Barret as overseer of his collections was in print for all to see. Bachaumont commented on it 11 March, and on 15 April gave the backstory of the testament a titillating spin, making it sound like Barret really fooled Commerson, and that when she showed herself to him after many months of disguising and dirtying her body with tar so that he would not recognize her "he could only be enchanted by such a mark of fidelity and attachment."[144] A more scientific acknowledgment of the important liaison with Barret came in Lalande's *Éloge de Commerson*, published in the *Journal de Physique* in February 1775 and read at the Académie des Sciences, where the naturalist was said to have chosen a smart, tireless, and brave female companion whose collaboration was invaluable on the voyage.[145] Commerson's friends in the Jardin must by now have begun to wonder where Barret was, where the fruits of the couple's collecting had ended up, and in particular when they might hope to receive the promised botanical bounty.

Basseporte's infirmities were increasingly distressing, A. L. de Jussieu telling his mother at one point that her asthma was so acute "I fear this might end badly."[146] Thouin would insist periodically, as for a spell in 1775 and again in 1777, that Basseporte come live in his house to spare her being sick alone.[147] She was tutoring one of his sisters and considered a member of the family. While she eventually moved back to her own apartment on the same grounds in the Jardin, she felt poorly enough to write the first of several wills and testaments on 7 March 1776. (She was to write another a year later with "my friend" Biheron added to her doctor Dubourg and his wife as another executrix.) Basseporte would live and continue working for several more years, but at junctures such as these that must have been impossible for her to imagine.

At some point in late 1775 Barret and her new husband returned to France.[148] Once in Paris she would have discovered the appalling state of the materials that were brought over a year before by Jossigny and the complete neglect into which the manuscripts and herbaria had fallen. Already much of what they sent had been lost, or cannibalized by other men of science, as Commerson had not been there to defend himself against such predation; it seemed the recurring nightmare that haunted him since the start of the trip was playing out. On 6 February 1770 he had warned Bernard de Jussieu about the importance of protecting his shipments, saying that he had sent with his trusted astronomer friend Rochon three months earlier two *coco de mer* specimens—coconut palms—but only because Rochon had promised to keep the plants upright next to him in the carriage after landing, so that the "violent hands of the indiscrete curious" did not "ravish the discoveries that I hold most dear."[149] He told Lemonnier, "There are in the republic of letters, as in honey hives, heavy and lazy drones who live at the expense of busy and industrious bees. I have already felt several times their half-starved and perfidious bite."[150] In yet another letter to Lemonnier a few months before his death, Commerson insisted that his mailings travel only under his own direct supervision or with very explicit orders for their safety provided by the ministers in Paris. "I can predict that I would die of chagrin if I saw, without this favor, [the loss of] these precious spoils from so many countries."[151]

Well, he *had* died—of other causes than chagrin, although that might indeed have contributed—and exactly what he predicted had come to pass. In summer 1774 Beau's and Vachier's attempts to secure for Commerson's son those materials that actually reached France had fallen flat because Buffon had claimed them all for himself and then forgotten about them.[152]

Barret might at this point have approached Turgot, Daubenton, Thouin, and others for whom, out of friendship, Commerson had named plants. Although she had no formal rank in scientific or governmental circles, she had known these people before the trip. And those who had not known her before Bougainville's voyage certainly knew of her now, from the captain's 1771 book in which he attested to her botanical expertise. The 1774 publication of Commerson's testament and Lalande's *éloge* also highlighted Barret's important involvement with their joint work. This may have sufficed to give her some influence or at least allow her to be heard.

There is no record of this, but timing suggests the possibility that Barret's return to France might have galvanized important men into action: there was suddenly new eagerness to locate and attribute properly the great wealth of material Commerson had accumulated over the course of many years, that which had already arrived in France and that which had not yet been sent. Barret could have reported detailed information about all the treasures still unrecovered from Isle de France, and by doing so she may have spurred people in high places to act.

We know that Barret had returned to France by late 1775, as her husband signed as witness to his brother's wedding southeast of Bordeaux on 27 November.[153] Turgot and Malesherbes, recently appointed ministers by the young King Louis XVI, took an interest in Commerson's collections at that time, abruptly eager to safeguard his name and reputation. The prevailing indifference and neglect by Paris naturalists might simply not have been brought to their attention until Barret alerted them. Turgot had known Commerson well enough to have a plant named after him on the voyage, and as we saw Malesherbes had extremely close connections with the Jussieu family and the Jardin, still deprived of and awaiting all the new species promised from Commerson's harvest. Who but Barret could describe accurately the botanical wonders that remained in boxes across the sea? Turgot, despite having received an impassioned entreaty from Beau in 1774,[154] had not pursued Commerson's affairs at that time, but in late 1775 he asked Thouin to put the gifted botanist Joseph Dombey in charge of gathering, organizing, and cataloguing what could be found of Commerson's collections in Paris.[155] And Thouin wrote to Sonnerat in December 1775 that Malesherbes, freshly aware of the rich untapped lode that never made it to France, wanted to find all of Commerson's missing material. Sonnerat, sailing at the time in the Indian Ocean with stops on Isle de France, was ordered to locate all the boxes remaining there so that they could immediately be sent for.[156]

Wheels had been set in motion. On 11 January 1776 A. L. de Jussieu received from Daubenton at Buffon's request some *herbiers* of Commerson's that had not been turned over initially,[157] but by then the twenty-four boxes and forty-five cartons were already all mixed up with Sonnerat's collection. Jussieu was not up to the task of sorting and organizing these, or was too busy with other duties, or simply did not care. It is even possible that he deliberately downplayed the importance and value of Commerson's work because Beau continued to request money for the dead man's son based on the significance of the father's discoveries.[158] To his shame, A. L. de Jussieu did nothing for over a decade, until 13 March 1789, by which time it was far too late to assemble, catalog, and properly attribute items, or to determine just what Commerson's original contribution really was.[159]

In the spring of 1776 Barret came again to Paris with her husband, from Dordogne where she had settled, appearing on 3 April at the office of Vachier's notary near the Place des Victoires, to claim the inheritance due her in her former capacity as "gouvernante," from Commerson's testament. It was a long trip, and why she did not take care of this business when she first returned to France the previous year is unclear. The notary reported that she and Dubernat lodged at the Hôtel Dauphin on the rue du Bac in the Faubourg Saint Germain, across the river. Beau was not present at this meeting, probably did not want to face Barret, for aside from his disapproval of her and perhaps because of it he had helped himself to some of her rightful bequest. Barret received 1,282 livres, 8 sols, and 8 deniers. This was the sum of the funds from Vachier that Commerson had left to her in his testament—600 + 100/year of wages calculated from 6 September 1764 to 6 December 1766 (after which she was presumably a fugitive who turned up at his boat), plus 457, 8, 8, from the notary who had sold the things in the apartment (furniture, linen, clothes) that had been left exclusively to her, minus questionable fees to Beau and some others involved in preparing the inventory of Commerson's effects.[160] Satisfied or not with this outcome, Barret accepted it.

From her hotel on the rue du Bac, did nostalgia perhaps propel her to walk a short half hour east along the quays of the Seine toward the Jardin to see her old friend Basseporte, now seventy-five years old, one more time? We have no way of knowing if they had stayed in touch as no letters between them survive. A month earlier, on 7 March, Basseporte had written her first will, for she had never been robust and was then feeling especially poorly.[161] She was living intermittently with the family of Thouin, for whom Commerson had fondly named a plant, and whose sister Marie-Jeanne was one of Basseporte's students.[162] Bernard de Jussieu himself, though failing and nearly eighty years old, was still going daily to the garden.[163] Barret might have stopped in to see if Commerson's materials were being treated any better after the ministers' intervention of the previous year.

Nothing, not even Barret's likely advocacy, could prevent the plunder without scruple of Commerson's material, which he was never able to claim for his own glory.[164] Some of it was eventually recovered, but the magnitude of the job of separating and properly attributing the specimens was stupefying and any efforts were too little too late. Rochon, a caring friend of Commerson's through two brief overlaps on Isle de France who spoke of him with admiration, would lament during the Revolution when he wrote up an account of his own travels, that over the intervening twenty years only a small fraction of the botanist's enormous collection had been recovered.[165] And one Macé, apparently a correspondent of the Académie, also took up the cause of tracking down Commerson's unretrieved treasures.[166] The covetous who raided Commerson's available materials included Dumeril, Lacepede, Lamarck, Lemonnier, and A. L. de Jussieu himself.[167] "Commerson," Georges Cuvier wrote, "would hold one of the premier ranks among naturalists. Unfortunately, . . . those to whom his manuscripts and herbarium were entrusted were guilty of neglecting them in a shameful manner. . . . We cannot regret enough the abandon in which his collections were left, because had they been used immediately France would have had one of the most distinguished rankings among those nations that contributed to the progress of natural history. Commerson's . . . ardor [was] without example."[168]

It was later estimated, based on what naturalists in the next century were able to untangle, that Commerson's stockpile included roughly 30,000 specimens, 5,000 new species, 3,000 of which were previously unknown to Europe, and 1,500 drawings.[169] A good part of this collection, of course, was joint work which could never have been assembled without Barret's vigorous collaboration.

In 1778, when Basseporte was seventy-seven years old, she began to put her age on her botanical illustrations right next to her signature, in proud defiance of infirmity and passing time, and she continued to do this until her eightieth year when she died (figure 9).[170] She had written a second will on 18 May 1777, adding her intimate friend Biheron as another executor should her doctor Dubourg and his wife predecease her, and she had lost her dear colleague Bernard de Jussieu on 6 November of that same year, which was a huge blow. But despite her own bouts with severe illness the demands on her to produce the *vélins* never decreased, and she was determined to keep her job until her last breath. She would not countenance the naming of a successor although one had been designated on 1 July 1774 by the newly minted king, Louis XVI, who had not yet been informed of the protective arrangement and had succumbed to pressure. But everyone involved was then sworn to the strictest secrecy, and they kept amazingly quiet.[171] The successor was an ambitious Dutchman, Gerard van Spaendonck. He was to get all the customary "honors, privileges, franchises, exemptions, gages, rights, fruits, profits, revenues, and emoluments" enjoyed by Basseporte.[172] But he also had to agree to give 400 livres of his yearly pension to another man, Jacques Desève, a favorite of Buffon's who was equally deserving of the post.[173] Van Spaendonck in any case was relegated to the wings until Basseporte's death. She never relinquished her position nor did anyone suggest that she do so.

At least her old friend Franklin had returned for a long stay in Paris, informing Dubourg on 4 December 1776 that he was back and settling in Passy. Basseporte resumed contact with him directly, but because his residence was far from the Jardin she also wrote to him through her neighbor Dubourg, as when she asked him to deliver two *mémoirs* on 12 February 1778, one to Franklin and another to his landlord Chaumont who had been since 1776 a great upholder of the American cause, supplying gunpowder and clothing for the war effort. Presumably these communications had to do with America, for Basseporte, along with Biheron and Dubourg, staunchly supported the revolution in the colonies. Franklin at the time was deluged by droves of young Frenchmen volunteering to fight but without uniforms or money. On 15 April 1778 Marie-Charlotte

Fig. 9. Basseporte's painting of "Ixia," one of her last. On the bottom right, next to her signature, she recorded that she was "80 years old" (detail). [Muséum national d'Histoire naturelle, Paris, France. ©RMN–Grand Palais/ Art Resource, N.Y.]

Leullier L'Enfant, wife of a painter at the Gobelins tapestry works, asked Franklin to mail a letter to her only son because hers had not reached him. She wrote that Basseporte, whom she was "fortunate enough to know," joined her in imploring Franklin's protection for her boy. (This son, Pierre Charles L'Enfant, remained in the U.S. and became the designer/urban planner of the city of Washington, D.C., which was arranged according to the now famous L'Enfant Plan.) A year later, after one of Basseporte's illnesses, she recovered sufficiently, as Dubourg told Franklin on 17 April 1779, to have "found the legs" to come seeking information on one more French soldier who appeared to be lost.[174]

Buffon was another friend Basseporte could still count on, and the two corresponded when he was away in his *terres* in Montbard. He was filled with wonder that she continued her exertions and her painting into old age, a phenomenon noticed by others too, one of whom wrote: "The works that come to life under her brush are regarded by all the connoisseurs as masterpieces, where art challenged nature for truth of expression, delicacy, and the precision of the coloring. Although presently seventy-eight years old, we still see this indefatigable artist exposed, throughout whole days, to the ardors of the sun, in the most uncomfortable positions, copying for the Cabinet du Roi all that nature offers of the most magnificent, the most precious and the most rare in the plants gathered in the royal gardens."[175] On 12 January 1780 Buffon, aware of her frailty, answered one of Basseporte's letters, saying, "I was enchanted, mademoiselle, to receive a quite long letter all by your hand, and as well written as thought out. I hope that in ten years we will exchange others like these, and that you will have for me then the same sentiments that you have the goodness to accord me today."[176]

But on 29 June of that year Basseporte updated her will for a third time. Things must have deteriorated through the summer, for on 26 August her niece Mlle Aillaud wrote a grasping letter to Franklin trying to get Basseporte's pension commuted to her as the old lady lay dying, reminding him that she had seen her aunt entertain him and Buffon in her home and hoping Franklin would intervene for her with Buffon. He did nothing of the kind, of course, and the government refused the greedy request.

On 4 September 1780 Basseporte died, reportedly with a brush in her hand. A. L. de Jussieu wrote a letter lamenting to his wife that this would be the funeral of yet "one more dear friend gone."[177] The Jardin was just not the same without her. Thouin, by now an internationally known botanist in the garden, mourned Basseporte's passing so profoundly that those who knew him worried. Laurent of Brest feared he would make himself ill, and wrote on 28 September 1780:

> You have experienced, my dear friend, some days ago, a moment of grief very natural for someone with your kind of attachment to her. Permit me . . . to tell you how I see things. Mlle Basseporte deserves to be mourned, I also was very obliged to her, and I will not hide that the news of her loss affected me greatly. But in this case you must absolutely come to terms with it. The most sensitive people will be the first to point out to us that she did nothing more, after having lived a beautiful career, than to pay the tribute she owed to nature. This reason alone can, I believe, suffice to have you acknowledge that I make a good point, and to promise me that you will not feed a chagrin that not only would be completely useless, but could again put in jeopardy the health of a friend to whom I owe infinitely more and whose loss would be a much greater wrong to society.[178]

Lalande had said of Lepaute's death that it was worse for him than for astronomy, and here was Laurent saying Basseporte's death did not compare to the possible loss of Thouin. And this from men who sincerely respected the work of these two women.

In her will Basseporte's left linens, furniture, and a gold watch to her faithful housekeeper, wife of one of the gardeners, who had stayed loyally with her throughout the decades despite the modesty of her situation and opportunities to earn more money elsewhere; art supplies to her students; some money for the poor of her parish; and a set of silver candlesticks to her most famous female student and special friend Biheron.[179] The inventory of Basseporte's possessions done after her death on 5 October 1780 under Thouin's supervision revealed that she had numerous artworks, including copies of Correggio from Sery's time in Italy that he had given her as thanks for her kindness to him in his final illness. There was also a shrine, and many religious paintings by her and others. Relating to her life in science: works by Sybilla Merian on the plants and insects of Surinam; numerous cartons of natural history drawings and paintings mostly by Basseporte herself, of butterflies, birds, bugs, and studies of a pineapple in pastel; vestiges of the work of some of her protégés—a sketch model by Larchevêque, a wax ear under glass on a pedestal surely by Biheron. There were maps of Paris and of Versailles, where she went regularly. And there were geographic globes, no doubt by Mentelle who wrote her lengthy and intimately detailed obituary in the *Nécrologe des hommes célèbres*.[180] Mentelle was a geographer, one of those who had benefited from Basseporte's network of science colleagues and other savants among whom she had found tutors for him in his formative years. He became extremely prolific, writing thirty-six geography works, producing maps, globes, and atlases, and teaching until his death in 1815. "I owe to Mlle Basseporte the happiness of my life," he wrote. "I recommend the memory and veneration of her to my son, his children and his children's children."[181]

Basseporte painted a self-portrait, now lost, which found a loving home with Biheron who cherished this souvenir for fifteen years until her own death in 1795. Both women earned their living through science with their handy minds and mindful hands, having that unique bond in common and much else. Both, for example, beyond their other talents were also proficient healers, although this skill was necessarily practiced off the record. Naturally Basseporte knew the medicinal virtues of the plants she studied and cured individuals whose prognosis from the doctors of the Sorbonne medical faculty had been dire; A. L. de Jussieu repeatedly urged his mother in Lyon to try, and even market, some of Basseporte's effective remedies in that city.[182] Yet despite her many and varied achievements, Basseporte remained genuinely modest. Her eulogist asserted that she was "equally admired by painters and naturalists" but that "she was unaware of the celebrity she enjoyed; no foreign prince came to Paris who did not wish to see her works and know their creator."[183] Her protegée and partner Biheron attracted that very same kind of following.

Barret was never prosecuted for her infractions. Quite the contrary, she was to receive a royal surprise.

But first, after seeing to her inheritance in Paris in 1776, she settled into her husband's homeland in Périgord, in the little village of Saint Aulaye (today the commune of Saint-Antoine-de-Breuilh) along the banks of the river Dronne. They resided in a neighborhood called Les Graves within the parish, and local researchers believe it is possible to identify her house at *parcelle* 496–73, a low stone building which now sports bright blue shutters (figure 10). Barret also brought members of her own family to live close by, first a nephew from Bourgogne who became a hosier and established himself and his large brood in Gensac just southwest of Saint Aulaye, then a niece who produced another four children, so Barret was very occupied as a great aunt.[184] Jean Dubernat had

Fig. 10. Jeanne Barret's last known house, in the village of Saint Aulaye, Dordogne. [Photo by author]

many relatives as well, and their domestic scene must have been busy and their marriage solid, attested to by the mutual care given to each other's families in their respective testaments.

It was a simple country life, Barret surrounded by *agriculteurs,* many of whom were illiterate like the ones who witnessed her death, and like her own parents and godparents who decades before had also been unable to sign her birth certificate.[185] Such a contrast to the company of the erudite and eloquent Commerson with his Latin and Greek dedications and purplish French prose. How long did it take her to readjust to a bucolic existence among modest, plain-spoken folk? It is notoriously hard for those who have lived daringly and taken huge risks to resume normal behavior, and those who have willingly put their lives in jeopardy often find peace and quiet difficult to accept and may seem "puzzling creatures" to outsiders.[186] Commerson was regarded as a martyr to science who died too soon, but Barret lived for several decades after their world travels and would never again experience that buzz.

Did she dwell in the past, reliving her adventure in her memories? Did she speak frequently about it, regaling those around her with stories? Did she stay in touch with Bougainville? Might she have reminded the powers that be of her past contributions and accomplishments?

Suddenly, in 1785 when Bougainville was consulting on the Lapérouse expedition that was gearing up to leave just then from Brest, almost exactly ten years after Barret's return to France, King Louis XVI awarded her a pension of 200 livres a year to be paid from the treasury of the Invalides for her equal participation in Commerson's "work and perils." Bougainville had predicted in his ship's log that she would be forgiven for her infractions and, perhaps reminiscing about his own voyage, was likely instrumental in this gift being granted to her at just this moment. "Cette femme extraordinaire" was honored 13 November 1785, as follows:

> The woman named Jeanne Barre, with the help of a disguise, made the voyage around the world on one of the ships commanded by de Bougainville. She devoted herself particularly to the service of M de Commerson, doctor botanist, and shared the work and perils of this savant with the greatest courage. Her conduct was exemplary as M de Bougainville noted. M de Commerson having died, Barre, whose sex had been discovered, married one Du Bernat, former petty officer in the royal regiment of Comtois. Today, the woman Du Bernat and her husband having arrived at the age of infirmities and unable to continue working, the minister therefore gladly awarded this extraordinary woman a pension of 200 livres on the fund of the Invalides, retroactive to 1 January 1785.[187]

The actual *brevet de pension* was dated a couple of weeks later, names spelled correctly, worded slightly differently and with details about it being personally delivered to her by a marshal of France:

> Today, 1 December 1785, the king being in Versailles and wanting to gratify and treat favorably Jeanne Barret, wife of Dubernat . . . His Majesty accorded to her and made the gift of the sum of 200 livres of annual pension to be paid to her for the rest of her life on the treasury of the invalids of the Navy. . . . And to show his good will, His Majesty ordered me to expedite this brevet that she will sign in her own hand and that I will countersign. Marquis de Castries.[188]

This pension, seemingly modest, was in range with those given to many workers in the royal palace by Marie Antoinette. The annual income of a family of weavers at this time was about 150 livres and sufficed for a couple with children to live independently. Lodgings could be rented in the provinces for about 40 livres a year.[189] While Commerson's orphaned son was awarded a much larger gratification of 1,000 livres, we must remember that Beau had lobbied ceaselessly to obtain some recompense for the boy based on his father's efforts in the service of the Crown. Barret, on the other hand, was not only off the royal payroll but was an illegitimate and even technically criminal participant in Bougainville's mission. That she received any pension at all was astonishing and entirely due to her contributions to science.

The reference to infirmities and old age is odd, and it suggests either that Barret might have requested help, or that someone in power was keeping track of her since her return to private life. She was only forty-five, Dubernat only forty-eight. Although both already far exceeded the average life expectancy, they could not have been gravely ill for they lived over two decades more and he rejoined the military during the Revolution.[190] In any case, Barret's annual pension of 200 livres seems to have been forthcoming for about five years, but as the Revolution advanced, the couple struggled to get this money. The National Assembly ruled that pensions granted under the Old Regime could only be reinstated through rigorous examination of each individual case, a bureaucratic morass in which Barret and her husband got hopelessly trapped. And then came the Reign of Terror.

The purge from the National Convention in June 1793 of the moderate Girondins, delegates from Barret's southwest part of France, brought the Revolution right to her door. On 7 July of that year she wrote the authorities in Paris that she had not received a payment for several years. The answer on 18 August told her to apply in Bordeaux and supply all manner of paperwork.[191] Then she was referred back to Paris. When Dubernat's battalion returned from Alsace the next year, he took over the pursuit of the elusive pension, now almost four years in arrears. He sent a letter, certificate of residency, and Barret's precious 1785 brevet to authorities in Paris who requested them, but when he followed up he was told on 26 July that neither his letter nor the enclosed documents had arrived.

Panic might well have ensued then, as the irreplaceable brevet was apparently lost. Together Barret and Dubernat sought the assistance of a notary in Sainte-Foy-la-Grande and also chose two men to represent her in Paris to claim the pension "that belonged or will belong to her on the republic or otherwise." This wording reflects the extreme confusion about the coffers from which her monies should come. They had been granted to her from the royal treasury by a king so reviled since then that he had been guillotined the previous year. The fugitive Girondins, in hiding throughout the Terror, were smoked out in her own backyard, so to speak, with the help of ferocious mastiffs from Sainte-Foy-la-Grande in the summer of 1794, and those ousted deputies who did not kill themselves were executed in Bordeaux. Danger was on Barret's doorstep. Dubernat took the risk of sending a last request for the current and retroactive payments to Paris in December 1794, signing with the requisite "salut et fraternité," but the answer, scribbled directly on his letter, referred Barret once again to Bordeaux.[192] It was a runaround, and we do not know if she ever saw the payments again.

During the following decade she must have begun to feel her age, writing three successive testaments in 1805, in January, then April, and finally in June, each simplified from the previous one. She left everything to her husband, and at his death some was to go to her servant, to the poor of the commune, and the rest to two nephews, a mason and a hosier, and to a niece.[193] Dubernat's testaments were the mirror image, providing for everyone in his and her families.[194] At eleven at night on 5 August 1807 Barret died.[195] Dubernat was absent, possibly at the wedding of a family member that occurred the same day, so the witnesses to her death were two *cultivateurs* who could not sign their names. She was buried in the churchyard of Saint Aulaye (figure 11).

Intriguingly, Barret left to her heirs a sizable sum for her social class, 10,300 livres. When Commerson set out on Bougainville's voyage he had between 10,000 and 12,000 livres, which he considered "a small fortune."[196] Her royal pension, even without the documented interruption, could account for only a small portion of it. How might she have accumulated that much money? Had her Isle de France tavern been lucrative? Once back home, did she stay in touch with other botanists, sell souvenirs of her voyage, keep some of Commerson's papers that later became valuable? She must in any case have been enterprising, organized, and frugal. It is further intriguing that the adult children of one of her nephews, despite their upbringing in the hinterlands among people who

Fig. 11. Barret's grave in the Saint Aulaye churchyard where she was buried.
Because the names on the stones can no longer be read, this one has been designated
as hers by the local historians. [Photo by author]

rarely strayed far from their own village, seem to have inherited their great-aunt's wanderlust, requesting passports in Bordeaux for travel to New Orleans, Buenos Aires, and Valparaiso.[197]

Prince Nassau-Siegen, who witnessed all of Barret's exploits on Bougainville's voyage, esteemed her enormously and, acknowledging her agency, declared that the trip had been her idea exclusively. "Rather than suspect the naturalist of having hired her for such a painful voyage, I prefer to accord to her alone all of the honor of such a courageous enterprise; abandoning the quiet occupations of her sex, she dared to face the fatigues, the dangers, and all the events that one can morally expect in a navigation of this kind. Her adventure, I believe, deserves a place in the history of famous women."[198]

Basseporte flourishes still, through her hundreds of magnificent botanical illustrations in the prized *Vélins du Roi* collection at the central library of the Muséum nationale d'Histoire naturelle. They are preserved like treasures, consultable only with special permission and handled only with white gloves. And Barret sails on through the streets recently named for her in such cities as Dijon, Strasbourg, Rochefort and countless villages throughout France, a mews in Nantes, a school in Montpellier, the Saint Aulaye churchyard where she is now commemorated, and in the attention her daring feat continues to attract today.[199]

Dear Jeanne,

Mostly, I wonder what really happened once Bougainville left you and Philibert off on Isle de France. Did he desert you? Did he expose you? Or was your story perhaps not spread about, making it possible to remain in disguise as if nothing had happened and to continue as his manservant? Hard to believe you would have agreed to prolong such painful sartorial constraints. Yet given your zeal for science you might have, and even taken a new name, hiding in plain sight to accompany him on further adventures, like the trip to the volcano on Isle Bourbon. On the other hand, when he dedicated an interesting shrub to you he seemed nostalgic, writing that your plant name had significance "autrefois," which suggests that you were in his past. But then what did you do? It was falsely reported that you married the owner of a forge on Isle de France; someone else spoke of your "personal papers" ending up in Strasbourg. Writers write the damnedest things.

Did you spend time with Aotourou after his visit to Paris, during his year on Isle de France from 23 October 1770 to 18 October 1771 while Poivre waited for favorable winds to return him to his Tahitian home? The Parisians who had examined him and heard him play his nose flute said he never learned French, yet you two had years earlier figured out ways to communicate with each other, when he first came onboard. Did he explain what it was like for him in the capital? Was he paraded around, exploited, or treated with dignity? Was he curious about the city, or terrified? They said he loved the opera . . .

And speaking of Paris, I would love to know when and why you used the name Bonnefoi there. What purpose did it serve?

The house where you lived before the trip, right in the elbow of the rue des Boulangers, has a modernized exterior now and is unrecognizable as an old building. But longtime denizens of the street remember the magnificent wood balustrade and banister on the stairway leading down to the court's well, the worn steps you trod countless times, living as you did on the third floor. From there to the

Jardin you walked east for a quarter of a mile along the old rue du
Faubourg Saint Victor—today the rue Jussieu and rue Linné—past
a cabaret called Le Buisson Ardent (a present-day restaurant on the
spot still bears that name). Did this give you the idea to open your
own saloon as you did years later?

I saw the village where you spent the last three decades of your
life with the sturdy, faithful Dubernat, in gentle rolling hills with a
beach on the river Dronne. Twelve hundred people lived there when
you died, and it has almost the same population now. The locals are
enamored of you, as are the more far-flung members of the Société
Botanique du Périgord. They have chosen a grave in the cemetery
adjoining the twelfth-century church and designated it as yours because, while they know you were buried there, the stones have all
been rubbed clean with the passage of time. They have named the
town meeting hall for you also. And I saw Sainte-Foy-la-Grande, a
thirteenth-century bastide town with half-timbered houses, unusual
because the river Dordogne runs right through it, where you and
your husband stood more than once before a notary trying to get
your royal pension restored.

Which raises the question: how well could you write? We have
several signatures of yours—from your pregnancy declaration, your
wedding, and the settlement of your inheritance claim. They look
pretty consistent. But there is no trace of anything longer in your
hand. You wrote to Vachier—we have only his reply; you wrote to
the minister of the navy to get your payments flowing again after an
interruption—we have only his reply. Even Dubernat, whose letters
we <u>do</u> have, seems to have dictated them; the pieces of his correspondence are all in different hands with different spellings. Perhaps neither of you had great facility with the pen? But regardless, we know
you made tickets for Commerson's herbaria, kept records of the
collections. And you were obviously organized and disciplined with
your own domestic financial affairs, amassing as you did a lot of
money by the time you died. Where did that sum come from?

Leaving a family of illiterate, small town peasants you traveled
the wide world in the company of a sophisticated scholar, then looped
back to the country life once more among simple *agriculteurs* who

could not sign their names. That, I suppose, was your comfort zone, but you willingly left it for a good long while, ready to be very uncomfortable and not knowing what the future held.

About your decision to cross-dress and morph from female to male, from the realm of submission to that of power. . . . Because you could not hide your small stature and hairless face, you discouraged the approaches and silenced the suspicions of the other sailors by claiming to be a eunuch! That was inspired. You did all this for a great adventure. Marie-Marguerite donned male garb also, as we shall see, in order to attend dissections while learning anatomy. Mme de Beaumer, editor of the *Journal des Dames,* drove the book trade authorities mad by always dressing like a man. Evidently you pulled this off with aplomb.

Yet there are costs. Identity lives in the brain and is not something to play around with lightly. I am thinking about the stress and strain of your passing onboard without being "read." The act you had to perform day after day required unaccustomed moves, blustery bearing, an assertive and unapologetic stance, different voice, demeanor, mannerisms, gestures. Binding your breasts with straps and bandages caused rib-cage compression and, as we learn from the surgeon Vivès who treated you, badly injured your skin, but you were lucky the tightness didn't asphyxiate you. Perhaps the sailor's outfit was baggy enough to somewhat conceal your form. Living in stealth may also have been a bit easier because the ship's population had never seen you before, had no sense of how you looked as a woman. Some say that seamen are men of few words, so maybe you didn't have to talk much.

Still, there were protocols, and you needed to appear familiar with them. Observing sailors in Rochefort while stuck there for a month and a half, picking up their vocal patterns and body language surely helped. But the pressure of not letting down your guard, the dread of discovery, the immense toll of the deception must have become increasingly painful. It could only have been a relief to be "clocked" and relieved of this difficult role which you had consummately mastered for the fourteen months until the Tahiti incident in April 1768, after which you lay low a bit longer until your confession to Bougainville. Then, once deposited on Isle de France in November, you could have been entirely released from that bondage. Is that what you chose?

I found it strangely touching to learn that just as you were disappearing into your country life in Périgord, the *flûte Etoile* on which you sailed for so many months was ebbing also. Sent to China in 1773, she was then retired and reduced to a flat-bottomed barge in 1777. The *Encyclopédie* has a picture of a *flûte,* and all I can think of is how impossibly cramped it must have been, even in its heyday, with over one hundred men on board and you in the midst of it all.

Commerson said you were "armed with a bow, like Diana, armed with intelligence and gravity, like Minerva." Diana, goddess of the hunt, the moon, the woods, and wild nature. Minerva, goddess of wisdom, medicine, war, craft, knowledge, and perspicacity. This from a savant well versed in the classics who knew his mythology. Wonderful.

Dear Madeleine Françoise,

Not long ago I sat under the cedar of Lebanon planted in 1734 by
Bernard de Jussieu while you and everyone in the garden looked on.
It is a huge, wide tree now, its lush needles and spreading branches
providing welcome shade to visitors at the Jardin des Plantes as your
beautiful park is now called. This gigantic evergreen, shaped like a
little pyramid when you saw it but luxuriously broad and open now,
outlived you and will long outlive me. Other trees planted centuries
ago are still here too: I recognized an Acacia, grown from seed orig-
inating in my part of the world, North America, which was already
here in your day, and there is the tall Sophora Japonica, transplanted
in 1747 by Bernard de Jussieu, again while you all watched, from
the Place Dauphine where it first took root. Next to the cedar is the
Labyrinth, a tall hill with rows of hedges in rising circular paths that
take you around and up to the gazebo at the top, one of the oldest
metal constructions in the world built at Buffon's orders and from
which one can see all of Paris. I strolled through the majestic avenues
of plane trees, for which we also have Buffon to thank, and enjoyed
the famous banks of roses, irises, and peonies, picturing you bent over
them as you sketched and painted. The Jardin Alpin, the materials for
which were accumulated during your day, is now a secluded space
for plants from mountain climates that you can only get to through a
tunnel passage. The big old pistachio tree, grown out of seeds from
China and still there, fascinated an earlier Jardin botanist, Sébastien
Vaillant, who figured out—by observing its sterility until he mingled
its flowers with those of another tree of the same species—that plants
reproduce sexually just like animals. He was the first to introduce
terms like male, female, and ovary into discussions of floral anatomy.
This nomenclature initially created a scandal but was soon picked up
by Linnaeus, whom you met in the garden in 1738 and for whom plant
sexuality was central. He wrote and spoke freely about it with you
and Bernard de Jussieu. It wasn't shocking anymore.

Did you walk often down through the newly expanded Jardin to

the Seine and then west to the rue des Bernardins where Bernard de
Jussieu lived? Was it there that you painted his portrait? And was it
the side view you did, with him peering through a magnifying glass
at a plant, or the front facing image? Both have been attributed to
you, and it says a great deal about his admiration and trust that he
allowed you this special closeness.

There is of course a big statue of Buffon in the Jardin. What
would you expect? It was made, exceptionally, during his (and your)
lifetime, in 1776, by the royal sculptor Pajou. Oddly, he has his back
to the gardens and instead faces the street where his house and the
cabinet of natural history, today the museum, still stands. Even more
oddly his cerebellum, in a crystal urn and said to have been preserved
according to "Egyptian methods," is encased in the statue's base,
which serves as a reliquary. It irks me that nothing around here is
named for you. Instead, this neighborhood's streets bear the names
of all your male colleagues in the Jardin. You situated yourself and
pulled your weight amidst these men, but women were not supposed
to have such ambition and, sadly, history hasn't paid tribute to you in
this visible way.

I am sorry to report, although not surprised, that some have
deemed your work inferior to the images made by the men who held
your position before you. I disagree. Yours are not flashy like some
of theirs, but they are meticulous. And some have seen vanity in your
determination to work until you died, finding evidence in your later
gouaches of a "trembling hand." Rather than admiring your stamina
one art historian called it "a bit too much perseverance." Even your
eulogist Mentelle saw your refusal to admit you were vanquished as
prideful, your "only concession to human frailty." I disagree with this
too. I see it instead as your way to give meaning to the experience of
being old. Were you inspired by your brilliant German/Dutch predecessor in scientific illustration, Maria Sibylla Merian, whose books
you owned and who worked until her dying day despite having suffered a stroke? We know that staying active with new projects is
healthy and keeps brain cells alive. Focus, continued social interactions, and a sense of purpose promote longevity. You must have sensed
this intuitively, writing proudly next to your signature on your late

botanical works that you were in your seventy-ninth, even eightieth year!

And you were known to be very beautiful. Your colleague and intimate companion, anatomist Marie-Marguerite, was often said to be plain, and even when young she was gratuitously described as an old maid. Not you. Did being pretty cause problems for you? You never relied on it, and clearly wanted other women to count on skills, not looks, in order to get through life. So you taught them flower painting, to give them marketable competence and a livelihood. You enjoyed and respected other women, including Marie-Marguerite's mysterious other "très bonne amie" Mlle Lainé, who was in the picture from around 1766, but about whom we know nothing. A. L. de Jussieu refers to your trio in his letters as "the three dowager demoiselles."

There surely was discrimination both subtle and overt at the Jardin, but you did not let it derail you. You broke boundaries and taught your most famous disciple, Marie-Marguerite, to do the same, to act as though anything is possible. Today, some high-achieving women in science suffer from "imposter syndrome," the belief that they are frauds who will soon be found out and who had no legitimate right to reach their level of success. They question their own abilities and therefore feel undeserving of what they consider to be luck rather than earned victories. This is a fairly common problem, the fear of exposure. But not you. Purposely setting up shop in that masculine enclave you were not plagued by doubts or you would have been unable to keep, and insist on keeping, that job for nearly fifty years.

Illness—yet you lived to be so old. A. L. de Jussieu's letters made it sound as though you were almost always sick. Elisabeth, Marie-Marguerite, and Geneviève persevered also despite serious infirmities. Some other scientific women—I'm thinking especially of Florence Nightingale—developed "creative maladies" to keep the world at bay so they could accomplish their work in peace. I wonder if some of that was going on here too, with you.

It's fitting that you were buried in Saint Médard with its Jansenist history. I have often sat in that church at the foot of the cobblestoned rue Mouffetard market, breathing the cool, stony, waxy air, watching the flickering candles, and thinking about your funeral there. I know A. L. de Jussieu attended, and the distraught Thouin too, and according to archivist A. Jal the artist Belle of the nearby Gobelins tapestry works was there too, but the rest of the attendees are literally lost in the ellipses of Jal's notice. . . . I wonder if Buffon came; he would have been in town as his unwavering schedule had him in Paris every fall and winter, in Montbard every spring and summer.

And Marie-Marguerite would of course have been there, heartbroken. You had shaped her life since girlhood, in so many vital ways. Not only did you choose vocation over family for yourself, but you chose her as your partner and instilled in her the same keen ambition to make an independent living based on her own unusual talent. You worked hard, seemed to revel in what you were doing, and never retired, a stellar example of engaged, creative aging.

FOUR

Anatomist and Inventor

Marie-Marguerite Biheron and Her Medical Museum (1719–1795)

> There is a young surgeon of great merit who wants to . . . profit from the lights
> of your artists of this kind. I presume to recommend him to Mr. Hewson.
> —MLLE BIHERON to Benjamin Franklin, 26 June 1773

> Biheron, destined to be an anatomist, will be such,
> whether a college of dissectors smile or frown.
> —CAROLINE H. DALL, *Historical Pictures Retouched* (1860)

MADEMOISELLE MARIE-MARGUERITE BIHERON, EXPERT ANAT-
omist, artist, and teacher, in 1762 sent the Danish king a collection of her perfectly accurate wax models, among them a cross-sectioned heart, a kidney, a
liver, an ear, an eye, some bladders, and a set of reproductive organs which included a "carefully confectioned" penis. King Frederick V welcomed the mailing and awarded her a handsome "gratification."[1] Already in 1759 the Jesuit
newspaper *Journal de Trévoux* had called her work, which she demonstrated
that year before the assembled members of the Académie des Sciences, an "anatomical marvel." Ten years later a disciple of Linnaeus visiting Paris referred
to Biheron's medical exhibit on the rue de la Vieille Estrapade, a *must-see* for
visitors to the capital, as an "anatomical miracle." And in 1771 Diderot, in a
letter to his friend the English firebrand John Wilkes, lauded her useful productions for their "marvelous truth and exactitude." Such admiration was echoed
by numerous other observers across a broad political, religious, and international spectrum, including the philosophe d'Alembert, Ben Franklin, physician

to the British monarchs Sir John Pringle, and the Swiss physiologist Albrecht von Haller. What was it about the "anatomie artificielle" that Biheron fashioned, displayed, and taught with for decades in her private home museum that elicited this kind of wonder? The fact that a woman had produced them? Her models, made from a formula of her own invention that neither melted nor broke, were said to imitate the human body with consummate precision. But of course these "marvels" and "miracles" were made possible only through her study and dissection of actual corpses, too numerous to count and ongoing over decades.

Why did Biheron choose and persevere in such an offbeat vocation? No one was forcing her. Au contraire, it was entirely counterintuitive, especially for an "infinitely devout" young woman.[2] Yet she was driven, and she dedicated her life to this work.

Anatomy experienced a scientific heyday in the eighteenth century, as more and more was learned about the structure and workings of the body. The very title of Giovanni Battista Morgagni's 1761 *On the Seats and Causes of Diseases* reveals the characteristic Enlightenment move toward solidism, the study of situated organs, and away from the age-old focus on fluid humors and vague discussions of phlegmatic or choleric constitutions.[3] The increased use of hospitals, and of postmortems as a follow-up teaching practice, made possible the confirmation of conditions only surmised while the patient was still alive. Many believed the study of anatomy, the analysis of the body's structural components, to be essential for all educated individuals. Physician and surgeon Jean-Joseph Sue wrote, "In dissecting, one sifts through the very entrails of Nature, which becomes a book for us, and the impressions that last from this are infinitely more pronounced than those acquired through other studies. The object of anatomy is the living person, its subject the human cadaver. . . . As those who study this science are not all heading to the same goal we easily see that it is not just for persons practicing the Art [of medicine]. It is good for physicists, theologians, jurists, and almost all those in liberal or mechanical professions."[4]

Convinced of the general importance of such knowledge, dedicated anatomists throughout Europe—but even individuals not specifically in that field, including most of the players in my story—bequeathed their own bodies as specimens to be used for scientific research when they died. Lalande, Commerson, and Buffon whom we have already met asked to be dissected, as did Diderot and Morand whom we will get to know in this chapter. Numerous anatomical amphitheaters cropped up and ads for public courses filled the papers, all attesting to their broadening popularity. Mme de Genlis who studied with Biheron reported that her friend, the young Comtesse de Coigny, was so enamored of the subject that she always traveled with a cadaver in the trunk of her carriage.[5] The status of surgeons was vastly elevated during this period because of their practical hands-on experience, now at least on a par with the book-learned theoretical doctors and even above them. Anatomy, clearly, was both a hot scientific field and a fashionable obsession.[6]

But there was a problem. Corpses rotted, the smell so vile and noxious in this age before effective preservatives or refrigeration that at the schools and public classes in the capital no dissections were permitted during the hot and humid months from May to September, when decay would be especially rapid. As the inimitable Rousseau put it, "What a frightful apparatus is an anatomical amphitheater: stinking cadavers, slavering and livid flesh, blood, disgusting intestines, dreadful skeletons, pestilential fumes! Upon my word, that is not where Jean-Jacques will go looking for his fun."[7] Biheron provided a way around this by crafting and teaching with her ingenious, entirely reliable anatomical models instead of actual bodies, and doing so all year round. Hers were lessons in splanchnology, the study of the viscera, but without the gore.

While the preparation of such uncannily lifelike "artificial anatomies" required that Biheron dissect real corpses in order to constantly hone her models, her final products with all of the internal organs available for removal and examination were not the least bit unpleasant to study and manipulate. On the contrary, Biheron's pupils and visitors were full of enthusiasm for these fascinating, tidy, orderly, reassuring models and for their knowledgeable female creator. Diderot would later assure Empress Catherine the Great of Russia that Biheron's profound understanding of human form and function was "rare even among men" who studied such things, d'Alembert confirmed her expertise with his claim that he learned more anatomy from her than from all his lessons at the Paris medical faculty, and the geographer Mentelle, speaking of her mastery, said simply that she "owned anatomy."[8]

Although Biheron burst into public view at age forty with her first of three widely and wildly heralded anatomical demonstrations before the illustrious Académie des Sciences on 23 June 1759, we can find traces of her much earlier, which may elucidate her motivation for taking on such a unique occupation. Was it to cater to the appetites of society ladies like the Comtesse de Coigny? To escape from boredom or from marriage? Was it to gain esteem within her family? To instruct and enlighten others, something for which she had a natural talent? For the sheer joy of it—Mme de Genlis said it was her passion, her *délice?*[9] To reinforce her faith by proving that nature, and in particular the human body, revealed God's infinite wisdom? To satisfy her own burning curiosity? To earn a living? Probably all of the above.

The youngest daughter of the well-established pharmacist Gilles Biheron on the rue Saint Paul, himself descended from a long line in that profession, who would almost surely have betrothed her to a fellow apothecary had he not died when she was four years old, she had a sister who took nun's vows, an older brother who became a pharmacist, and another who became a military surgeon. She, unable to follow in their footsteps because of her sex, grew up in the same healing culture and forged her own independent but related medical path. Biheron's anatomy teacher, Sauveur-François Morand, believed he was molded by his upbringing in his father's surgical practice, and a similar kind of imprinting may have given her a natural scientific bent, surrounded during her early formative years in her father's shop by the tools of that trade.[10]

Pharmacy was much closer in those days to medicine and chemistry than drug stores today would have us believe. A great deal more was involved than the dispensing of remedies. Although apothecaries were often associated with spice merchants, they were in fact the respected and knowledgeable collaborators of doctors and men of science, joining them for experiments and demonstrations, and many of them had their own gardens where they cultivated medicinal plants in an effort to come up with new cures.[11] Shelves, cupboards, and drawers held earthenware jugs, mortars and pestles, brass balances with weights and measures, translucent horns and glass vessels for weighing liquids and inspecting urine, decorated faiences for storage, dispensing pots, ointment containers, leech jars, leaves, seeds, roots, mushrooms, tinctures, extracts and infusions of all sorts, spatulas and slabs for mixing creams and thick syrups, powders in their folded paper sheaths, tablets, capsules, lozenges, and pastilles. There were mustards and vinegars (some of which during the eighteenth century were even reputed by their maker, Maille, to prevent plague and restore virginity), and of course the ever-present suppositories.[12]

Biheron's mother, widowed in 1723 and left with four children aged four, thirteen, sixteen, and seventeen—did she dote on her baby Marie-Marguerite, born almost a decade after her next older sibling, or resent her for coming along when the hard child-rearing was supposed to be over?—seems to have kept up the shop with the help of her sons until she remarried in 1734. Biheron was then almost fifteen, and about that time she was sent to learn painting from Mlle Basseporte, eighteen years her senior, the gifted botanist and botanical illustrator (see chapter 3). Basseporte had recently moved from her home on the rue Vieille du Temple in the Marais, where she ran an art school for young girls, to the Jardin du Roi, where she would become an indispensable permanent fixture. She quickly assessed Biheron's particular artistic talent to be sculpture. Building on this special pupil's upbringing in the medical world of pharmacy, Basseporte urged her toward the increasingly popular study of human anatomy but with a distinctive twist, the emphasis on modeling, in this way nurturing her mindful hands in three dimensions as she had cultivated her own in two. She secured for Biheron the mentorship of two savants in her network, the renowned surgeon Sauveur Morand and the eclectic doctor Jacques Barbeu Dubourg.

Basseporte's choice of advisors for her protegée was significant, revealing that her sphere of friends reached far beyond the Jardin itself into other professional milieux, and that she wished to harness these broader connections for this exceptional student. Morand was a member of the Académie des Sciences, a founder of the Académie de chirurgie, a royal censor, and chief surgeon at the Hôpital de la Charité and the Hôpital des Invalides, so among other things he could easily supply bodies for the budding anatomist. Dubourg was a member of the Faculté de médecine, although much more progressive than most of his colleagues for he favored inoculation and condemned the practice of bleeding. He was also a botanist and author who taught pharmacy at various times and an active medical journalist who could make Biheron known in the pages of his paper. Between these two influential men, she would have access to immense knowledge and to wide circles of assistance and visibility should she need them.

Meanwhile a great bond developed between the two women, whose interests steadily overlapped, botanist Basseporte displaying her own considerable skill in anatomy by contributing a masterful drawing of the insides of a human head for the third volume of Buffon's monumental *Histoire Naturelle*, and anatomist Biheron botanizing with Basseporte and Dubourg, who named a species of mushroom after her in his *Le Botaniste Français*, a handbook admired by Rousseau.[13] Neither woman married, and they seem to have been each other's intimate companion for some time, although they did not share a home. Basseporte named Biheron as an executor of her will, and Biheron, who outlived Basse-porte by fifteen years, cherished the older woman's self-portrait as a treasure. Many saw them as a couple, Franklin for example sending a single gift to the two of them.[14]

In 1744, once her artistic apprenticeship was over and she reached age twenty-five, her legal majority, Biheron could no longer live in the Jardin du Roi with her teacher for she had no official connection there, but she found a place a short walk from one of the entrances to the garden. Already deeply involved in her anatomical studies at surgical amphitheaters (where she had to disguise herself in male attire to be admitted), in private sessions with her mentors, and through numerous dissections of her own, she took up lodging on the rue de la Vieille Estrapade, at the corner of the rue des Poules (today the rue Laromiguière), in a house that would soon be occupied also by the philosophe Denis Diderot when he moved his family into a third-floor apartment there in 1747.[15] This neighborhood with its evocative street names—the rue du Puits-qui-Parle (a man was said to have drowned his gossipy wife in the well there), the rue de l'Arbalette (named for its crossbow shooting range), the rue du Cheval-Vert (today the rue des Irlandais)—was certainly not fashionable. The medieval torture method of the strappado, for which the Place de l'Estrapade was named and where this practice had for centuries taken place, may no longer have been current, although it is unclear exactly when it was abandoned.[16] But this was a bustling, popular quarter, and Biheron was soon well known there. Nearby the construction of the new Église Sainte Genevieve, later to become the Pantheon, was underway, and not far in the other direction was the Église Saint Jacques, whose priest Jean-Denys Cochin, a friend of Biheron's, would later purchase four houses that she owned on the Faubourg Saint Jacques to form the seed buildings of the Hôpital Cochin, a hospital that offered public assistance for the poor of the quarter and the wounded from the nearby quarries and is still functioning and well known today.[17]

When Diderot became Biheron's immediate neighbor he joined Basseporte, Morand, and Dubourg in her close group. Sharing her great interest in medicine and anatomy, he took lessons from her on several occasions. She was routinely working with corpses to perfect her models, and one Diderot biographer even supposed he might have chosen that apartment expressly to be closer to Biheron and her anatomical research.[18] Since 1744 he had been translating Robert James's *A Medical Dictionary*, and his six-volume French version, called *Dictionnaire uni-*

versel de médecine, appeared between 1746 and 1748. He next wrote a work ostensibly on visual perception and blindness—we saw that his then-friend Condillac and Mlle Ferrand were exploring similar subjects—although Diderot's book touched on subversive topics, enough so that this *Lettre sur les aveugles* quickly got him into trouble and into jail, the police labeling him "dangerous" and "impious" for speaking irreverently of, among other things, the "holy mysteries."

Biheron was very probably present when Diderot was arrested in their building at 7:30 in the morning on 24 July 1749, and thence escorted to the tower prison of Vincennes, and also when he returned home upon his release 3 November. Despite the seemingly irreconcilable differences separating her Jansenist piety from Diderot's free thinking, a real friendship developed between the two which lasted many decades, and he would freely admit how much he and his friends learned from her during these years.[19] Biheron may also have witnessed, a few years later, the strange police "raid" orchestrated by the progressive director of the book trade Malesherbes, a secret friend of the philosophes who had actually rescued Diderot's most incriminating *Encyclopédie* papers beforehand for safekeeping in his own home. Biheron became and remained an admirer of Malesherbes for that and other reasons which will become clear later.

Diderot respected Morand's medical integrity, urging him to use his influence for the public good to end the age-old feud between surgeons and doctors and recruiting him for consultation on some articles in the fledgling *Encyclopédie*.[20] He agreed with Morand's insistence that giving one's body to science was "the best thing that the dead can do for the living."[21] In 1751 Diderot added to d'Alembert's *Encyclopédie* article "Cadavre," which already stated that "dead bodies are the only books where one can study anatomy properly," a section going still further: "For the preservation of men and the progress of the art of healing . . . there should be a law that forbids interment of a body before it is opened. What a wealth of knowledge we would gain in this way!"[22] Around this time Diderot began to insist to his family that his own corpse be given to science. As his laconic daughter would later write, "My father believed it was wise to open up those who were no more; he believed this operation useful to the living. He requested of me more than once [that he be opened]. And so he was."[23]

Diderot and his wife had already lost two children in infancy, but on 2 September 1753 his daughter Marie Angelique was born, baptized at the nearby Saint Etienne du Mont, and Biheron who surely knew her as a baby would teach her anatomy as a young woman on the eve of her marriage. Diderot's family moved away to the rue Taranne in 1754, so he and Biheron would no longer have had the same daily contact after that. And already he had become absorbed by his herculean labors as editor of the *Encyclopédie* as he would be until the last text volumes appeared in 1765, and the eleven volumes of plates were finally finished in 1772. But even during this time, once Biheron opened her *Cabinet d'Anatomie artificielle*, he loyally took foreign visitors to see it—even complaining at one point, in 1768, that Baron Grimm was pushing him to do still more of this and that it was wearing him out.[24] Once the *Encyclopédie* was done, Diderot would again champion Biheron and more vigorously than ever.

She was part of another scientific circle as well, the electricity crowd. Her immediate neighborhood was the scene of some of the earliest experiments with lightning, inspired by Benjamin Franklin's first French translator Dalibard, a botanist and physicist friend of Biheron's teacher Dubourg who would himself be Franklin's second and more complete translator. Dalibard performed Franklin's kite experiment in June 1752 in Marly for King Louis XV, using a forty-foot iron rod grounded in bottles, and that same year translated into French Franklin's theory that lightning is electric. After the initial success of this royal demonstration, extracting lightning from a cloud, which thrilled the monarch beyond measure, crowds flocked to an encore performance in Paris where Biheron's neighbor, a physicist named De Lor, repeated what he referred to as "the Philadephia experiment" with a rod more than twice as high (ninety-nine feet) on the roof of his building, right on the Place de l'Estrapade. Dalibard would be among the first, in 1762, to urge Franklin to come to Paris to meet all the "électriciens," and later Franklin would thank him profusely for being so kind to him, "a Stranger at Paris" during his first brief trips in 1767 and 1769.[25] It was through Dubourg and Dalibard that Biheron met and began her long friendship with Franklin.

But even earlier her network of scientific friends had become increasingly intertwined. In 1753 she helped distribute Dubourg's unrolling timeline, what he called his *Carte Chronographique*, a scrolled invention about which Diderot raved in the *Encyclopédie* article "Chronologique (machine)."[26] By 1755 she was preparing anatomical models for Morand as he put together a collection for then-empress Elisabeth of Russia, to be expedited to her in 1759 (by way of Poland) in order to bring up to speed the study of anatomy and medicine in Saint Petersburg. All of this, then, preceded Biheron's famed demonstration to the Académie

in 1759; Basseporte, Morand, Dubourg, Dalibard, and Diderot formed her growing *bande*.

In order to do her anatomical work Biheron became of necessity a fearless
handler of dead flesh. Her calling was to show and teach the beauty and complexity of our bodily organization. She was working with death in order to
understand life and help the living. For her, a person of faith, demystifying the
body was a pious act, revealing the intricacy and wonder of Creation. Unapologetically assuming the medical gaze, she went way beyond gazing; she cut,
dug, delved, extracted, and took apart in order to reassemble the accurate whole.
Seeing the hand of God in all that she discovered, she was said to have argued
with a professed atheist that the complexity of our organization proves a higher
power at work, challenging said materialist as they looked together at the ram-
ifying structures she was demonstrating: "Well then, professor of chance, do you
have enough brains to convince us that chance is equally intelligent?"[27] It was
this sensibility that animated Biheron as both worshipper and researcher.

Earlier anatomists wrote graphically about their furtive studies, as most did
not get bodies legitimately, and beyond that they had to do their gruesome work
in extreme cold to avoid quick putrefaction. So they froze as they labored alone
on cadavers—"flayed, hideous, reeking and horrible to behold" had been Leo-
nardo da Vinci's earlier characterization—the vapors ghastly, the display har-
rowing. Corpses in rigor mortis would be stiff, but soon the bodies became more
pliable as the cells degraded, easier to work with but already fast decaying. The
remains would have to be wrapped at night in paper and cloths dipped in vinegar to avoid dehydration and spoilage, but despite such precautions they would
dry out, then liquefy and rot. Biheron, a woman of literally penetrating vision,
dealt over many decades in this macabre space. Her end-product was aesthetic
and sanitized, but she had to pass through deep zones of morbidity to create it.[28]
She surmounted, as a pair of contemporary admirers marveled, the "difficulties
there were for a girl to insinuate herself into the hospitals, and into the surgical
amphitheaters, at hours when she would not risk encounters with students; one
could see her there, the scalpel in hand, applied with an incredible ardor in discovering the most hidden secrets of the human structure. Devoted without rest
to these repulsive occupations, she suspended them only to apply at her home,
in the most ingenious models, the knowledge and the discoveries with which she
had enriched herself."[29]

Time and again, working alone, Biheron would open "les trois ventres," the three main body cavities: first the thoracic with the heart and lungs, next the abdominal, overcrowded with viscera and covered by its epiploon or omentum which appeared to float on the surface of the digestive system, and finally the pelvic with its bladder, reproductive organs, and rectum. Historically the *trois ventres* were the abdominal, chest, and head, but Biheron defined them differently because of her interest in the generative parts as an independent system. She modeled separately the dorsal cavity, including the brain and spinal cord covered by their meninges. In a perpetual race against time and the ravages of putrefaction, she would peel away the ubiquitous connective tissue that holds together but hides the internal layout she was trying to study, working in the glass enclosure of her garden, a dissecting chamber that Mme de Genlis tells us Biheron referred to as her "petit boudoir."[30] The carcasses she studied, at least several hundred over her lifetime, some of dubious provenance—it was said she occasionally hired people to steal them from the military—would have been in various stages of decomposition when she received them, a very few perhaps fresh from execution or snatched immediately after death, but the vast majority either unearthed and already rotting or emaciated, deformed, even desiccated from long illness if obtained straight from the hospitals.[31] She concocted some sort of wetting and preserving solution and then toiled as fast as she could until decay made necessary the procurement of another body. Always in a hurry to take molds of the organs, she would later make casts to produce her waxen copies.

Cutting a line through the skin, her scalpel's blade would proceed through the yellow, loose fascia layer, the sometimes fatty, sometimes webby connective tissue, to the gray, more clearly defined muscles, probing gently with closed scissors, or forceps, or most likely her hands themselves to spread organs and vessels apart without damaging them. In the thorax she would need to saw through ribs and cartilage—bone particles must have flown all around—to expose the viscera. There lay fragile-looking veins, muscular arteries, strong whitish fibrous cords of nerves, shiny tendons and everywhere, even in thin bodies, globular, messy fat obscuring her view. Inserting her thumbs at the midline above the sternum, she could then open the body like a book, the ribcage spread like a butterfly, and she would behold the two asymmetric grayish brown lungs and the fist-sized heart filled with black hardened blood. Proceeding through the omentum, the apron-like membrane supporting the abdominal structures, she would dig into the cavity and find the flat, pale pink sac of the stomach, the greenish gall bladder, the broad portal vein, and the liver.[32]

In the pelvic region of the females she'd see the apricot-sized womb, a tiny irregular ball that could stretch with the developing fetus, the little bluish egg-shaped ovaries, the small but sturdy fallopian tubes. She studied the male generative organs as well, making cross sections of the testes and the penis. These models of the reproductive system, along with others of embryos at various stages of development in utero, were used in the lessons she provided to women. We will return to this, as it was a singular, progressive, indeed modern service that she provided, her female students said to number well over a hundred.[33] Nobody else in Paris was known to offer this kind of education.

Making her wax models was also time-consuming. Biheron would grease the surface of the specimen she was reproducing in order to prevent sticking, put down some thread or string to aid in taking off the hardened mold, and then brush and eventually apply with a spatula layers of plaster onto the prepared organ. The front and back of the organ would each need their separate treatment. Once the molds set and were gingerly removed, their internal sides would be coated with oil or soap to fill in any small pores, the two halves of the mold reattached to each other, the wax poured into the cavity of the assembled whole and rotated around slowly and patiently in successive coats. Several additives would have been mixed with the wax—olive oil, turpentine, suet, and Biheron's mystery ingredients, perhaps silk, wool, threads, feathers, vegetable resins, tints, resulting in her secret alloy which famously neither melted nor broke. Once the wax hardened and the mold was removed there followed the painstaking corrections, painting and finishing of each body part.[34] Biheron might have been particularly drawn to the malleable medium of wax, associated as it was with church candles and votive offerings; her work could be said to blend the sacred and the scientific.[35]

In 1759 the Académie des Sciences invited Biheron to make the first of three unprecedented presentations, bringing her into the limelight. Morand had arranged this extraordinary event—women were not allowed to participate directly in the academy—and spoke proudly of her triumphant imitation of the human body before actually introducing her to the assembled savants. Praising the great superiority of her models, he said that previous ones, like those of Guillaume Desnoues's traveling show earlier in the century, were made of hard wax and presented only the position, form, and color of the organs, any parts with lightness or suppleness simply left out. Biheron's, he boasted, showed hollow viscera and membranes "in a way that fools the beholder." Soft organs like the stomach, intestines, and lungs could even be inflated. Solid elements like the liver, kidneys, brain, while made mostly of wax, were composed of a mixture which seemed indestructible. "A real skin covers this model . . . , which will neither yellow over time, crack from dryness, nor break in transport. . . . Mlle Biheron has succeeded in copying and imitating nature in this way with a precision and faithfulness that nobody has yet approached. If we are astonished by her talents . . . , we should be as much so by her taste for anatomy, which enabled her to surmount the almost invincible repugnance of persons of her sex for objects of this kind."[36] After these preliminaries Morand ushered her in; she discussed and manipulated the various artificial parts, which then underwent close scrutiny by the spectator-academicians. Her models along with a collection of surgical instruments were about to be sent to Russia for the school of surgery, as was a cloth obstetrical model, at the time not demonstrated to the academy but also perfected by Biheron.[37]

The periodicals of the day enthused about her ingenuity, performance, and the brilliance of the pieces she fashioned. The *Journal de médecine, chirurgie et pharmacie,* edited by the Montpellier-trained doctor Charles-Augustin Vandermonde, described his trip to her workshop to see these wonders for himself, and there he was stunned by their "exactness and truthfulness. . . . It seems to us that Mlle. Biheron redoubled her efforts every time she encountered obstacles, and that her talents are stimulated [*irrités*] to fight against difficulties. It is impossible to better render, than she has, the internal surface of the cranium, the different substances of the brain, and the ropey cordons of the nerves. The choroid plexus seemed to us a thing of beauty; we equally admired the transversal sections of the ribs, the texture of the sternum and especially the membranes that cover the chest and lower abdomen; the omentum makes a complete illusion. There are countless beauties of detail that we cannot adequately describe here; we must freely admit that there are parts that we have never seen as well in a cadaver." If Biheron continued in this way, Vandermonde added, her entire set of models would allow anatomy students "to thoroughly learn this science, without the disgust and without the horror," better and with less wasted time.[38] The Jesuit *Journal de Trévoux* raved about innards that were transparent and observed that wax, though used throughout in the models, was "never used alone" but rather mixed in a unique blend to trick the eye and seem entirely life-like. "Thanks to this anatomical marvel, we are spared having to do dissections ourselves," but the reporter went on to recognize that Biheron was only able to produce these artificial bodies by repeatedly and scrupulously studying real ones.[39]

Her colleagues involved Biheron in other activities as well, some surprising and even bizarre. During 1759 and 1760 on several occasions both Morand and Dubourg had her participate in monitoring a curious recurrence of the Jansenist *convulsionnaires* phenomenon that had caused widespread hysteria in Paris in the 1730s and forced the closing of the Église Saint Médard at the bottom of the rue Mouffetard. Miracles said to be occurring there on the tomb of a Jansenist deacon led to crowds writhing in fits of ecstasy and derangement. The move-ment had been driven underground for disturbing the peace, but remaining cells of believers held occasional séances in Biheron's neighborhood, so the govern-ment wished to round up enough information to suppress the fanatical cult de-finitively. Dubourg was sent by the medical faculty and by the police chief Ber-tin to infiltrate and observe the proceedings, during which women and girls were said to be "crucified" and, claiming that they felt no pain, were happily eating *riz au lait* and oysters when they descended from the cross. Morand attended and took notes on similar sessions.[40]

From their descriptions it is impossible to know exactly what went on at these weird gatherings, but Biheron was present at them, reportedly washing the hands and feet of the women, who then were said to leave and go back to work. Whether she was there as a reassuring female presence, as a scientific skeptic, or as a student of just how much the flesh could endure is not clear. She, Basseporte, and Morand had Jansenist leanings themselves.[41] But their Jansenism was as much philosophical as religious, and they were certainly not possessed. A new kind of "enlightened" Jansenism had evolved from a theological to a political and economic position, supporting agriculture, free trade, Gallic liberties, even republicanism. Both Basseporte and Biheron later sympathized with Malesher-bes in his *patriote* support of the *parlements* against despotic royal ministers, a *frondeur*, subversive position associated with the nobles of the robe, many of whom were Jansenist magistrates.[42] Biheron, level-headed and scientific, was an observer, not a participant, in these cultish frenzies. Something of a medical phenomenon, these rituals had a physical component as well and probably in-terested her on that score.

The official opening of Biheron's medical museum in her home on 13 May

1761 was a big day. She was for the first time displaying her models for all to see. The four-page pamphlet announcing her *Anatomie artificielle* had appeared a few weeks earlier, endorsed in late April, predictably, by none other than Morand, this time in his capacity as royal censor, explaining that while her work already had the nod from the Académie des Sciences, the Faculté de médecine, and the Académie de chirurgie, she now wanted another vote of confidence, the "approval of the public." The pamphlet could even have been written by Morand, for the text suspiciously resembled his words two years earlier for her Académie appearance. "The public is alerted to the display of an artificial anatomy on a body truncated at the extremities with the unfolding of the viscera contained in the three cavities." There followed almost verbatim the favorable comparison of Biheron's models to all previous ones and praise for their precision and astounding realism.

We must remember of course that although this little brochure is repetitious of the report to the Académie, that earlier account was seen only by a select few, whereas the advertisement circulated to a broad public and was intended for an entirely different audience. Biheron's exhibit would be open every day except Sundays and holidays from eleven to one and from four to six; for those who wished "a more detailed displaying of the viscera in the three bodily cavities" or wanted lessons on each separately, such arrangements were available.[43] This pamphlet reveals the anatomist's readiness to launch herself into the wider world, to demonstrate for all comers, to try to make a living from her science and her teaching. Interestingly, no fee was mentioned for admission; she was probably planning to determine that later based on the size and enthusiasm of the crowds that came.

Once again, as after her academic appearance, the press applauded Biheron's wondrous collection, in *l'Avant Coureur*,[44] in the *Gazette de France* which called it an "astonishing" exhibit absolutely unique of its kind,[45] and in the *Journal des Dames*, whose radical female editor Mme de Beaumer was especially thrilled that this was the work of a scientific woman.[46] The *Gazette d'Epidaure*—a medical journal edited by none other than Dubourg himself—claimed that even before her cabinet opened she was already "so celebrated in all of Europe" that many foreign monarchs had ordered pieces from her assemblage.[47] All reported that people flocked to her.

But Biheron was much more than a popularizer or "montreuse," for she reached a double public.[48] She was certainly spectacular for spectators and entertained as she educated. Many loved her "please DO touch" policy and felt unthreatened, amused, and probably heartened by the strange but undeniable beauty of the body's parts. In that respect her studio was a crowd-pleaser for the curious. But the other half of her audience, the serious students, natural philosophers, and medical men she knew, learned from, and in turn instructed, understood full well that Biheron was a woman of science whose work entailed hard and gruesome labors of a kind not recognized by casual visitors, the beholders and holders who saw and handled only the odorless final result. Those who had experienced what dissection entailed, like the famed English doctor Sir John Pringle, admired the extreme faithfulness of her models yet immediately exclaimed, "Mademoiselle, all that is missing is the stench."[49] This compliment showed his own identification with the ghastliness (and even ghostliness) of working repeatedly with decomposing bodies.

We might well wonder what allowed Biheron to maintain her emotional balance in this transgressive occupation for nearly half a century. Although as a person of strong faith she surely regarded the religious rituals of proper church burial as sacrosanct, she also considered her work essential, which, for the sake of the living, involved disturbing the dead. In this context she did not experience her studies of the body as a violation but instead as an educational necessity. Focusing on her practical goal, she addressed the shortage of cadavers for medical research by providing an innovative pedagogical tool, a substitute that could be used to teach anatomy. Her lifelike but imperishable models with all their removable parts were made in the service of science, allowing her to offer her classes continuously, even during the months that the police prohibited dissections of real corpses in the all-male schools. And that is precisely what she did.

Overall, the 1760s were good times for Biheron. Bachaumont's newsletter the *Mémoires secrets* lauded her in 1763, arguing that she deserved official compensation for what she was contributing to medical science: "Mlle Biheron gives us one of the most curious and interesting spectacles. This girl, as active as she is industrious, has for several years applied herself to anatomy in such an intelligent way that she executes models of the greatest perfection. She uses all sorts of material, whatever is most appropriate for the illusion and rendering in all their faithfulness to nature the different parts that she wants to show. Such works can be very useful for many operations, and this able worker should be encouraged by the government."[50] In 1766 Dubourg asked Biheron to model a baby girl born with only one eye, although she was not primarily focused on deformities and pathologies. This "Cyclops" would later attract the attention of Buffon, who included an image from her wax model, along with great praise for her, in a supplementary section of the *Histoire Naturelle* called "Sur les monstres."[51] In 1767 she met Franklin on his first trip to Paris, which led to a sustained relationship with him and with his traveling companion Pringle. The two men stayed barely over a month, from 28 August to 8 October, and Franklin, who had been apprehensive about what kind of reception he would get in France, received a warm welcome especially by the electricity group. Dalibard, Dubourg, Basseporte, and Biheron entertained and hosted the visitors from England. Due to the brevity of their stay in the capital—Dalibard called it a mere fleeting "apparition"—their encounters with men of science were limited.[52] Yet they made time to see Biheron's exhibit and encouraged her to come to England with her teaching demonstrations, an idea she took under serious consideration.

All the "Franklinistes" awaited with impatience another visit from the two travelers,[53] and they did return in 1769, once again for a quick summer appearance. They followed the latest scientific activities and curiosities, including the presence of Aotourou, the Tahitian prince and Jeanne Barret's friend whom Bougainville had brought back from his voyage. This "noble savage" was being examined by various doctors, academicians, and linguists—La Condamine, Péreire—and he played the nose pipe for Pringle, creating much interest. But Franklin was also most thrilled and surprised to find great keenness among his French friends for the idea of American independence. Once back in London he wrote to a Boston friend, "I found our dispute much attended to. . . . In short, all of Europe except Britain, appears to be on our side of the Question."[54] Biheron was awakened to this political struggle against despotism across the sea, and both Dubourg and Dalibard would soon become major supporters and suppliers of rifles, ammunition, woolens, and more for the colonists in their fight for independence, well near bankrupting themselves in devotion to the American cause.

Delighted as Biheron was with her new scientific colleagues from across the Channel, trouble was brewing for her by the end of the decade; this may be the moment that the patriarchal medical institutions began to actively block her efforts. In 1769, when Antoine Ferrein, who had taught anatomy conscientiously but without particular distinction at the Jardin du Roi, suddenly died of a stroke on 28 February, the charismatic doctor Antoine Petit who succeeded him at this job immediately began to draw enormous crowds. These public anatomy lectures, "almost deserted previously" according to Bachaumont, were now the scenes of veritable stampedes, and police had to be called in to maintain order. One student compared Petit to Jesus Christ for the gentleness, charity, and shining spirit of his instruction. This compliment reportedly made the professor weep, the spectators fittingly moved by the touching moment. Petit found it nearly impossible to press through the assembled multitudes, it was said; students and admirers almost suffocated him and rent his garments in adoration.[55]

Petit's theatrical popularity and his uncanny ability to make real dissections attractive surely lured some anatomy fans into his audience and away from Biheron's. The university faculty evidently hardened against her too in what later writers called "medical despotism," barring students from attending her group classes.[56] At this time she may have had to renounce them and turn more to individual visits and lessons, for example, the one she provided for a young Swedish traveler, Jacob Jonas Bjornstahl, a disciple of Linnaeus who was in Paris trying to procure for his master some rare seeds and while there sought out scientific activity around town. After a visit to Biheron's cabinet, dazzled by her models and her knowledge of all body part names in both Latin and Greek—Mercier later remarked on this too, perhaps patronizingly although both men genuinely admired her—he reported that he had witnessed an "anatomical miracle." Through him Biheron sent regards to the famous naturalist of the North, whom she must have met when he visited the Jardin du Roi in 1738 while she was studying there with Basseporte.[57] But such private appointments were not going to keep Biheron's Paris museum afloat.

Morand, as impressed with her as ever, arranged at this delicate juncture another demonstration at the Académie in 1770, now of a different model entirely, her flexible obstetrical mannequin of a pelvis with a fetus in utero. He elaborated that this model had a movable coccyx, an anus, labia majora, and a cervix that could be tightened or dilated at will to show the effects of the baby descending and crowning. By placing the model of the infant in different positions, one could do maneuvers too risky to practice on live mothers and thus become more proficient at assisting childbirth.[58] Even the hard-nosed Antoine Louis, secretary of the surgical academy, said this model was far superior to the already-praised childbirth "machines" of Mme du Coudray.[59] Of course du Coudray's mannequins were meant to be handled and practiced on by her hundreds of students, were hastily produced in large numbers for wide distribution, and showed only the broad contours of the mother's body in sturdy materials that could withstand wear and tear.[60] Biheron's, in contrast, were made with an exacting eye to detail and were part of her admirably precise group of gynecological models of all kinds. In 1806 a doctor in Strasbourg credited her, retroactively, for having produced the very first *hystero-plasmata* models showing the vagina in a state of virginity, although others had wrongly claimed the priority.[61]

Morand's stimulus may have turned things around for Biheron, at least temporarily. In attendance for this second academic demonstration were Jussieu, Cassini de Thury, d'Alembert, Vaucanson, Lemonnier, La Condamine, Adanson, Buffon, Nollet, Maraldi, Duhamel, Pingré, Macquer, Lavoisier, Condorcet, and Portal, some of the most influential luminaries of her age, and she again displayed her skill to dazzling effect. Portal would now mention her in his *Histoire de l'anatomie* and even the great Albrecht von Haller expressed enthusiasm for her obstetrical mastery.[62] Haller set up the medical school in Göttingen, and although he had stopped teaching there in 1753 he remained involved enough to acquire one of Biheron's models; reports confirmed that, indeed, one of her "anatomies artificielles" was on display in that city.[63] Her "Représentations Anatomiques en Cire Colorée" were mentioned in an English guidebook of 1770, alongside sights like Notre Dame, the Luxembourg, the Invalides, and other outstanding venues that "merit a stranger's notice."[64]

On 6 March 1771, Biheron made yet a third appearance at the Académie, this time for the new young king of Sweden, Gustav III. Grimm, in his newssheet the *Correspondance Littéraire*, now gave her some of his most fervent praise, comparing her presentation favorably to those delivered by d'Alembert, Macquer, and Lavoisier in the same academic session. He discussed the perfection and accuracy of her model, how impressed Pringle had been with her invention, the extraordinary "odorless" and "incorruptible" material of which it was made, the exactitude of the tiniest detail and the most exquisite nuance, so that observers "will have trouble distinguishing the limit between art and nature." Calling Biheron "unique in all of Europe," he then scolded that the government should long ago have acquired this collection for the Cabinet d'histoire naturelle in the Jardin du Roi and compensated her in a way that did justice to and encouraged her inventiveness. Here he was echoing what Bachaumont had urged almost a decade earlier.

But, Grimm went on to lament, instead of official recognition and support Biheron's situation was actually quite different; she is poor—this mentioned twice—living on a small pension of 1,200–1,500 livres, "infinitely devout besides . . . has never been pretty, has never had either protection or ploy" and so, with no scheme up her sleeve, "remains neglected and forgotten on a corner of the Estrapade in a house formerly occupied by Diderot the philosophe."[65]

Really? Were things in fact that bleak, or was this said for dramatic effect to encourage financial backing for such a deserving woman? Grimm was right that the earnest, hard-working Biheron had never had the benefit of a patron. She did have loyal friends, however, and Diderot arranged for his daughter Angelique to take a course with her that same month, interrupting the young girl's plans to go to the countryside and sending her instead for *diner,* the meal usually consumed in the early or mid-afternoon, chez Biheron, and then what could only be called a sex-education lesson, deeming this essential as she was reaching marriageable age.[66] Adamant that his daughter be a pure and modest wife when the time came, Diderot believed the best way to assure this was education about her conjugal duties presented in a matter-of-fact, scientific manner so that all seductive mystery, all whims and desires, be entirely removed from the subject of procreation. Now that an actual wedding might soon occur, some personal sessions were in order with the "talented and sage" Mlle Biheron, whose anatomical models of the genital and reproductive organs would be used to explain what was in store for the bride and mother Angelique would eventually become.[67] Diderot was convinced that when his daughter inevitably came upon "indecent" books, Biheron's instruction would be an antidote to any arousal or temptation.[68]

The philosophe was certainly giving the anatomy teacher a lot of credit. Yet however much she was appreciated by particular clients, she was finding Paris increasingly inhospitable. Attendance at her museum was diminishing because the medical and surgical faculties, both of which had originally approved her models, had not foreseen her popularity, disliked her competition, and found more and more ways to sabotage her, interfering with her classes, forbidding her to offer any advice on healing, and excluding her from exhibiting some of her pieces in an obvious place at the Cabinet du Roi.[69] With her growing success through the 1760s, Biheron had come to be viewed by these all-male institutions as a threat and a serious rival.

Anyone else might simply have quit at this point. Never one to surrender, however, she decided to take Pringle and Franklin up on their invitation to teach in London, to associate with William Hunter and William Hewson, famed anatomists and surgeons of their acquaintance, and to display her models to new, potentially more appreciative savants and audiences there. Although men did it all the time, it was unheard of for women to travel for scientific collaboration, but that did not stop Biheron, who felt naturally propelled to do so.

In late fall 1771 she made the first of two trips to London, carrying a letter of introduction from Diderot to John Wilkes, in which he greeted his friend, the "honored Gracchus," extolled him for rousing the patriotic demon in England, and then lamented the current state of France, a seemingly great edifice that was in fact crumbling, and from which smart rodents, with excellent instinct, were fleeing before the collapse. The philosophes, like himself, less wise than those with pointed muzzles and long tails, stayed until the rubble fell on their heads. "Mlle Biheron, who will give you this extravagant letter, is an alarmed mouse who comes out of her hole and goes to seek some security with you. This mouse is a mouse distinguished in her species. She justifies the consideration she enjoys here, by a quantity of very beautiful works. They are anatomical pieces of a marvelous truth and exactitude. I urge you to welcome her and give her all the good help that you can. My daughter did, easily and without disgust, an anatomy course with her. If you believe me about this, you'll engage Mlle Wilkes to take some of her lessons. While this is not the object of her visit, because she joins to her knowledge a great generous character[,] I'm certain she will be pleased to oblige you for your child."[70] Diderot made clear his distress that talents like Biheron's were not sufficiently appreciated in France, and that people gifted in science and art needed to search for refuge and safety elsewhere. Less than a month later he sent Wilkes another letter introducing the miniaturist Pierre Pasquier, a painter in the Académie de peinture who was also quitting France, and referring once again to Biheron. The French empire, it seemed to Diderot, antagonistic as it was to those with any useful skill, was disintegrating, and by scaring such worthy people away was "returning to stupidity and misery."[71]

Although Diderot would thank Wilkes for welcoming Biheron, and she herself would speak kindly of his attention, the radical politician may have been too busy with his own personal slander issues, for he was of little tangible assistance.[72] Franklin and his circle, however, did try to introduce Biheron to some men of science in London on this first trip, and she stayed for about six months. Soon after her arrival, Pringle invited her to dine at his house in December along with Franklin and the distinguished Dutch chemist and physiologist Jan Ingenhousz, who would later discover photosynthesis but was now passing through doing research and performing smallpox inoculations on the rich and famous.

This company shows that Biheron was regarded as a scientific person rather than just a visitor they felt obligated to entertain. She would be eternally grateful to Franklin, to his scientific friends, to his landlady Mrs Stevenson, and to her daughter Mary, known as "Polly" and now Mrs William Hewson, for their help.

Biheron was also an important go-between, as Polly would translate Dubourg's *Petit Code de la Raison*, a deistic work he was afraid to publish in France, while Dubourg would translate Franklin's work into French. The anatomist carried letters, corrections, and printers' proofs to and fro on both of her London trips. Disappointingly, Franklin concealed Polly's role as a translator, believing it would be "vanity" to call attention to her part in putting Dubourg's work into English. He left her name off the published book, explaining that "modesty" was to be preferred.[73] This conservative decision belittled the importance of the translators' task, a matter that Mme d'Arconville would have set him straight on (see chapter 5). But the fact remains that Franklin, eager associate of anatomist Biheron, was unenlightened with respect to translator Polly's different kind of contribution to the propagation of ideas. Sadly, the famed Dr William Hunter also failed to appreciate her. He and William Hewson were in a stormy partnership, and confirmed bachelor Hunter disapproved of his younger colleague's 1770 marriage to Polly, disregarding her keen intelligence and fearing that domestic entanglements would distract Hewson from their shared work. Although they taught anatomy together, the feud between these two men was increasingly corrosive and hindered Biheron's visit.

Notwithstanding A. L. de Jussieu's report to his mother in a letter of 24 February 1772 that Biheron was having a wonderful, productive time in London and was greatly admired by both intellectuals and *gens de condition*, Dubourg's letters to Franklin suggest that her first visit to London was a financial bust.[74] For example, the *Public Advertiser* of Saturday 18 January 1772 announced an anatomy course she might well have been involved in—"On Monday next, the 20th instant, at 2 o'clock Dr Hunter's and Mr Hewson's course of anatomical lectures will begin at their Theatre in Windmill St"—but there was no mention of the special Parisian guest and no suggestion that she was one of the instructors for the fee-paying attendees. The only notice of her in the periodical press during this first visit was a rather odd one in the *Gazetteer and New Daily Advertiser* in March, very late in her stay. "We hear that the celebrated Mademoiselle Bieron [*sic*] from Paris is now in the Metropolis, and has brought her curious Cabinet of Anatomical Figures, which has been so long the admiration of all the connoisseurs at Paris and of all Europe. If Mademoiselle Bieron would be so kind as to render her direction public in this paper, she would oblige an admirer of her talents."[75] This suggests that she had worked behind the scenes instead of pushing to show her collection, the publicity for which was in any case woefully inadequate. Apparently the time was just not ripe.

By mid-May 1772 Biheron was back home in Paris to take advantage of better audiences at her cabinet in the warm spring, summer, and early fall months when others were prohibited from doing dissections and people came to her in larger numbers for instruction. Dubourg wrote to Franklin on 31 May 1772: "Mlle Biheron has not let me forget your repeated goodness to her, for which I am as appreciative as I would be if they had been directly to me."[76] But France was now in turmoil. The royalist coup by minister Maupeou that had so displeased the Jussieu family had resulted in the rout of the legitimate *parlements*, the country's law courts, the installation of yes-men in those judicial bodies, and also in excessively strict censorship of the press. In September Biheron, already sensitized by Franklin to issues of oppression and the grievances in the American colonies, sent him a manuscript copy of Malesherbes's eloquent remonstrances condemning the suppression of the *parlements*. Like him, she sided with the *patriote* magistrates, the ousted defenders of the fundamental laws of the realm who were protesting ministerial despotism. Her chatty letter to Franklin ranged over many subjects but we see that she was becoming politicized:

Paris 10 September 1772
Monsieur,

I am very appreciative for the honor of your memory and thank you for your gracious letter. I am taking immediate advantage of Mr Walsh [returning to England] to send you the discourse at our Cour des Aides by our celebrated magistrate M Malherbe [Malesherbes]. I would have preferred to send a printed version but as it is I desire that it be agreeable to you. I am sorry that Madame Stevenson didn't yet receive her thread given to Captain Guilbée master of the ship called *l'Aigle* by Mme Bosse at Calais. I have the honor of thanking her and implore her to receive my particular compliments and for her dear family she will please forgive me for sending her something [which] can be of use to her and accept it from the hands of friendship that I pledged with all my heart to all of you. I do not know if Mr Walsh will be pleased with us, but our savants valued him very much for his merit and great knowledge. I was so flattered to be able to speak of you with him having had the honor of hosting him several times may I have this pleasure more often or go reiterate to you in person in the month of November how much I have the honor of to be, with the greatest consideration, your humble and obedient servant,

Biheron

Dare I request that Madame Stevenson remit the little package of powder to a cloth merchant who will come get it from her, he's a poor unfortunate whom I had the happiness to relieve of a pain in his eye and I hope that he will see well enough in a few months. I know the excellence of your heart and the love you have for the public good as does Madame Stevenson it is what engaged me to ask you this service.[77]

Phonetically spelled, repetitive, and erratically punctuated as was the writing of many autodidacts, this letter reveals that Biheron was busy, daring, and politically aware. Malesherbes's impassioned plea had not been publishable in France, and she was perhaps hinting that Franklin could get it printed in London. Confident about a remedy as the daughter of a pharmacist would be, her mailing of a prescription to a London patient suggests that she furtively engaged in healing work in Paris too, as did Basseporte.[78] John Walsh, a Fellow of the Royal Society who traveled in lofty scientific and social circles while in France, had sought Biheron's enlightened company more than once during his visit, and she casually conveyed how much other savants, with whom she was clearly in contact, also appreciated him.

While back in Paris Biheron continued her teaching of both groups and individuals, notably giving more lessons to Angelique immediately before her marriage on 13 September 1772. In November Diderot wrote a letter to his priest brother, after which the final rupture with this sanctimonious sibling occurred, telling him that he had relied on Mlle Biheron to educate his daughter: "It will soon be two months since she married, and she has kept and I hope will keep all of her life the simplicity, the gentleness, the modesty of a young girl. Modesty, decency, do you hear me, sir? And do you know to what she owes this so rare privilege that *gens du monde* notice in her? To three courses in anatomy that she took from one Mlle Biheron recommended as much for her talent as for her wisdom [*sagesse*], before passing into the nuptial bed. Obscene words are as insipid for her as religious disputes are for me."[79] Diderot's obsession went beyond the call of duty for the anatomist, but literate girls were indeed in danger of reading suggestive material. Whether Biheron was really neutralizing this menace or not, she was surely and candidly teaching adolescents and young women about their bodies.

It cannot have been accidental that her second trip to London was again timed to coincide with the cold months when dissections were permitted in the Paris faculties and her business slowed. Franklin said she was certainly not choosing an ideal moment by venturing to London in the dead of winter. Dubourg, who in addition to being Biheron's close colleague was also her doctor, had written a sweetly protective letter to Franklin begging him to speak frankly if he felt her prospects in London would be no more favorable than the first time. He wanted to save her unnecessary expense and exertion as she had still not entirely recovered, either financially or physically, from her previous journey. "In God's name, if you do not see better success for this voyage than the last one, tell me the truth confidentially and I'll insist that she stay home for her health."[80] At just this time Franklin was trying but failing to get Hunter and Hewson to reconcile; their feud would not enhance Biheron's reception.[81] But in early November he reassured Dubourg that if she could survive in "this Foggy Climate and Smoaky [sic] City without too much hazard, I will do my utmost when she comes to promote her interest."[82]

When Biheron returned to London Hewson had broken with Hunter and set up his own anatomical research on Craven Street in Franklin's residence, the older man having temporarily moved out and up the block. Biheron passed much of the Christmas holiday with Franklin and his company. Together they wrote to Dubourg on 26 December, Franklin explaining that she had been unwell but was on the mend now, and closing with, "We join in wishes of the season for you and Mme Dubourg." On 13 January 1773 Dubourg wrote back that news of Biheron's illness had made Basseporte sick with worry, but Franklin, trusting in the healing ministrations of the Stevenson women, sent a reply on different subjects—his design for a new stove, his views on music. On 24 February Dubourg thanked him for his goodness to their beloved traveler, and letters followed regarding his French translation of Franklin's works. On 11 April, still unsure as was Basseporte about Biheron's recuperation, Dubourg wished she were safely home and asked Franklin to escort her personally back to France.

He mentioned that they had received a few letters from her but had not gotten many others that they knew she sent.

We see from this that Biheron wrote frequently during her time abroad, but most of this precious correspondence is forever lost. It probably touched on some of the subjects—medical, political, and even fanciful that preoccupied the two men—such topics as restoring life to apparent drowning victims and those struck by lightning. Late that April Franklin described flies apparently dead in Virginia Madeira but that reportedly dried out and flew off when the bottle was opened in London. Please embalm me in Madeira, he begged Dubourg, so that I may wake to see "the condition of America 100 years from now."[83] No archivist thought to save letters from a female anatomist who was right in the thick of this relationship, and who wrote about her views on a regular basis. Letters between Franklin and Dubourg are numerous, and preserved of course. We have only a handful of Biheron's.

This time her arrival in London was promptly reported in the press on 7 January 1773, through both the *Gazetteer and New Daily Advertiser* and the *Public Advertiser,* and she was already showing her collection, including new pieces, in two sessions, from eleven to one and from one to three.[84] This was a still more exhausting schedule than the one she had proposed at the opening of her museum in Paris twelve years earlier, for now there was no break for a midday meal or rest. It was soon shortened slightly to noon until three, but during this second stay in London Biheron was determined to make her presence known and to be readily available. Long notices about her were well placed in the middle of the classified ads sections. Franklin was friendly with the publisher of the *Public Advertiser,* Henry Sampson "Sam" Woodfall, and may well have arranged and even paid for Biheron to be featured prominently. The announcements, probably written by Franklin, stressed the scientific nature of her work (figure 12):

> ARTIFICIAL ANATOMY—To the curious in general, and particularly to the Lovers of Natural History.
>
> MADEM. BIHERON, whose Anatomical Figures have received such deserved encomiums from the Faculty both of this and of other countries, is returned to this Capital with some considerable additions, particularly an accurate imitation of the whole Body, calculated to give an

exact idea of the structure of the Human Frame, and to teach that most necessary part of science, the knowledge of ourselves. This Exhibition having the same advantages over those that have so lately had the countenance of the Public, as the works of Nature have over those of Art, cannot fail to interest the English Nation, distinguished equally for their love of science and for their generosity. And as Anatomy in the hands of Madem. Biheron is divested of the disagreeable circumstances attending the examination of a real corpse, by her delicate and accurate imitations in wax, &c. the Public may now in the most agreeable manner see how curiously and wonderfully we are made. Madem. Biheron, in order to make her Exhibition the more agreeable, has engaged an English Gentleman to explain her figures in his own language instead of the French.

The hours of exhibition are once every day (Sundays excepted) viz. from Twelve in the morning to Three in the afternoon, at her lodgings, at Mr Williams's, Villars-street, opposite Duke-street, York Buildings.

Ladies and Persons of Distinction may have other hours, on giving timely notice.

Admittance Five Shillings[85]

The notice was frequently repeated in both papers throughout February and March, emphasizing that whereas Biheron's previous demonstrations had been in French, this liability was now remedied by the presence of a translator. She was living near the Strand, on Villars Street, just a few minutes from Franklin's house, and her landlord was none other than Jonathan Williams, Franklin's grand-nephew and financial clerk who was away in Boston on business at that very time. Biheron would later reciprocate by receiving and hosting young Williams in 1775 when he journeyed to Paris.

Then on 13 April, about three months after she first displayed her models, there appeared her first farewell:

MADEMOISELLE BIHERON returns her most grateful Acknowledgements to the Nobility and Gentry for the many Favours they have been pleased to confer on her during her Residence in this Capital, and begs Leave to acquaint them she will continue to exhibit her Anatomical

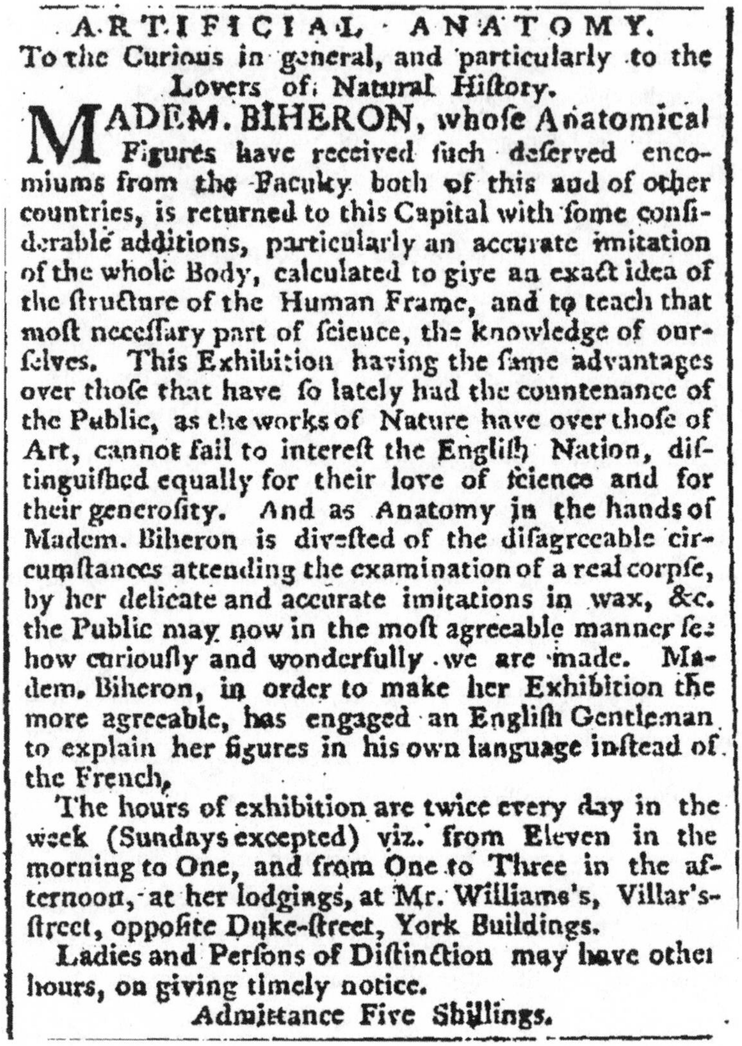

A·R·T·I·F·I·C·I·A·L · A·N·A·T·O·M·Y.
To the Curious in general, and particularly to the
Lovers of; Natural Hiftory.

MADEM. BIHERON, whofe Anatomical
Figures have received fuch deferved encomiums from the Faculty both of this and of other
countries, is returned to this Capital with fome confi-
derable additions, particularly an accurate imitation
of the whole Body, calculated to give an exact idea of
the ftructure of the Human Frame, and to teach that
moft neceffary part of fcience, the knowledge of our-
felves. This Exhibition having the fame advantages
over thofe that have fo lately had the countenance of
the Public, as the works of Nature have over thofe of
Art, cannot fail to intereft the Englifh Nation, dif-
tinguifhed equally for their love of fcience and for
their generofity. And as Anatomy in the hands of
Madem. Biheron is divefted of the difagreeable cir-
cumftances attending the examination of a real corpfe,
by her delicate and accurate imitations in wax, &c.
the Public may now in the moft agreeable manner fee
how curioufly and wonderfully we are made. Madem. Biheron, in order to make her Exhibition the
more agreeable, has engaged an Englifh Gentleman
to explain her figures in his own language inftead of
the French,
  The hours of exhibition are twice every day in the
week (Sundays excepted) viz. from Eleven in the
morning to One, and from One to Three in the afternoon, at her lodgings, at Mr. Williams's, Villar's-
ftreet, oppofite Duke-ftreet, York Buildings.
  Ladies and Perfons of Diftinction may have other
hours, on giving timely notice.
Admittance Five Shillings.

Fig. 12. Notice for Marie-Marguerite Biheron's London exhibit of her
anatomical collection, from the *Public Advertiser* of 26 January 1773.
[British Library, © British Library Board, Burney 599.b]

Curiosities every Day at Twelve o'Clock at her Lodgings, at Mr Williams's. Villers-street, York Buildings, in the Strand, till the 20th Instant, when she proposes leaving London.

Admittance Five Shillings

A notice on 28 April explained that her departure was postponed several weeks "by particular Desire of Several Persons of Quality" until 6 May, and on that day another notice read, "Sensible of the Honor conferred upon her by the late numerous Appearance of the Nobility and Gentry at her Exhibition . . . she has deferred her departure on that account til the 4th of June when she will positively leave England." Franklin must have made a big effort for her trip to yield fruit toward the end. Ever proud, Biheron kept her hours and her fee steady during this extra month, simply stated her intention to close up shop, and did it, sailing home on 6 June.[86]

Who exactly were her clientele? Five shillings was a fairly steep fee, so her cabinet was not as readily accessible to the broad public as her Paris exhibit had always been. A single shilling in 1720 had the buying power to pay for a dinner of beef, bread, and beer plus a tip, admission to the Vauxhall or Ranelagh pleasure gardens, or a pound of perfumed soap. In the mid-1700s, workmen at the Wedgwood factory, who were paid comparatively well, earned four to five shillings a week.[87] The average person, therefore, could not have afforded Biheron's attraction; it seems indeed to have been targeting the nobility and gentry precisely as the advertisement noted.

During this second visit to London, when she was not busy showing her collection she probably helped Hewson in his new Craven Street lodgings perform dissections on animals and humans, these last on cadavers illegally procured. Although by this time in England human dissection was no longer a crime punishable by death or deportation as it had once been, but rather a misdemeanor for which a whipping would be in order, it was still condemned by public opinion, and so Hewson would have had to dispose of these corpses secretively. Biheron was well-practiced in the art of obtaining and discarding bodies, and we know that she worked closely with him during this time.[88] Buried bones were recently discovered at this site during excavations.[89] Hewson's untimely death on 1 May 1774 was caused by septicemia, resulting from dissecting a putrid body while having an open wound.[90] It is a wonder that Biheron did not succumb to the same deadly infection.

And what about the esteemed Dr. William Hunter? Did he learn from Biheron or vice versa? The two were virtually the same age. Because she, a true specialist on female anatomy and childbirth, visited London twice between fall 1771 and spring 1773 and he did not publish his *The Human Gravid Uterus* with its starkly detailed illustrations until 1774, there has been speculation that she had a hand in the preparation of that famous book, or at least brought with her some demonstration pieces that he must have used as models.[91] Letters between Pringle and Albrecht von Haller show that various images for Hunter's work were completed much earlier, yet oddly he did not publish until after Biheron's trip.[92] Other factors, including his feud with Hewson, make his appreciation of her unlikely. Franklin as middle man tried to be neutral, referring to this as "the unpleasant Time in which, as a common friend, I was obliged to hear your . . . mutual complaints."[93] But he surely favored Mrs Stevenson's new son-in-law. Pringle too admired Hewson, and he just then got the young man elected as a Fellow of the Royal Society. Under these circumstances how could Biheron, close colleague of Franklin and confidante of the Hewson household, have had much sway with, or even attention from, Hunter? In 1784 he would write that he had little patience with human models and minced no words: "Many of the waxworks I have seen are so tawdry . . . and so very incorrect in the circumstances of the figure, situation and the like, that, though they strike a vulgar eye with admiration, they must appear ridiculous to the anatomist."[94] Here he was referring to some of the later shows targeting the commoners on Fleet Street, and he must have thought more highly of Biheron's rigorously accurate obstetrical productions. But we just do not know if anything transpired between them.

As soon as she returned to Paris Biheron wrote to Franklin taking care of business, expressing her continued interest in advancing medicine, her eagerness to learn new fields of science, and her anxiety about a shifty adventuress, taking on the role of self-styled spy protecting the American colonies:

> I gave to our friend [Dubourg] the depots of papers you entrusted to me concerning the translation of your excellent works of physics. I hope to be among the first to learn from them. This study will be all the more agreeable because it is by you and our friend Dubourg. He and his dear wife, Mlle Basseporte, M Dalibard, and all those who have the happiness to know you present their compliments. I beg you not to forget to greet M Pringle and to think sometimes of she who across the sea will keep eternally the memory of your attentions to her. Madams Stevenson, Hewson, and Mr Hewson must please receive my compliments and thanks for the friendship they showed me for which I am most grateful.

My bundles crossed the sea fortunately and left Calais for Paris the

18th of this month to arrive tomorrow Sunday. M Dalibard seems enchanted with the acquisition you made for him without having seen it yet. In thanking you he will seize the next chance to reimburse you for this mirror.

There is a young surgeon with great merit who wants to leave for London [at] the beginning of next week to profit from the lights of your artists of this kind. I presume to recommend him to Mr Hewson. I have the honor of warning you in confidence that Mlle Guion de Saint Marie is little known in Paris and that someone who is interested in you came to find me to urge you to mistrust her. She seems to have duped father Berthier with her words. One of our ministers has, they say, written to the Comte de Guine to send back an *homme de condition* thirty years old who I saw with her who she, since a while ago, has turned against his family to whom he seems very dear to hear them tell it. As we don't need to know more, you'll do with this what your prudence dictates. I'm too attached to you not to inform you of this while reiterating the assurance of my devotion. Believe me for life, Monsieur, your humble and obedient servant Biheron. Allow me to greet M Fevre [Franklin's secretary/clerk].[95]

Father Joseph-Etienne Berthier, or Bertier, another friend of Biheron's, was a priest and a physicist, a foreign member of the Royal Society but eager to support the revolutionaries. As early as 27 February 1769 he had written to Franklin, "I was a Frankliniste without knowing it, now that I know I won't fail to cite the author of my sect," and he would later recommend many men who wished to go fight for America.[96] Franklin was trying all this while to find a peaceful solution to the conflict between England and the restive colonies across the sea, and he was not beyond using humor for his cause. He had fought for the repeal of the Stamp Act with satire. In the *Public Advertiser* in September 1773 he impersonated the arch conservative QED proposing "Rules by Which a Great Empire May Be Reduced to a Small One." Later that month the paper carried one of Franklin's most famous hoaxes, "An Edict of the King of Prussia," proclaiming the island of Britain a Prussian colony, settled as it was by Angles and Saxons, which must therefore submit to taxation by Prussia identical to that imposed on the American colonies by the British.

Quite probably Franklin's now lost letters to Biheron concerned these matters as much as they did her scientific work. We know that the two continued to write directly to each other since Dubourg explained to Franklin, "I will tell you nothing about Mademoiselle Biheron as she has the honor of corresponding with you herself, and loves such exchanges with you almost as much as I do."[97] Franklin soon indicated that no diplomatic solution seemed possible for the colonies and that the rumbles of war were getting louder. In October 1774, Biheron addressed him in a more informal and personal way; owing to a long illness, she had not replied to a letter he had sent a full year earlier. Together with Basseporte, she acknowledged a single gift he had sent to the two of them as a couple:

> Monsieur and dear friend,
>
> It is with the greatest satisfaction that I received the honor of yours dated 13 October of last year. If I did not answer sooner my poor health was the unique cause, I grab therefore with urgency a young man who returns to your city to assure you of the vivid interest I take in the present affairs of North America. I make sincere prayers to God that you are rendered justice and that the tranquility of you and your family will be reestablished as in the past. . . . Not a day passes without my having the pleasure of speaking about you with our friends. I am very grateful as is Mlle Basseporte for the beautiful present M Dubourg gave us [from you].[98]

Franklin sailed from London in 1775 as revolution broke out, the next year signed the Declaration of Independence, and then returned to France for a long stay as representative of the new United States. Biheron had followed these events attentively, and she might even have influenced Diderot toward the American side, for while he hardly wrote about this matter he did mention in a letter to Wilkes in 1776 that he disapproved of, indeed dreaded, the British repression against the "insurgents" of America.[99] Franklin saw Biheron immediately when he returned to Paris, writing to Polly Hewson on 12 January 1777 that "on my arrival Mlle Biheron gave me great pleasure in the perusal of a letter from you to her so I learned you were all well in August last." The following year, in April 1778, Polly wrote a long letter to Franklin at the end of which she thanked Biheron for the elegant present she sent to her daughter. Her letter was carried by Ingenhousz, who remained a friend although he had earlier written to

Franklin from Vienna where he served the royal court deploring the horror and bloodshed of the American "civil war" and advocating respect for the mother country.[100]

Biheron stayed in touch as Franklin got increasingly busy with political affairs during his long residency at the Hôtel de Valentinois in Passy. Basseporte supported him too, writing during this time to both Franklin and his landlord, Jacques-Donatien Le Ray de Chaumont, who was himself another major force in persuading France to support the colonies against Britain. The two women could not afford to provide financial backing, but with Dubourg and Dalibard they could round up soldiers and donate uniforms.[101] It took a long time before the French government officially joined the war and went on to help the Americans clinch a definitive victory. Biheron and her crowd had been early drivers, far in the lead with vision and enthusiasm for this military alliance.

And what of her anatomy museum during this time? When she returned from the second London trip and reopened her cabinet in late 1773, Biheron had advertised in the papers that her exhibit included many additional pieces not seen before in Paris.[102] But she demonstrated for the public during much more restricted hours now, only Wednesdays and only from eleven to one. That she was losing steam as a teacher and medical demonstrator seemed clear to Diderot. Unwilling to let her languish, he had hustled throughout 1774 to arrange an expedition for Biheron to Saint Petersburg. Happily free of his own *Encyclopédie* duties and his daughter securely married, he had finally traveled to Russia to meet in person his benefactress Catherine the Great, an empress he described as having the charms of Cleopatra and the soul of Caesar. The magnanimous ruler had bought Diderot's library back in 1765, when he was broke, but allowed him to retain the three thousand books until he died and even paid him a salary as keeper of his collection. This was his opportunity to thank and spend some months with Catherine in person at her court. He was especially eager to visit and comment on her prize educational project, the Smolny Institute for Girls, one of the first, if not the first, female boarding schools for young women. He stayed from October 1773 to March 1774, observing and advising the enlightened and energetic monarch. Sixteen years older than she, he believed she was genuinely impressed with his wisdom and experience, eager to receive his suggestions.

Diderot urged Catherine in the most persuasive terms—no doubt sharing a bit too much information and telling her more than she cared to hear about his own obsessions—to incorporate Biheron's anatomical expertise into the boarding school curriculum, pointing out that any scandal at the school would irreparably ruin its reputation. Were a seduction to take place, no parent would ever again want to send a child there.[103] Referring to the scientific lessons Biheron had provided his own daughter on sexual anatomy, he described how her teachings would guarantee that such a catastrophe not occur at Smolny because the students would be properly enlightened about conjugal duties, pregnancy, and motherhood. "This is how I cut the root of my daughter's curiosity. Once she knew everything, she no longer craved more. Her imagination was calmed and her morals only grew more pure." Such exposure to the facts of life fortified her, making her laugh at the advances of men instead of falling for them.

Diderot went on to explain how Biheron's teaching with models of the baby in the womb prepared his daughter for her first delivery, which she endured with a resolve and fortitude never seen, he argued, in "ignorant women." She also now knew how to keep herself, her family, and her domestics healthy. "But, you ask, where could she have acquired this knowledge of anatomy without consequences? From a demoiselle, very able and very upstanding, where I took my anatomy courses, me and my friends, twenty girls from the finest houses, one hundred society ladies, a science that she made quite widespread among us. Fathers took their sons and their daughters separately. . . . Your imperial majesty will ask me perhaps at what age my daughter and the other children took these lessons. At 16, 17, 18, one or two years before marriage. M Grimm, who went to this school can speak to your majesty about it . . . and tell you what he thought of it. Pringle, Petit, and our most celebrated anatomists all agree that her pieces are very perfect. D'Alembert who was a student of this demoiselle told me that he learned more from her in eight days than from our esteemed Ferrein in six months."

Hardly any foreigners, Diderot continued, came to Paris without visiting Biheron's museum and learning from her. This "singular girl" was in large measure responsible for spreading the taste for anatomy throughout the world. She would not be able to stay permanently in Russia because all her hard work had ruined her health, but she could come, bring her models, and train others to demonstrate; she might even be persuaded to leave many of her pieces. The empress, he promised, would be astounded by the effect of Biheron's teachings on her students, by their resultant modesty and reserve. They would know how to distinguish the fine man from the vulgar, the suitable book from the salacious, and remain untainted.

Diderot then gave the example of his own daughter reading Voltaire's *Candide* where Pangloss gives Paquette a "lesson in experimental physics" in the bushes, but thanks to her sessions with Biheron such reading could not corrupt her. His daughter and girls similarly educated would see through the idle flirtations of men to the sinister desires they disguise, the real intention of such overtures: "Miss, if you would like to forget your honorable principles, sacrifice to me your morals and reputation, dishonor yourself in your own eyes and those of others, change your name from decent girl to courtesan and lost girl, renounce all consideration in society and other establishments, blush the rest of your life, cause your father and mother to die of pain [just] to grant me a quarter of an hour of amusement, I'd be infinitely obliged to you." His daughter, Diderot boasted, unsusceptible to such ruses, would let men carry on and then burst out laughing.[104] He agreed to urge Biheron to make the voyage to Saint Petersburg with her pieces which, if properly maintained, could stay fresh for decades. Beyond demonstrating, she would gladly teach dissection to motivated students, enabling them to make models of their own. This rare talent could also generate some money for the Institute.[105]

While in The Hague in June 1774 on his way back to France from Russia, Diderot continued this negotiation with Catherine's factotum Ivan Betzky who was seventy at the time, the empress only forty-five. Assuming that the plan had been approved, Diderot explained that Biheron would be accompanied by a chambermaid and a valet. She would send her things by boat, he wrote, but she needed to travel on land because two sea voyages to England nearly killed her.[106] Betzky was Catherine's trusted advisor, and rumors persisted that the empress was in fact his biological daughter. He had helped her found Smolny and together they wrote pedagogical manuals. But apparently Diderot's way of presenting what Biheron could provide to the young ladies of Russia did not fit in with the imperial curricular plan. In September 1774 a refusal came from Russia and, realizing he may have presumed too much, Diderot wrote to Catherine that he sincerely hoped his suggestion had not offended her.[107] He himself became seriously sick on this trip to Russia—his daughter later said the cold and the waters of the Neva shortened his life[108]—so it seems strange that he proposed such an expedition for the fragile Biheron. Perhaps it was best that it fell through. But at least one of her full models with all its removable parts was ordered from Paris, and a Russian doctor was soon using it in demonstrations in Saint Petersburg.[109] That was the best Diderot had been able to do for his friend. In December 1774, despite having been so unwell that year, she again advertised her cabinet in Paris.[110]

A new devotee then appeared on the scene, the twenty-five-year-old classicist and linguist D'Ansse de Villoison. In 1775 he tried to help Biheron sell her entire collection to a German principality for 36,000 livres. When Charles August, the young Duc de Saxe-Weimar, was sent by his mother to visit Paris, Villoison escorted him to see "the so renowned museum of Mlle Biheron," where the duke enjoyed anatomy lessons for 96 livres. He then wished to purchase the collection. Meanwhile one Abbé Hemmer, secretary of the academy of Mannheim, wanted to buy it for Charles Theodore, Elector Palatine, who was eager to see the sciences make progress in his realm. Villoison campaigned strenuously for Saxe-Weimar, determined to outbid the other offer: "So I went immediately to see Mlle Biheron to intercept this sale, and . . . she would be glad to sell it for the same price to you instead." He injected a note of urgency, not wanting Saxe-Weimar to miss this chance, "I'd so much rather see this beautiful collection in your hands than in [Hemmer's], although it is sad that it has to leave France. In addition to the pieces you saw, there are many more that this *savante* did not have the time to show you." Then Villoison sent a list of the entire collection, again urging that it not get snatched up for Mannheim.[111] In the end neither sale was consummated.

Biheron had never been taken seriously by the court of Louis XV, despite Bachaumont's pointed plea in 1763, shortly after she opened her cabinet, that the government acknowledge and reward her, and then again the recommendation in 1770 by Morand, Tenon, and Delasonne, after her second appearance at the Académie, that her work be financially supported.[112] In 1771 Grimm's *Correspondance Littéraire* reiterated how the research of France's remarkable "anatomiste femmelle" "is without a doubt that which is most deserving of His Majesty's attention . . . and that she should, ever since her efforts began, have been properly honored, encouraged and recompensed by the king.[113] But all of this fell on deaf ears. In general, the attitude of the latter part of the reign of the sarcastically nicknamed Louis le Bien Aimé was well summed up by a courtier who mentioned Biheron's 1771 academy demonstration but, reflecting then-favorite Mme Du Barry's feeling that such things were a bore, dismissed her as a "poor girl fifty years old who spends her life in acts of devotion and studies of anatomy" and ended his report with, "Ugh! I don't know why I even write about such a thing!"[114]

The advent of the new king Louis XVI and his teenage bride Marie Antoinette in 1774 did not initially change Biheron's situation, but she kept working, seemingly undeterred by the fizzling of plans with Russia and the German princes. The 1776 *État de médecine* listed her as active and raved about her pieces "of a composition all her own."[115] In the same year Buffon published an image of a wax monstrosity by her in his *Histoire Naturelle*.[116] When in 1777 Emperor Joseph II of Austria, brother to Queen Marie Antoinette, traveled to Paris incognito as Count Falkenstein, he expressed interest in visiting the legendary anatomical models.[117] Abbé Riballier and his female co-author Cosson de la Cressonière described in their 1779 book on women's education how "after 30 years of this laborious study and of a multitude of particular experiments, done in her house and at her own expense, on human bodies, [Biheron] has at least today the satisfaction of enjoying the just praises that savants and the curious of all states and conditions insistently bestow on the masterpieces from her knowledgeable hand." The authors opined that women should be allowed to pursue science and learn from a teacher like the smart, diligent Biheron. "How many midwives, how many of those who serve the sick, would be able to glean essential illumination for their interesting functions."[118] Louis-Sébastien Mercier, in his 1781 *Tableau de Paris*, spoke of her as still exhibiting, and in 1788 the Marquis de Chastellux, a friend of Franklin's who had fought valiantly under Rochambeau in the French expeditionary forces of the American Revolution, must have known Biheron or recently seen her cabinet because he hailed her models as far superior to both the ones in Bologna and those of the anatomist "Shovel" in America.[119] A 1788 manuscript at the École de Pharmacie reported, she "lives still across from the walls of the Sainte Genevieve garden," although it did not explicitly mention her museum as being currently open.[120] But the military doctor R.-N. Des Genettes seemed to imply that it was, even in 1791.[121]

Yet there is no question that these were also difficult times for Biheron, as friends and significant supporters near and far died within a fairly short period, leaving her bereft. The first such loss was Morand on 21 July 1773, just after her return from her second trip to England. Hewson expired in 1774, Dalibard in 1778, Dubourg in 1779, and the all-important Basseporte in 1780. Pringle died in 1782 and Diderot in 1784. That year, at such a gloomy point in her life, Biheron wrote her own last will and testament, certainly unable to imagine that she would soldier on another eleven years. Franklin lived until 1790 but left Passy for America in 1785. Personally, she was increasingly isolated.

On other fronts, however, things improved. Her finances got a lift in March 1780 when she sold four houses inherited decades earlier from her mother to the visionary philanthropic curé Jean-Denys Cochin of the Église Saint Jacques, so that he could start the hospital for the neighborhood poor which still bears his name.[122] He paid her modest asking price of 17,000 livres partly in a lump sum and the rest, as the contract stated, in a guaranteed retirement income for her. Did she perhaps make this sale just then to raise money for the medical care of Basseporte who was dying? Cochin himself died prematurely in 1783, but Biheron had the satisfaction of seeing the hospital, designed by the architect C.-F. Viel, rise on the grounds of her former properties and get off to a good start thanks to other charitable contributions directly from the priest's parish.[123] Mercier's fierce attack on the barbarous behavior at the anatomy classes of the medical faculty and his high praise for Biheron in the widely read 1781 *Tableau de Paris* probably boosted attendance at her lessons around that time.[124]

And finally, in 1786, the Crown bought her collection. We do not know the reason for this change of policy at Versailles, but the French monarchs made the purchase 28 May, awarding Biheron an ongoing pension to teach the royal children and to continue producing additional anatomical pieces. She was allowed to keep the models and show them during her lifetime, an arrangement similar to what Empress Catherine had earlier engineered for Diderot and his library.[125] Biheron's pieces were inventoried by Vicq d'Azyr.[126] Villoison had hoped to claim the price of 36,000 livres but it seems to have gone to the French monarchs for an initial 6,000, supplemented by the ongoing pension, the total amount unclear. A gutsy note purportedly written by Biheron to a royal minister—she must have had some help with this as the style is much tidier than her spontaneous and phonetic letters to Franklin—suggests that in fact the overall sum awarded to her was considerably greater. It shows that, at sixty-seven years old, Biheron's confidence in herself and her work was undiminished:

My lord,
    Your greatness permitted me to give you to read the object of my request that the Queen deigned to accept and that her majesty recommended to your particular attention.
    I have the honor of offering to the King my cabinet of artificial

anatomy for the education of the children of France. As your greatness was good enough to propose that I determine what I hope for from the Government, I would desire the sum of 30,000 livres, of which 6,000 would be paid to me now, and of which the rest would be a fund for a life annuity. In addition, as recompense for a work unique in Europe, and to put me in a position to complete my cabinet with some rare pieces that I am considering, I desire an ongoing pension of 1,500 livres. Please give orders so that the Queen's wish may be accomplished. I await this favor to me from a minister who is the most enlightened protector of the fine arts.[127]

Apparently the underscoring in this note was Biheron's own, for emphasis. She was addressing Le Tonnelier de Breteuil, the head of the royal household, a close confidant of Marie Antoinette and an advisor known to appreciate useful intellectual efforts. The bookkeeping on this acquisition is very confusing as the documents are incomplete, disrupted, among other things, by the upheavals of 1789 and beyond. But Biheron must have been immensely relieved that the sale went through, and she certainly counted on the money.

Of the six women discussed in this book, only Barret, Biheron, and Mme d'Arconville (chapter 5) lived until, and through, the Revolution, which of course deeply affected their lives. But as it was not immediately clear how long disruptions would continue, Biheron complied by making her required patriotic "gifts," payments expected of all citizens, from 1790 to 1792. Then successive revolutionary governments accelerated the pace of new, invasive, and frightening laws, and when on 5 September 1793 the National Convention voted to declare "Terror the order of the Day" the atmosphere darkened dramatically.

This ominous development shifted Biheron into a different gear. She had not received her payments for some time, and it was easy to figure out why they had stopped. The monarchs who had bought her collection were gone, the king already guillotined and the queen, recently indicted, about to meet the same fate. France had declared itself a republic a year earlier, and the royal treasury from which Biheron had been paid no longer existed. Barret, it will be remembered, was having the same trouble at the same time receiving her pension. Both women realized they needed to speak up, remind the new leaders of past promises, and defend themselves although this was risky, calling attention as it did to roles they had played in the hated Old Regime. On 12 September, just a week after the official beginning of the Terror, Biheron summoned up her courage, approached the Comité d'instruction publique, one of the committees of the Convention, and asked that her *rente* be restored.[128] Whether she demanded or implored we cannot know, but she set things in motion.

Her request was supposed to have been handled by the physician Lanthenas, but he left the committee on 6 October, so it was dropped and ignored until 18 May 1794. The Commission temporaire des arts addressed it sooner, however, eager in January 1794 to establish the value of Biheron's collection before giving her an answer. That group studied the papers she provided about its 1786 sale to the Crown and descriptions of its contents.[129] Most of all, the new revolutionary École de Santé wanted to wrest from her the proprietary formula she used for her models that she had so carefully guarded all these decades, the celebrated indestructible wax alloy with its entirely natural look and feel that neither melted near high heat nor broke when dropped. And now, deprived of her pension and in need of money, Biheron was ready to consider disclosing the recipe for her invention. When in June 1794 the arts commission voted to continue and even make retroactive her payments in exchange for her secret, she finally capitulated.

It had been nine hard months since her initial request, during which the quality of daily life in Paris had deteriorated. The Law of Suspects led to neighbors denouncing neighbors, there were incessant, relentless executions, and food was scarce. Lavoisier had been guillotined the month before, and Biheron might justifiably have worried that the climate had grown hostile to science. The doctor Jean Nicolas Corvisart and Honoré Fragonard, brother of the painter, whose own anatomical preparations were becoming popular, went to her home amidst the paranoid hysteria of Robespierre and the crescendo of the Great Terror. On 8 July they gave their official report praising Biheron for "repugnance vanquished, difficulties surmounted, patience in trying times, and skill acquired by long work." A very elderly woman now, she impressed and moved them, and they deemed her "most deserving of praise." She had, we can imagine with what reluctance, written out the process for preparing and conserving her models, but this precious note, dutifully turned over to the committee, was unaccountably filed away in "the anatomy carton" and seems never to have surfaced again. Some pieces of Biheron's collection were purchased, enabling possible chemical analyses to determine their composition.[130]

We do not know if, over the last year of her life, she received any of the funds promised in exchange for vouchsafing her treasured procedure, but she survived the Terror, dying during its somewhat calmer aftermath, on 18 June 1795 (30 Prairial III), having worked and persevered despite infirmity into her seventy-sixth year, an example, like Basseporte's, of aging bravely. By a stroke of poetic justice, Biheron's cherished secret formula, acquired but then misplaced during the madness of the revolutionary month of Thermidor, accompanied her to the grave.

Her testament was opened by her notary a few days later, a tiny envelope with a red wax seal containing the square page folded in fourths that she had composed eleven years before, on 1 April 1784. "The hour of our death being uncertain I thought it appropriate to do this now," she had written. She asked in the name of the Father, Son, and Holy Ghost that her burial cost be no more than 250 livres. Among her bequests were to the poor of Saint Etienne du Mont parish, 600 livres; to Mme Aillard, Basseporte's niece described by Biheron as a friend, two gold bracelets; and to Mme Barrois, widow of the bookseller Barrois, the prized pastel self-portrait of Basseporte in its gold frame.[131] To her faithful maid Marie Elisabeth Vibart, who had been with her for more than thirty years and had accompanied her on both trips to England, Biheron bequeathed 2,000 livres and 200 more as a steady pension in addition to what she would get from the contract with the late M Cochin, which provided for the *rente viagere* to go to her domestic if she herself was no longer alive. Vibart was also to get all kitchen utensils, faience, clothes, laces, sheets, and table cloths, this because of her "unfailing attachment."

After the testament was read, seals were placed on the house so that the collection inside would remain undisturbed.[132] Biheron had died in her home on the rue des Postes (today rue Lhomond), where she must have moved in her very last years, just a few short streets from the rue des Poules house she had shared with Diderot. Vibart, mentioned in this will of 1784 and still alive in 1795, was tasked with the melancholy and lugubrious responsibility of watching over the whole anatomical assemblage until its disposition could be decided upon.

Biheron's models had a poignant posthumous life. The collection or at least some of it should have been preserved with care. Instead, the pieces were allowed to dry out while government officials debated whether to claim them or declare them worthless and even demand that Mlle Biheron's heirs return the money originally paid her by the Crown.[133] Eventually, in 1796, the founding director of the École de Santé, Michel-Augustin Thouret, concluded that the once-famed cabinet had been too long neglected to be pedagogically useful. But Biheron had certainly deserved the acclaim she received both in France and in England in her day. When Louis XVI acquired the collection ten years earlier "none could compete" with her wonderful invention. Now, unfortunately, it had only historical worth, representing the beginning of this scientific art, instructive to contemplate as "the first rudiments of a new branch of human industry," a completely original way to teach anatomy. The compassionate Thouret concluded that the government should acknowledge Biheron's zeal and the pioneering role that she played in "spread[ing] the taste for anatomy, and acquit its debt for such useful labors."[134]

Right here we witness the fateful moment, the disappearance of irreplaceable physical evidence and how it hampers the historian. The few distant cousins who materialized as Biheron's only heirs had no interest in the collection, so the notary Boulard who handled the estate put it up for sale. Despite Thouret's denial of its value for teaching it was advertised as being of use to amateurs, professors in the schools, or to anyone who gave anatomy lessons.[135] There seem to have been no takers because the next year, 1797, another notice offered citoyenne Biheron's models for sale. What transpired at that point is unclear, but by 1799 a showman named Bertrand displayed at least some of her pieces along with his own at his cabinet on the rue Hautefeuille and then later at his museum in the Palais Royal.[136] Then they scattered and vanished forever.

Fragonard, who benefited from knowing Biheron's formula, divulged to him during his official visit with her, was able to set up his own display at the École vétérinaire d'Alfort, where his injected anatomical preparations, using a somewhat different technique, can still be seen today.[137] He was, of course, a man and a member of the establishment. Biheron's pieces found no such permanent dwelling place and were lost. So was her own body. The churches with which she was associated during her lifetime could not hold traditional burials during the Revolution, and beginning in 1791 the dead on the Left Bank were interred in either Sainte Catherine or Vaugirard cemeteries, layers of bodies superimposed in large common ditches. These graveyards and others were closed and emptied in the early to mid 1800s, and any unclaimed remains were transferred to the catacombs. That is where Biheron's bones lie now, mixed in with many thousands of others under the streets of Paris, perhaps intermingled with those of some individuals she dissected. Given their contribution to her scientific life and teaching, this innovative educator would likely be grateful to be in their company.

Dear Marie-Marguerite,

I visit your ghost in several places as I stroll around the 5th arrondissement. You had two addresses on the rue de l'Estrapade behind
and south of what is now the Pantheon, the first on the corner of the
rue des Poules, today rue Laromiguière—although there's some am-
biguity because the building adjoins the one next to it on the corner
of the parallel rue Tournefort. Of course there is a plaque about
Diderot having lived there, but no mention of you. The courtyards
of those back-to-back buildings, where you did your experimental
work in a glass enclosure, are filled now with trees and garbage can
depots but back then that space was all yours. It was the scene of your
hundreds of dissections, a research lab. Your neighbors evidently
tolerated these unusual efforts, the constant traffic of bodies in and
out, and the crowds lining up to see your exhibit. You must have had
exceptional people skills to placate everyone around you, but they
probably recognized the deep reverence with which you approached
your work.

From your home to that of Madeleine Françoise in the Jardin du
Roi it was a short ten-minute walk along the rue Copeau, where your
mutual doctor friend Dubourg lived, a street named for a mill on the
lively river Bièvre, which is today hidden under the pavements but
then flowed freely through your neighborhood before emptying into
the Seine. That street is now named rue Lacépède, for a man of science at the Jardin, of course, and one who, incidentally, appropriated
and obscured Jeanne's and Philibert's contributions. But I digress.

Later you moved, it is unclear when and why, to the nearby
corner of the rue des Postes, today rue Lhomond, and the Place de
l'Estrapade. It was here that your anatomical models were inspected
by the revolutionary *officiers de santé*, who coveted your formula.
And it was here that you died at seventy-six years old. Your devoted
maid then had to guard all your models, rather eerie company but
familiar to her, while their value was appraised and their fate decided.
The pension arrangements you negotiated, first with Cochin, then

with the Crown, provided generously for this servant who stuck with you through everything.

You should know that the Hôpital Cochin on the rue du Faubourg Saint-Jacques in the nearby 14th arrondissement, that grew out of your vision for the care of neighborhood laborers and as such is also part of your legacy, has become a world-famous facility for medical research, public assistance, and Paris's main burn center.

I believe the poet Charlotte Catherine Cosson de la Cressonière, who wrote a short entry on you in a book on women's education, knew you personally. She was interested in things medical, composing an ode about the terrible 1772 hospital fire of the Hôtel Dieu. She seemed to know a lot about you—for example, your youthful talent in music—details that are repeated nowhere else. How would she have learned that except directly from you? It is fun to think of you being interviewed by a female biographer. She saw the value of teachers like you in preventing young girls from being discouraged and turned away from science. Cosson's view has been called "radically feminist" for her time; it is still relevant today and gets us back to the role-model issue. She also saw your lifelong anatomical experience as indispensable to potential students. So, far from being "ageist," Cosson was forward-looking in that respect as well.

Your main exhibit piece was described as a body of a female about twenty-five years old; so were the famous anatomical Venuses in Florence's La Specola museum, but those were blatantly sexual with their glass eyes, real hair, jewelry, their faces bearing ecstatic expressions. Your models were headless, sober, serious scientific studies of the body, not at all suggestive, just as Madeleine Françoise's botanical images were not at all showy. There was an artistic decorousness in both of you, because you were educators, aiming for minds rather than emotions. The purpose of your exhibit was to teach, not to titillate. By the way, some medical schools today use models instead of cadavers. And it is recognized that such replicas must include females, that universalizing the male as the anatomical norm is misleading, even dangerous.

Your assemblage resembles somewhat the current Body Worlds exhibitions that have been popular since the 1990s, created by two

German doctors, Gunther von Hagens and Angelina Whalley. Like you they give dead humans a "post-mortal" use of great benefit to the living. Like your cabinet they attract crowds. Like yours their show goes on tour, which is really what your collection did when you took it to England and almost to Russia. You caught the body between demise and decay in order to study it, model it, and then teach it.

Times have changed since then and we have become prudish over the intervening two and a half centuries. Body Worlds has been a huge hit, but it has run into ethical objections and fierce accusations ranging from bad taste to violation and desecration of the dead. Your museum seems not to have evoked such a response. Objections from the Faculty of Medicine, yes, but on different grounds. They were not offended, just threatened by your success. You showed that a woman could do this kind of work, had entrepreneurial smarts, and made a lucrative business out of it. This was more than they could bear. You were simply attracting too much public interest, rewarded for doing what you were good at. It made the Sorbonne jealous.

And demonstrating at the Académie three times! Did any other women get invited to appear and hold forth there? I don't think so. By the way, the image Buffon published of your model of the cyclopia congenital disorder is now on the internet Science Photo Library. The internet, another thing to explain and that you would appreciate.

Mlle Lainé, named frequently in A. L. de Jussieu's letters to his mother as your "very good friend," will probably remain a mystery Was she perhaps a love interest in the middle of your relationship with Madeleine Françoise, in the 1760s and 1770s? She is never mentioned by your scientific circle, who instead link you with Basseporte. It is something we may never know.

Despite obvious class differences, you had much in common with Geneviève, and your overlapping scientific fields could have been a great leveler. You were also the same age. Did you approach her when her *Traité d'ostéologie* appeared in 1759, when both of you were working on anatomy, she with her book and you with your first visit to the Académie, the two of you supported enthusiastically by Morand? You defied putrefaction by creating representations that would never perish, Geneviève instead sought antiseptic substances that would prevent or arrest decay of the flesh. You went twice to England, she taught herself and translated from English, another shared enthusaism. But mostly, both of you had working laboratories.

The Revolution. You had been political earlier, in your support of the *parlements* against the Crown, and of America against England. Were you still? What did you think of Olympe de Gouges's push for gender equality in her remarkable 1791 *Declaration of the Rights of Women?* You had in fact already lived as she advocated, exercising your independence and talents long before she directed women to do so. Did you lament the news of her execution in 1793? Then came the Terror, then Thermidor, then the alarming Days of Germinal in April 1795 when the Convention abruptly removed price caps on food, the starving poor rioting around your quarter. You yourself might have gone hungry.

Jeanne lived up to and through the Revolution in the southwest countryside, but you and Geneviève experienced it in Paris, the very heart of the upheaval. Both elderly by then, weathering the worst of it when in your seventies, you outlived many who were dear to you but coped, survived, and prevailed.

FIVE

Chemist and Experimentalist

Marie Geneviève Charlotte Thiroux d'Arconville and Her Choice of Anonymity (1720–1805)

I always had the desire, indeed the *need*, to *do* . . . and to acquire useful knowledge.
—MME D'ARCONVILLE

Study being, so to say, her principal interest, the greatest pleasure of her life . . . she was
bored with social functions unless she met learned men who spoke to her intelligence,
who suggested new ideas to her, who gave her subjects to work on.
—HIPPOLYTE DE LA PORTE

DETERMINED TO BE USEFUL TO SCIENCE EVEN AFTER SHE
died, polymath Mme Thiroux d'Arconville donated her body, stipulating in her
testament that she be left for twenty-four hours exactly where found and then
"opened," but only after a cross-shaped incision on the sole of her foot verified
that she was no longer alive. This last precaution was probably inspired by her
eldest son Louis Thiroux de Crosne who, as intendant of Rouen and then police
chief of Paris, had instituted guidelines to revive victims of drowning, asphyx-
iation, or seizures and others declared dead prematurely. The enlightened Crosne,
tainted by his association with the Old Regime, was guillotined to his mother's
horror during the Reign of Terror. Outliving him in unspeakable sadness by
more than ten years and thus "purged by the flames of purgatory," d'Arconville
hoped her sins would be forgiven, and that after a burial of the utmost simplicity
she might be admitted to paradise.[1] But before that her corpse could do some
good on earth. It was common among men to insist on a postmortem examination in the service of science, as did many in our story; Lalande, Commerson,

Buffon, Morand, and Diderot all demanded this, and d'Arconville's chemistry colleague Pierre-Joseph Macquer told his family as he was dying in 1784 that he too "desired very much to be opened after death so that the cause could be known."[2] Now, almost twenty years later when she wrote her own last will in 1803, d'Arconville made the same request for an autopsy to advance medical understanding.

This expressed wish for how her death should be handled was extremely unusual for a woman, almost unheard of. But then, so was much of what she decided to do with her life.

Marie Geneviève Charlotte Darlus was born into a rich tax farmer family in 1720, a smart, eager, and restless girl.[3] Her mother died when she was only four, but her father, André Gillaume Darlus, a resourceful fellow whose talents had made possible his remarkable rise, seems to have been a warm and kind man to whom she was quite attached.[4] Distraught after his wife's death, however, he failed to give his daughter the education she so craved. As she explained in the memoirs she dictated at the end of her life, she had only her little sister, whom she adored yet who had little stimulation to offer, for company. "I had no books to read, and was born with a very lively head and imagination. I was reduced to raising myself, without help, without advice. . . . I was completely ignorant."[5]

And yet, significantly, one of the things she did to relieve her boredom as a child was to perform rudimentary experiments. When she pricked her finger, she concocted a kind of ointment and tried to ascertain whether this salve had any healing properties, concluding that at least it hadn't made things worse. She next tried to make wine, paying a servant to buy some grapes, which she then strained and put into a corked jar, examining it eagerly each morning. One day the mixture exploded and stained her new dress, for which she was severely reprimanded; as she later commented, she was then unaware of fermentation, the subject that was to become her specialty. Finally, she created a kind of nectar from apricot pits, sugar, and water designed to be saved until the cold season to moisten the hard, dry winter bread that she and her sister would dip into it. To her dismay the paper-covered pot became overgrown with mold so foul-smelling that the hoped-for ambrosia needed to be thrown away.[6] Unmistakable signs, these albeit puerile investigations, of things to come.

Young Mlle Darlus actually asked to marry early and escape a life that numbed and saddened her for another, unknown but perhaps more interesting and eventful. Receiving a huge dowry of 350,000 livres, in 1735 at age fourteen, she wed the even wealthier landowner and magistrate Louis-Lazare Thiroux d'Arconville, eight years older than she and on his way to becoming a president of the *parlement* of Paris.[7] Within the next four years and while still a teenager she gave birth to an heir, a spare, and a third son for good measure, the family then spending much of each year in the chateau of the village of Crosne, which her father purchased after the birth of her children and where one could travel easily by river coach.[8] The extensive grounds included a park and floral garden that Darlus opened to the public, and the chateau, surrounded by a moat, was immense.[9] It was on the ground floor of this residence that Mme d'Arconville would later construct her chemistry laboratory.[10]

At twenty-two she contracted smallpox, narrowly escaping death but left so scarred both physically and emotionally that she made a radical change in her life, withdrawing from society as much as possible given the demands of her social station and turning instead to intellectual pursuits. This story, like Florence Nightingale's famously creative invalidism, also happened to be true, and it conveniently provided an acceptable reason for her subsequent unconventional occupations, handy for d'Arconville's family members of the elite *noblesse parlementaire* with reputations to uphold. According to one dramatic biographer, the disease marked "cruel traces" on her face, and "despite her youth, she renounced rouge, powerless and ridiculous palliative against irreparable ravages."[11] She gave up going to the theater which she had loved, and where she had reportedly attended more than a dozen performances of Voltaire's tragedy *Mérope,* a play whose appeal to her might reveal some of the struggles that raged within her regarding imbalances in domestic life and young motherhood to three sons.[12] Turning her energies now instead to learning—"I was at least twenty years old when I began my education; how much time had been lost!"[13]—d'Arconville taught herself languages starting with English and Italian, translating over the next many decades literary and scientific works that she admired and publishing many original ones of her own on morality, history, fiction, and of course science. Her command of these diverse realms was breathtaking.

Every book she published was anonymous, but while hiding her authorial identity she steadily gained confidence. Her family, friends, and the "erudites" of her circle knew of some of her activities, respected her wish for invisibility, but in fact were not themselves aware of many things she wrote; that it was she who produced the more controversial writings was kept even from them. More than anything she meant to be of service, a goal that propelled her throughout her life. As she said, "I always had the desire, indeed the *need*, to *do* . . . and to acquire useful knowledge."[14] Her use in this sentence of the French word *faire*, which means both "to do" and "to make," is telling, because she wanted to do the studying necessary to learn, but she also intended to make books in order to circulate her ideas, to communicate and get her thoughts out to readers. Upon receiving a copy of her first publication d'Arconville enthused, "I had such pleasure in seeing it, and especially in reading it, that I couldn't stop myself from kissing it with all my heart."[15] Remaining hidden emancipated her. As feminist scholar Carolyn Heilbrun explained about her own use of a pseudonym, "Secrecy is power. . . . I do not care for publicity. . . . Secrecy gave me a sense of control over my destiny that nothing else in my life . . . afforded."[16] Anonymity provided d'Arconville with autonomy, authority, and freedom.

Those few who knew of her intellectual interests—not necessarily that she was publishing but that she was always researching something seriously—admired her strong personality, domineering yet also generous and kind, and her mastery of "the art of obliging" so essential for her social role.[17] They were struck by how she dealt with her divided loyalties, particularly by her flexibility, reporting that she could be playful with young children for whom she gave parties, not far from a room where she studied anatomical specimens that she kept under her bed. She could plan a grand ball as befit her rank and station and then pivot to the intensity of chemistry experiments. Though extremely wealthy, d'Arconville ministered to the poor in both the country region of Meudon, where she founded a medical hospice tended by local nuns who lived on an estate she bought, and in her parish in Paris, where she gained a reputation as a "force" albeit a philanthropic and protective one. There was a glaring contrast between her roles. During her work on putrefaction she was said to have put rotting meat in "crystal vases placed of necessity on the mantel piece, like an ornament," these "monuments of destruction" jarring wildly with the freshness and vivacity she displayed when entertaining. She took no offence when someone "sent to her friend a scarf but to her a little ivory skeleton."[18]

Yet however gracefully d'Arconville seemed to navigate these disparate parts of her life, the push-and-pull took its toll. As she explained in personal letters and in her twelve-volume memoirs, which included her reflections not just on melancholy in general but on *her* melancholy in particular, she was plagued by depression.[19] A prize possession of hers was a life-size statue of *La Mélancolie* by the sculptor Étienne Maurice Falconet.[20] She called this immobilizing sadness a dangerous "sickness of the soul, or the heart" for which the only cure, and one that should be implemented immediately, was "serious and reflective occupation, like study."[21] She had what she called a "dog's hunger," a voracious need to constantly nourish her active mind, and it was this that kept her going.[22] In one of her very earliest books she had declared: "Work is the universal specific for all the ills to which our soul is necessarily subject: fear, chagrin and boredom. Pleasure distracts us, but it does not keep us busy. . . . We feel the need of a serious occupation that can fill the void in our soul and help us endure life . . . a daily remedy that can palliate . . . and prevent our being crushed. . . . That remedy is work."[23] D'Arconville cleaved to this belief until her dying day.

With her own children she was correct but not effusive or sentimental; maternal love did not come naturally, for she stressed that pregnancy and childbirth were painful and spoke of the danger of bad conduct and ingratitude from progeny.[24] Her marriage was clearly a disappointment, the comments she made on that institution withering, and these views remained unchanged from her earliest writings to her late-life memoirs. Wives, she wrote, might at best love their husbands for the first six months, never longer, and happiness in matrimony was only an idea, rarely a reality.[25] She considered marriage unnatural, demanding a "constancy that nature never established" although religion imposed it. Only phlegmatic women could possibly tolerate that kind of dependency, servitude, and inequality, the notion of belonging to someone like a dog or other animal, of wearing a yoke. She went on that while some believed the bedroom fixed everything, in truth it could do much more harm than good, and stated that the only time a wife enjoyed any deliverance was when her husband died.[26]

At the end of her life, far from waxing nostalgic or softening her stance on this matter she reiterated that remaining unattached was the healthiest thing, even if virtually impossible for women in her class. "Liberty," she wrote dramatically, "is nature's wish for all who breathe."[27] It had a charm for her that nothing could replace. Aside from a description of some early travels with her husband when they were newlyweds, he hardly appeared in her memoirs, but she wrote numerous essays praising solitude and celibacy.[28] The single life, she believed, was the happiest state. "I return to my *refrain*, that freedom is the paramount good, and one *prostitutes* it, so to speak, when one abandons it to chance by marrying."[29] Friendship, on the other hand, because freely chosen and cultivated, was something altogether to be cherished, and in her early book *De l'Amitié* she explored this topic in minute detail, stating that men of science, authentic, erudite, serious and devoted to their work, stimulated her brain, knew how to "feed her head," and were the best, most trustworthy people to befriend. So d'Arconville did precisely that.

By the 1750s, if not earlier, she was taking classes at the Jardin du Roi and in its amphitheater, where she got to know some of the leading men of science and philosophes of the day who gave or attended lectures and demonstrations there—chemist and apothecary Guillaume-François Rouelle, botanist Bernard de Jussieu, chemist Pierre-Joseph Macquer, agronomist Louis-Paul Abeille, and eager botany students Lamoignon de Malesherbes and Denis Diderot, among others. It was one of the only places where women were actually permitted to join the audience for lessons on botany, chemistry, even anatomy and a thriving center for science. It was also, of course, Mlle Basseporte's home turf, and d'Arconville must have met both her and the learned head gardener Thouin, who later provided plants for the hothouses she maintained on her estates. Eventually she turned away from botany—she often called it "agriculture"—recognizing that she was too "lazy" to go out on strenuous rambles gathering plants, that exertion was alien to her nature, and that she was basically a sedentary person. "One cannot do botany in one's room," she realized, speaking frequently about her plumpness—the "torment of my life"—and her basic physical lethargy and cowardice.[30] Activity, she said, all physical movement, was inimical to her being.[31] She far preferred the tranquility of an indoor laboratory and the relatively stationary work of chemistry experiments.[32]

The intellectuals with whom d'Arconville developed a close kinship were quite like her, serious, austere, and committed to the welfare of humanity with little interest in competition or self-aggrandizement. This could certainly not be said about most men of science in her day who jockeyed egotistically for position, priority, and fame, but she chose as her colleagues the ones whose integrity and honesty measured up to her standards. Macquer and the doctor-chemist François Paul Lyon Poulletier de la Salle helped her set up her labs at Crosne and in the city, although she then functioned quite independently. They were almost exactly her age and her letters show that she socialized with them, consulting both on matters of writing style as well as science, insisting that they be frank in evaluating her solo projects rather than indulgent.[33] Duties kept Macquer extremely occupied, but d'Arconville appreciated how well he stayed grounded. Wearing many hats, he excelled as royal censor, prolific author, consultant for the king's tapestry works at Gobelins and porcelain factories at Sèvres, and teacher of chemistry in his own private courses as well as at the Jardin du Roi. He was also an active lab researcher, editor of the official *Journal des sçavans*, member of the Académie des Sciences in Paris which he valued unreservedly as a "republic always at war with error,"[34] and a foreign member of those in Stockholm, Turin, Philadelphia, and Madrid. Macquer was sought after as a smart, enlightened guest at numerous functions where he mingled with *le monde*.[35] Although he met these social obligations with unfailing politeness, he wrote to his wife, "my dear great friend, I am neither gay nor happy, I am collapsed," far preferring to be quiet "in my element." Resigned to his busy pace, however, he saw home as "a port where I will find peace and tranquility in the last part of my life."[36]

The doctor and chemist Poulletier de la Salle was distantly related to d'Arconville on the Darlus side. He assisted Macquer on some of his writings without wanting, in fact refusing, any public acknowledgment.[37] His experiments on bile, which would eventually lead to nothing less than his discovery of cholesterol in 1769, were also done without any desire for recognition; on the contrary, he was to ask d'Arconville to put the early ones, those completed before 1766, into her book on putrefaction but without attribution, and she obliged.[38] Poulletier seemed to emulate d'Arconville in many ways. He set up several hospices for the poor outside Paris, just as she did in Meudon, and he translated Pemberton's *London Pharmacopoeia of the Royal College of Physicians*, adding many scholarly notes, just as she was teaching herself English and undertaking richly annotated critical editions of several scientific translations. His work appeared slightly after hers and it seems probable that she was the influence on him rather than vice versa.[39]

Botanist Bernard de Jussieu, Mlle Basseporte's famously modest colleague, was another whose extraordinary wisdom and humility attracted d'Arconville from the time she first met him when taking his courses at the Jardin. He was, as she put it, "made to bring illumination to every subject he touched."[40] This man, "to whom we are indebted for so many discoveries because of his command of so many different fields," taught her much, including the preservative power of plant juices and sap in connection with her interest in antiseptics.[41] His generosity speaks to his immense respect for her, because as we saw he found most women unworthy of his time. He also gave her keys to all the hot houses and gardens in the Jardin du Roi and helped her set up her own collection of exotic trees in her country properties, visiting and advising her there several times for stays of a few days.[42] To her he seemed the very model of the masterful but retiring sage.

Guillaume-Chrétien de Lamoignon de Malesherbes, a dedicated amateur botanist, liberal scholar, and statesman who, even as director of the book trade believed in press freedom and loosened the prevailing strict censorship, was greatly admired by both Basseporte and Biheron, as we have seen, and was especially close to both d'Arconville and her sister. They had met through Jussieu at the Jardin and remained close until Malesherbes was beheaded for defending Louis XVI before the Revolutionary Tribunal. As a magistrate he had been outspokenly critical of ministerial despotism and was a staunch defender of the law, but like d'Arconville he regarded monarchy as more stable than republics, especially the increasingly violent and unruly French republic of the early 1790s. She saw him as a person of keen intellect, probity, discernment, tenderness, and authenticity, and he would entrust some of his manuscripts to her care.[43]

Scientifically, d'Arconville's bond with Macquer, the preeminent chemist of the time before Lavoisier completely revolutionized the field, was perhaps the most important.[44] When they met he had already written a highly regarded *Éléments de la chymie théorique* in 1749, its historical preface fiercely attacking alchemy as "vain reasoning," a "folly of the human mind," an "illusion," a kind of "leprosy which disfigured [science] and opposed its progress." His *Éléments de la chymie pratique* soon followed in 1750. In these books he showed his determination to legitimize chemistry as a hard, rational science, separate from vain speculations of course but also separate from pharmacy, and he talked about chemical affinities in Newtonian terms as quantifiable qualities.[45]

However much d'Arconville savored interacting with her colleagues, she truly valued time alone, to read, think, and learn languages so that she could relish works in their original form. This aligned with her exacting standards but also with her desire to make important ideas available to the French-speaking world. Her interests spanned literature and history as well as science, and after English, Italian, and Latin, she tackled German and Spanish. Her ex libris, a personalized bookplate engraved for her by Louise Le Daulceur, one of the few women in this line of art, was of the classical allegorical variety, showing Minerva, goddess of wisdom, craft, and war, helmeted and holding sword and shield, floating on some clouds near two volumes by d'Arconville's favorite poets at the time, Milton and Tasso.[46] Her very first publication, which appeared in 1756, was *Avis d'un père à sa fille,* a French translation of Lord Halifax's *Advice to a Daughter.* Had Le Daulceur not died that same year, d'Arconville might well have commissioned a quite different bookplate, because her next two translations, also from the English, were not literary but scientific. Appearing together in 1759, Alexander Monro's treatise *Anatomy of the Human Bones,* of which she translated the 1741 edition, became *Traité d'ostéologie* in her hands, and Peter Shaw's 1734 *Chemical Lectures* became *Leçons de chymie, propres à perfectionner la physique, le commerce et les arts.*

Translation was (and is) not the passive activity we might mistakenly tend to consider it. Especially for eighteenth-century women, it was a way of participating in the intellectual life of the day, and a few of them actually shaped the dissemination of ideas, blurring the lines between themselves and the author, altering, shortening, expanding, and judging their material. Mme Du Châtelet had done so in her translation of and commentary on Newton's *Principia,* a partial edition of which was published posthumously in 1756. D'Arconville may well have been familiar with this work, and she too was exceptionally aggressive in this regard, appending "para-texts" to all of her translations, introductions, prefaces, preliminary discourses, "avertissement," afterwords, and other front and back matter that framed them, as well as numerous critical notes within the text itself. She also took liberties with the actual words.[47] For example, in her first translation, the Halifax book, she explained: "I thought myself obliged to change and even suppress sometimes certain portions of my original; some because they were so contrary to our ways that they would have been shocking, and others because the expressions used by the author could not pass into our language without becoming completely ridiculous. I flatter myself that my changes and omissions will be approved. With the exception of the passages just mentioned, I tried to render my translation as faithfully as possible."[48] D'Arconville, like Du Châtelet, adopted an almost editorial approach to modification and what she deemed to be improvement of the texts on which she worked.

As a scientific translator she was still more forceful because during the 1750s she had begun to do her own experiments, testing those mentioned in the originals and finding fault with them when she could not replicate the reported results. She was not afraid to criticize the likes of Bacon, Boerhaave, and Pringle, believing that such forthright critique was necessary for the progress of science. As usual she did not identify herself in these publications, but her collaborators and the royal censors assigned to her writings—Biheron's friend and colleague Sauveur Morand for the Monro, Macquer himself for the Shaw—were of course aware of her masquerade. And their enthusiastic approbations emphasized the value of d'Arconville's extensive, knowledgeable notes, prefaces, and annotations. These translations made lasting contributions to scientific culture throughout Europe. Decisions about what to translate for successful communication took into account the target recipients, the pulse of interest, what subjects would be eagerly welcomed and by whom. French, much more than the Latin of previous centuries, was becoming the language of science in the Enlightenment, and d'Arconville's efforts not only made these texts accessible for the first time in France but also enriched the flow of knowledge across many frontiers.[49]

Her 1759 translations of Monro's book on bones and Shaw's chemical lectures were reviewed in enthusiastic detail in the periodical press, at least as much as the originals had been, and they became standard references throughout continental Europe over the next centuries. Monro himself in a later edition of his own work even gave a citation from its French translation. Clarifying, fleshing out, updating, and correcting the base texts, supplying additional features like tables, glossaries, indexes, and illustrations, providing scholarly preliminary discourses full of historical and geographical background, d'Arconville's voice could be plainly heard as she balanced fidelity to the texts with challenges, striving to plumb their depths and at the same time to enrich and enhance their appeal. Not a popularizer—she assumed she was reaching an educated readership for she felt no need to translate passages in Latin—she nonetheless meant to spread about information she regarded as beneficial.[50] And these translations were also the crucibles in which d'Arconville's confidence and scientific autonomy formed, laying the groundwork for her own independent research which would culminate seven years later in her ambitious treatise on putrefaction.

The *Traité d'ostéologie*, d'Arconville's anatomical translation, was published in two huge folio volumes, one of which had magnificent full-page illustrations of complete skeletons and separate bones, and received praise in three successive approbations by the censor Morand, in March, May, and July of 1759.[51] It was a collaboration with the surgeon Jean-Joseph Sue, but as she kept her characteristic anonymity and only his name appeared on the title page the work was thought to be exclusively his. I have explored elsewhere, by a close examination of the notes, the ways that he and d'Arconville shared and fulfilled different roles.[52] Monro, author of the English original, who was not aware of her involvement at all, arranged in appreciation of this magnificent effort to have Sue elected into the Edinburgh academy.[53] Interestingly, Monro himself had never found illustrations necessary, in fact had considered them a dangerous distraction. Countering this in her unsigned preface d'Arconville insisted that, if made with scrupulous accuracy, such images were essential in the teaching of anatomy for those who found actual dissection repulsive. The contrast between the English and the French version could not have been more dramatic. Monro's was a compact pocket-sized handbook for students at the medical school, her translation a thing of imposing magnitude and visual splendor meant to persuade readers of the beauty of the human form and the usefulness of anatomy (figure 13).

Beyond the skeleton pictures, which were of course the main focus, d'Arconville's work was embellished with allegorical illustrations, including the frontispiece, some vignettes, and cul-de-lampes by the painter J. B. M. Pierre, a sought-after artist whom she hired at considerable expense. One sumptuous image shows the god of Time withdrawing his scythe and, instead of cutting life short, pulling back a heavy drape to let bright light shine in. This beam illuminates a woman, symbolizing Anatomy, who looks boldly back at Time and holds a knife in her hand, ready to start a dissection. A foreshortened cadaver whose feet we see lies on a marble slab, and a winged cherub gets ready to remove the sheet hiding the rest of the body so that the work can begin. Anatomy's feet rest on the great books in the field, and a second image pays homage to such authors as Vesalius, Paré, Monro, Winslow, Haller, Cheselden, and other "illustrious men" who labored to "allay the ills to which we are condemned and which are only too real and too numerous"(figure 14). There are depictions of cherubs playing with bones, of death charmed by the pursuit of truth, of everyone greeting this "positive science" with enthusiasm.[54] D'Arconville meant to produce something aesthetically pleasing and grand, and she succeeded (figure 15).

She and Sue had a division of labor. He selected and articulated the skeletons, she did the introduction, translation, financing, and hiring of the draftsmen. Her grand-nephew Pierre Bodard explained: "She had made for this purpose paper of the greatest beauty, ordered the character fonts, and had thirty-one plates engraved by the most able artists."[55] Morand's approbation of 27 March 1759 raved, "These images, precious for their beauty, surpass all those that we have known up to the present time." At this same moment Morand was praising Biheron's anatomical models as superior to any that preceded them and arranging her demonstration before the Académie des Sciences. He was also

TRAITÉ
D'OSTÉOLOGIE,

TRADUIT DE L'ANGLOIS DE M. MONRO,
Professeur d'Anatomie, et de la Société Royale
d'Edimbourg:

Où L'ON A AJOUTÉ DES Planches en TAILLE-DOUCE,
qui repréſentent au naturel tous les Os de l'Adulte
& du Fœtus, avec leurs explications.

Par M. S U E, Profeſſeur & Démonſtrateur d'Anatomie aux Ecoles Royales
de Chirurgie, de l'Académie Royale de Peinture & de Sculpture, Cenſeur
Royal, & Conſeiller du Comité de l'Académie Royale de Chirurgie.

PREMIÈRE PARTIE.

A PARIS,
Chez GUILLAUME CAVELIER, Libraire, rue Saint Jacques,
au Lys d'or.

M. DCC. LIX.
AVEC APPROBATION ET PRIVILÉGE DU ROI.

Fig. 13. Title page of Madame d'Arconville's 1759 anonymous translation, *Traité d'ostéologie,* on which only the name of her collaborator Sue appears. [History & Special Collections for the Sciences, UCLA Library Special Collections]

Fig. 14. Frontispiece of the *Traité d'ostéologie* showing Father Time encouraging the dissection of the corpse whose feet we see. [History & Special Collections for the Sciences, UCLA Library Special Collections]

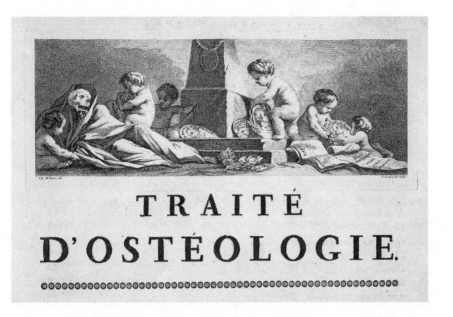

TRAITÉ
D'OSTÉOLOGIE.

Fig. 15. Vignette from the *Traité d'ostéologie*, showing the names of all the illustrious medical authors whom d'Arconville admired etched on discs assembled by putti. [History & Special Collections for the Sciences, UCLA Library Special Collections]

that year supporting the midwife Mme du Coudray's remarkable obstetrical text-book, which he claimed exceeded all previous ones in usefulness.[56] He was gen-uinely enthusiastic about women's scientific contributions to the advancement of knowledge.

What of the argument that d'Arconville was sexist and misogynistic, be-cause the female skeleton depicted in the *Traité d'ostéologie* had a markedly large pelvis and a small head?[57] This image did do substantial subsequent damage, used repeatedly to show that women were only fit to bear children and that their intellect was blighted. But it was Sue, not d'Arconville, who made and chose the image. The differences between her preface as it appeared in 1759 and a 1775 reprint of it in her many-volume *Mélanges de littérature, de morale, et de physique,* where references to operations and specimens do not appear, show that the notes about the female body based on measurements of the illustrated skeleton were not hers, but Sue's. He used the identical artwork and wording almost thirty years later in his *Éléments d'anatomie à l'usage des peintres, des sculpteurs et des amateurs,* with which d'Arconville had no association whatsoever, and he added still more about the uniqueness of the female skeleton.[58] D'Arconville had indeed financed the pictures for the *Traité d'ostéologie* and could have intercepted the female image, which she did not. But that it would have such uncanny staying power and be used as biological proof against women she could not have envisioned. Sue, on the other hand, perpetuated it in later works when he could easily have eliminated or revised it had it not been his own idea.[59]

With the *Traité d'ostéologie*, d'Arconville succeeded in creating something monumental, lavish, a "superb . . . typographic masterpiece," as she said herself.[60] Her second scientific translation of 1759 was a different enterprise entirely, though equally idiosyncratic. The *Leçons de chymie, propres à perfectionner la physique, le commerce, et les arts. Par M Pierre Shaw, premier médecin du Roi d'Angleterre, Traduites de l'Anglais,* received from Macquer the following approbation: "I believe that this book cannot help but be very useful and agreeable for the public as much for the importance of its subject as for the instructive notes, and the interesting Discourse with which the translator has enriched it." Once again here, d'Arconville amplified and reimagined the original work, expanding and making it more broadly accessible and user-friendly, though not in the same way. Shaw was an entrepreneur who had given these lectures in London and then in Scarborough, a resort town where he attracted a high society crowd. He even sold a "portable laboratory" along with his lectures to broaden interest and participation in chemistry, and his success was related to his ability as a performer who enticed audiences at these staged spectacles.

D'Arconville took on the challenge of making a book of written lessons as exciting as Shaw's live enactments. Criticizing a number of his conclusions in order to get a conversation going with her readers, she had already tried to reproduce many of his experiments and her results occasionally did not match his. Long detailed footnotes contained running commentaries on problems she found, that Shaw's method had been "unsure" in one case, his procedure "unfaithful" in another, that there was "good reason to doubt" his findings in yet another. She brought in the opposing views of "several [other] chemists" in some notes, of articles from the proceedings of various scientific academies, and of her own experimental results. Of course the vast majority of Shaw's lectures presented helpful, valuable information, but d'Arconville believed that controversies of this kind would involve the readers of her translation, encourage them to try some of the more elementary experiments for themselves at home, and convince them that questioning the status quo was healthy in the pursuit of new knowledge.[61] Chemistry was a dynamic, exciting science—she used the word "science" over forty times in her preface—that became and remained d'Arconville's true love, and here she was embarking on her own adventures in that field, unafraid of challenging male authorities.

This Shaw translation was preceded by her *Discours préliminaire* which was nothing less than a masterful ninety-four-page history of applied chemistry, her astonishing erudition displayed throughout in long footnote commentaries on procedures and products. Dyeing, painting, tanning, distilling, metallurgy, glassworks, porcelain, brewing, enamel, precious stones, soaps, acids and alkalis, gums, resins, and hard water—all secured their place in her discussion. A thorough glossary and extensive remarks in the body of the text itself totaled almost two hundred instructional annotations, some references biblical, most to books about science, and a few to papers in proceedings of scientific academies both in and outside France. Dismissing the obscurantist charlatans who deluded people for centuries with promises of elixirs of immortality so as to enrich themselves, she showed how they were vanquished by the likes of Newton, Boerhaave, Stahl, Lemery, and Rouelle, who put an end to their dangerous fictions and established accurate, practical chemistry which now made beneficial products for industry, agriculture, pharmacy, and medicine. "The goal of chemistry," she wrote, "in general, is to render all the substances that compose the universe useful to mankind."[62]

On scientific method, d'Arconville favored an empiricist, inductive approach. "When we have only experience as our guide the progress of knowledge is always slow, but its very slowness contributes to making it more sure. It transforms the genius into an observer, and forces him . . . to calm his natural vivacity so as to ponder the smallest details. These details . . . gathered and compared . . . form the foundations of the most certain and invariable doctrines."[63] So beyond experiment, Lockean reflection on such details and analogical thinking were essential for arriving at broader principles in science. Thus while she emphasized the practical Baconian gathering of facts, d'Arconville also sought generalizations that discerned patterns in the workings of nature. She favored both experiment and the clarifying theory it might lead to.

There was a strong moralistic tone to her *Discours*, in which she stressed the uses of chemistry, not only for medicine and pharmacy but also for the fields of human industry, commerce, and "arts," by which she really meant technology. Even though she saw self-love (amour propre), ambition and the desire for fame, honor, and reputation as unattractive features of the human condition, she acknowledged their importance as motivation for discovery, although some discoveries, like gunpowder, and printing itself when abused to spread about evil ideas, were destructive.[64] Condemnations of greed were sprinkled into her mostly positive discussion of curiosity and inventiveness. Regarding mining, for example, she recognized the lust for luxury, the cupidity of our appetite for gold which led to atrocities committed against the Peruvians.[65] But then she went on to describe the ways in which other metals and minerals were found and their properties figured out: iron, lead, pewter, platinum, zinc, bismuth, antimony, pyrites, and strange mercury, always with their utility in mind, in the latter case to cure "the shameful malady, the result of licentiousness and debauchery" (syphilis).[66] Fire appeared mysterious, until it was tamed for cooking, baking, and eventually the making of metal tools. Yeast was known to Moses—she made many references to scripture—and the fermentation, rising, and swelling of dough must have amazed people of his time. This, like so many other things, seemed wondrous and almost magical until elucidated by the advances in chemistry.

D'Arconville's Shaw translation was impressive for many reasons, not the least of which was the way she had to adjust his lectures for readers. Shaw, staging his numerous experiments before live audiences, narrated as he went along. There were twenty lessons, some including as many as ten different demonstrations. She needed to make them appealing and sufficiently vivid for readers to become "virtual witnesses."[67] Providing in her text a detailed table showing the scope and thrust of each lesson, she explained why readers should be as keenly interested as Shaw's spectators in this most serviceable of all studies. Subjects ranged over earth and mining, the effects of fire, the atmosphere's air, water, different solvents, chemical analysis, synthesis, distillation, dyeing, mineralogy, stones and salts, pyrotechnics, and the way chemistry enhanced our daily lives. There were many experiments on fermentation and putrefaction in the vegetable, animal, and mineral realms—already of special interest to her—which included spoiling, rotting, and decomposing but also the positive applications of this natural process, for example in vinegar and in wine and beer making. Next came discussions of the healing powers of oils, salts, infusions, and syrups.

But it was in the preliminary *Discours* that d'Arconville's broad command of and admiration for chemistry shone through. She also promoted the forthcoming work of her two chemist colleagues, Macquer's study on tinctures of silk and Poulletier's imminent French translation of Pemberton's *London Phar-*

macopoeia. Nonetheless, she wrote, lovers of truth still have much to discover. "Nature invites us, the Art gives us the means; let us seize the chance to augment further, if that is possible, the treasures that are the heritage of all men, and that all good Citizens should seek to render useful for everyone."[68] Reviewing her Shaw translation in his *Journal de médecine,* the doctor C. A. Vandermonde, unaware of her identity, praised "the man of letters who undertook this translation" for having the "zeal . . . to give French chemists such a present." And he singled out especially "the eloquence of the diction and the poetics of style" in its long introductory essay.[69] Of this *Discours préliminaire* d'Arconville later wrote, "I added [to the Shaw translation] a discourse . . . on all the arts related to chemistry; it succeeded quite well, and in effect, it is the best thing I wrote in my whole life."[70]

Shaw's views on fermentation and putrefaction, mentioned countless times, obviously fascinated d'Arconville, not least because her own experiments led her to disagree with some of the author's findings. In addition, she brought up John Pringle's work on antiseptic substances meant to retard putrefaction, with which she also disagreed. "I repeated Dr. Pringle's experiments, and I found his results exact at least for some substances he used." She confirmed his conclusions on quinquina as being effectively anti-putrid, but "this was not the case for chamomile, which did not appear to me to stop putrefaction as M. Pringle maintains." She did not think it appropriate to discuss her own discoveries in this translation of Shaw, although a long footnote made clear that she would very much like to do so. "I would overstep the limits I prescribed for myself if I went into more detail now."[71] This was in essence a promise that the anonymous "*Traducteur des Leçons de Chymie de M. Shaw*" as she next styled herself, would soon be back to elucidate this matter further in an offering of her own.

We can only imagine the reaction of her husband and three teenage sons when she disappeared day after day throughout the 1750s into the laboratories she constructed at both her city and country homes, trying to reproduce results reported by other chemists. Beyond this scientific activity she was writing prolifically as well. How available she was for her family we cannot know, but she went about her work with enormous intensity.

During the 1760s, d'Arconville moved away from translation to writing original works, yet while she was still deeply involved with her putrefaction experiments and preparing to make them known, the things she chose to publish showed an additional kind of preoccupation. It is important to recognize how inextricably linked her science was with her general outlook on life, and therefore how relevant, even essential, her nonscientific writings of a moral and philosophical sort were for her and for our full appreciation of this brilliant, resolute, complicated, and many-splendored woman.

Her 1760 *Pensées et réflexions morales sur divers sujets,* which she later revealed was dedicated to her sister, contained essays on religion, women, marriage, and many other topics that concerned her. Notable were her wide range and her candor, particularly on delicate subjects like attraction in male-female relationships, and here her anonymity gave her extraordinary license, for much of what she wrote, unbeknownst to her readers, was autobiographical. "Women should stop being women at forty. It is enough, it seems to me, to have played with dolls and [then] played wife for twenty-five years."[72] This was a highly personal comment, for she had herself just reached the forty-year milestone. "It is necessary to wait until women stop being pretty to judge their merits and talents rationally."[73] If they do learn it is usually only to impress others, their comprehension remaining superficial, and as they all want to be flattered they are jealous and mistrustful of other women.[74] Such remarks have led one modern commentator to say that d'Arconville really wished to be a man, that she practiced a kind of literary transvestism. But it seems more likely that she believed in a higher, asexual state where erotic interactions would no longer interfere with study, where the body lost significance and only the intellect mattered.[75] Dissatisfied with her marriage, she wrote that while some may think the death of a loved one the worst sadness that could befall, "other ills" are far greater for they continue and only become more bitter with time. When reprinting this very passage in her 1775 *Mélanges,* she modified it to say more explicitly that "domestic sorrows" create the most unhappiness and make life unbearable.[76]

And d'Arconville provided here the first iteration of what would become a leitmotif, a succinct version of her life's credo, that work was the remedy for all calamities. She would always seek the company of intelligent scholars who understood the complexity of things and did not rush to conclusions, realizing that penetrating thought required application and patience. "The ignoramus assures, the learned man doubts, the sage reflects and suspends judgment."[77] She also made clear her reason for remaining nameless on her publications. Although the savants with whom she chose to surround herself appreciated her and her work unfailingly, the majority of men did not think women capable of profound intellectual activity: "Should [women] exhibit science or wit? If their works are bad, they are heckled; if they are good, they are appropriated by others; all they have left is the ridicule for having declared themselves authors."[78]

Many of these ideas were soon developed further in *De l'Amitié* (1761) and *Des Passions* (1764). The first was approved by Macquer in his capacity as censor on 12 April 1761, endorsing in glowing terms "the way the affections of the heart are developed in this work, the sentiments of virtue that are spread throughout it, and the graces of style with which it is adorned." D'Arconville dedicated it "A Mon Ami," most probably Thiroux d'Espersenne, her husband's youngest brother and often referred to as her very dearest friend. In this book she minced no words. Marriage is usually arranged by selfishly ambitious fathers, or sometimes the result of unchecked desire, and neither is the path to anything worthwhile. Instead we must seek virtue, esteem, and mutual respect. Women are too preoccupied with their looks and their hope to be alluring. In fact friendship, which requires "firmness in the soul, justice in the ideas, consequence in the principles, truth in the character, stability in conduct and discernment in choice, suits very little a sex feeble by nature, frivolous by education, featherbrained by pretension, coquettish by vanity and fickle by idleness." Friends of the opposite sex can only be confident of the feelings they have for one another "after age has deadened the fire of passions, their senses are muted, and the difference between the sexes has become null for them."[79] Intellectual associations are best, and men of *science* are the most honorable of all, having "preferred being useful to shining among those who think they can judge everything but know nothing." This represented her ideal—and idealistic— model for sociability. "Relations based on compatibility of tastes and occupations are the most agreeable of all and the most durable, and of all preferences there is none that provides as much resource for friendship as study."[80]

Des Passions soon followed. Her cousin Poulletier might have been the one to whom she namelessly dedicated this book, for on 21 September of that year he sent Macquer a copy of it with a little wink.[81] She spoke of how much more friendship, a "superior sentiment," meant than her blood ties to this person, how she had freely chosen him despite their being related.[82] In this work she upheld friendship against passion, the latter leading always to misery. As in her *Traité d'ostéologie*, allegorical illustrations adorn the two sections, the first on *l'Amour*, the second on *l'Ambition*. In the latter part she discussed Julius Caesar and others who ruthlessly, crushingly impose their will on others.[83] Jealousy, unbridled desire, and tyrants who aspire to saturate the world with their name spread poison everywhere. With keen insight d'Arconville viewed such behaviors, whether by leaders or ordinary men, as attempts to fill a void within, like the smoking, drinking, and consuming of ecstatic substances used throughout the world as "remedies for the pain of living."[84] She lamented that even girls, deprived of education and so reduced to pitiful, idle, trivial creatures, formed ambitions based on their charms, relying on their beauty to attract men through whom they might play some role in the world.[85] Was d'Arconville perhaps revealing her own tactics here? Did she decide to marry her husband because he was besotted enough with her when she was young and pretty to leave her alone in her unladylike scholarly pursuits once she bore him three sons? She had, as we saw, satisfied him in this regard before she turned twenty, and at least one contemporary suspected that the real reason for her publishing anonymously was to not embarrass her mate in exchange for the considerable leeway she enjoyed.[86]

De l'Amitié and Des Passions went through multiple editions and were generally well received, although the journalist Baron von Grimm called them pedestrian, the works of a carpenter who just threw chips to the public, and he expressed no wish to learn the identity of their author.[87] What must have been his surprise when these two works were later attributed, mistakenly, to his good comrade Diderot, whom he greatly admired, in two German editions of 1770.[88] Not to mention d'Arconville's own consternation at being confused with a daring philosophe whose ideas she repudiated.

While writing about passion, ambition, and the all-important friendship, she had been performing literally hundreds of experiments and was finally ready to make known her conclusions. Her 1766 Essai pour servir à l'histoire de la putréfaction marked her return to scientific publishing, this time with a work entirely her own. It was in some ways an outgrowth of her earlier translations which had given her authority, provided an opportunity to put forth her own views in the explanatory notes, and bolstered her confidence to embark on her own original treatise. The title page identified the author as "the translator of the Leçons de chymie de M. Shaw," visible yet invisible, creating a link to her previous publication and thus underscoring her legitimacy, her right to be contributing to the discipline of chemistry (figure 16).[89] Encouraged by Macquer, to whom she dedicated the Essai, calling him "Mon Ami, Mon Maître," she thanked him for his support and intellectual generosity.

Putrefaction had fascinated Francis Bacon, who devoted much attention to it in his 1627 Sylva Sylvarum. Johann Joachim Becher wrote at length about

ESSAI

POUR SERVIR A L'HISTOIRE

DE

LA PUTRÉFACTION.

Par le Traducteur des Leçons de Chymie de M. SHAW,
premier Médecin du Roi d'Angleterre.

A PARIS,

CHEZ P. FR. DIDOT LE JEUNE, Quai des Augustins,
près du Pont S. Michel, à S. Augustin.

M. DCC. LXVI.

AVEC APPROBATION, ET PRIVILEGE DU ROI.

Fig. 16. Title page of d'Arconville's anonymous 1766 *Essai* on putrefaction,
where she identifies herself only as the earlier translator of a chemical work
by Peter Shaw. [History & Special Collections for the Sciences, UCLA
Library Special Collections]

it in 1669, and the subject had certainly intrigued Newton, who wrote in *Opticks* Query #30 that "nature seems delighted with transmutations. . . . Eggs grow from insensible magnitudes, and change into animals, tadpoles into frogs, and worms into flies. All birds Beasts and fishes, insects, trees and other vegetables with their several parts, grow out of water and watry tinctures and salts and by putrefaction return again into watry substances." In Query #31 he reiterated this interest in the cycles of generation and decay but recognizing, as with gravity and both magnetic and electrical attraction, that the *cause* of fermentation was not yet known, and deferring to others to figure it out.[90]

The subject had not been addressed after that until John Pringle tackled it directly in a presentation to the Royal Society in 1750, which he then appended to a 1752 book on diseases of the British army.[91] D'Arconville, who probably read this work in the original English even though it was translated into French in 1755, saw Pringle's formulation as just a starting point that, in her words, "leaves several things to be desired."[92] With characteristic self-possession she plunged into the timely topic of putrefaction, positioning herself at the cutting edge with her own new research, ready to challenge prevailing but unsatisfactory views.

Putrefaction was an enormous problem in the days before refrigeration or effective preservatives. It made difficult the study of human anatomy, as we saw in Mlle Biheron's race against time with the cadavers she dissected, and it vastly complicated actual medical practice in dealing with wounds. D'Arconville, who was acquainted with anatomical specimens in connection with her 1759 *Traité d'ostéologie,* had made this very point in her introduction to that work, and she had referred to the subject many more times in her translation of Shaw. Now in 1766, in her thirty-five-page preface to her *Essai* she called putrefaction with all its destructiveness "the marvelous operation that nature executes on all organized bodies."[93] With somewhat disingenuous modesty she wrote that if her labors could shed "even a feeble glimmer on a matter as broad as the one I treat," she would consider all her time in the laboratory worthwhile.[94] The approbation was by Macquer as royal censor, who gave the now predictable rave: "I think that this work, filled with interesting experiments and useful views presented in the clearest and most exact manner, can only be very advantageous to the progress of medicine and chemistry and is most worthy of being printed" (figure 17).[95]

The boldness of d'Arconville's undertaking cannot be overemphasized. She

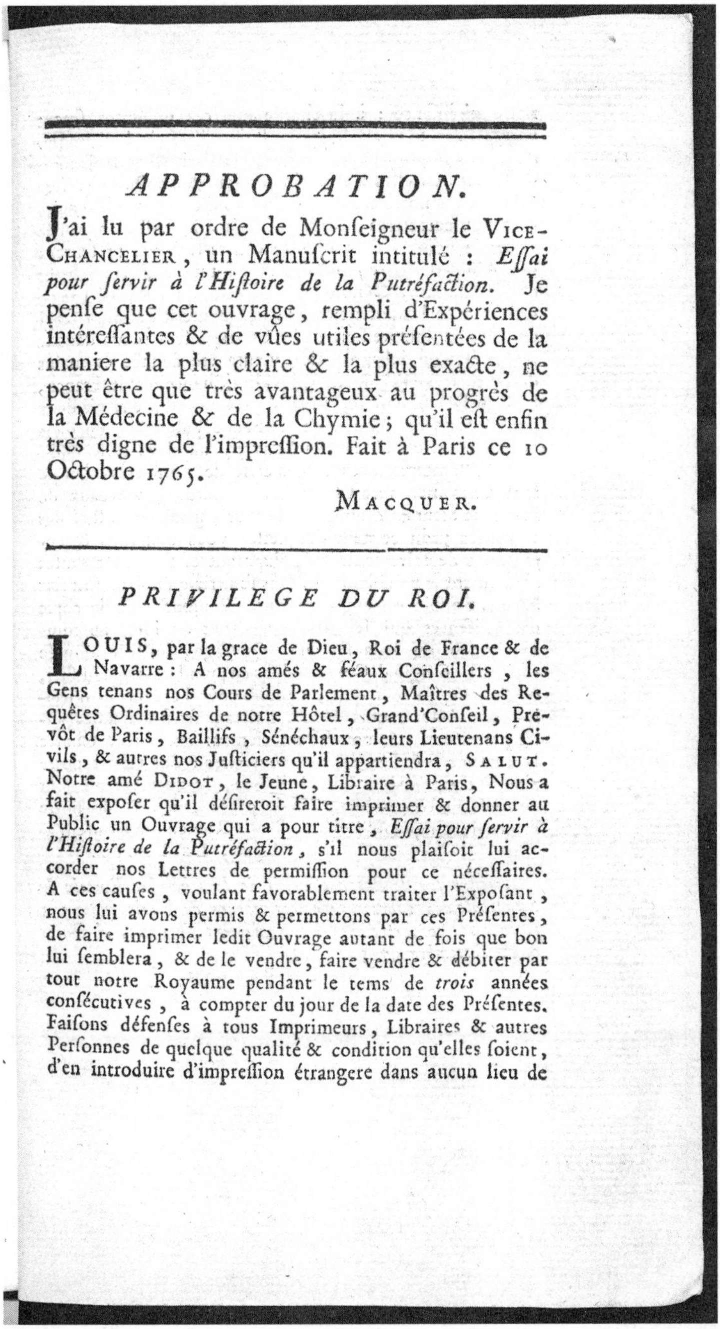

APPROBATION.

J'ai lu par ordre de Monseigneur le VICE-CHANCELIER, un Manuscrit intitulé : *Essai pour servir à l'Histoire de la Putréfaction.* Je pense que cet ouvrage, rempli d'Expériences intéressantes & de vûes utiles présentées de la maniere la plus claire & la plus exacte, ne peut être que très avantageux au progrès de la Médecine & de la Chymie ; qu'il est enfin très digne de l'impression. Fait à Paris ce 10 Octobre 1765.

MACQUER.

PRIVILEGE DU ROI.

LOUIS, par la grace de Dieu, Roi de France & de Navarre : A nos amés & féaux Conseillers, les Gens tenans nos Cours de Parlement, Maîtres des Requêtes Ordinaires de notre Hôtel, Grand'Conseil, Prévôt de Paris, Baillifs, Sénéchaux, leurs Lieutenans Civils, & autres nos Justiciers qu'il appartiendra, SALUT. Notre amé DIDOT, le Jeune, Libraire à Paris, Nous a fait exposer qu'il désireroit faire imprimer & donner au Public un Ouvrage qui a pour titre, *Essai pour servir à l'Histoire de la Putréfaction,* s'il nous plaisoit lui accorder nos Lettres de permission pour ce nécessaires. A ces causes, voulant favorablement traiter l'Exposant, nous lui avons permis & permettons par ces Présentes, de faire imprimer ledit Ouvrage autant de fois que bon lui semblera, & de le vendre, faire vendre & débiter par tout notre Royaume pendant le tems de *trois* années consécutives, à compter du jour de la date des Présentes. Faisons défenses à tous Imprimeurs, Libraires & autres Personnes de quelque qualité & condition qu'elles soient, d'en introduire d'impression étrangere dans aucun lieu de

Fig. 17. Pierre Joseph Macquer's approval for the printing of d'Arconville's *Essai* on putrefaction. [History & Special Collections for the Sciences, UCLA Library Special Collections]

was tackling a subject that she saw as "the key . . . and the story of all nature," and many agreed with her about its central importance. She was taking on Pringle, a famous man who would earn the sobriquet "father of military medicine" and whose star was on the ascendant in Britain. D'Arconville's goal, on the face of it, was to discover antiseptics that would retard putrefaction, and she had started her three hundred experiments in search of such substances over a decade earlier. The outcome of that quest itself would have been hugely useful. But she was also seeking something far more ambitious. Close reading of her text shows a relentless pursuit of an overarching *law* or *theory* that would explain the phenomenon itself, the *cause* of putrefaction that others had sought and failed to find.

Most men of science, she wrote, could not equal the genius of Newton, Stahl, Boerhaave, Winslow, and von Haller.[96] Nor did others have to be mere historians of nature talking about the ideas of the past. A middling group of researchers like herself could seek inspiration from the giants and elucidate new matters, and in that spirit she had spent more than ten years doing hundreds of precise experiments attempting to classify substances that would prevent decay in organic matter, in particular beef, lamb, rabbit, fish, and eggs. Just as she had repeated Shaw's experiments, tried to replicate his results for her translation of his lessons, and challenged him when she could not, she did the same for Pringle's and found many of them wanting.

Assuming that in true scientific spirit Pringle would not object if she pointed out what had escaped him and instances where he was plainly wrong, she still thought it prudent to deploy a clever tactic so as not to wound him. She praised his talents as "so superior" to her own and made excuses for him by suggesting that a man so busy tending to his other lofty occupations—with his many noble patrons he was well on his way to becoming physician to the royal court—must simply not have had time to follow through and check things more than once. There was no place for rivalry, she argued, when the fecundity of nature left open a "vast field for new discoveries," especially in this area of fermentation and putrefaction, the most interesting part of physical science, "the key to all the others . . . [for] every living thing is subject to its powers. . . . We can therefore regard putrefaction as the wish of nature" which "seems to destroy only to create anew."[97]

Displaying in this work a dazzling command of her discipline, d'Arconville nonetheless referred more than once to being on a path akin to a labyrinth—this a Baconian reference—for the subject of putrefaction was still quite new and she was sure to take false turns as her research proceeded and she tried to find her way. She was testing different liquids and solutions of minerals for their antiseptic properties by submerging organic matter in them. She aimed for impeccable accuracy, knowing that her experiments would need to prove entirely reproducible, just as she expected those done by others to be. For this reason she wrote up everything in a lab notebook, adhering to strict protocol, journaling each possible variable in her experiments, not only the kind of meat and the substance in which it was immersed—solutions of metallic salts, gums and resins, winey liquors, acids, fixed alkalis, volatile alkalis, plant juices and saps, neutral salts, earthy salts, quicklime, many different mineral waters—but also whether the work was done in her lab in the country or in the city, the exact time of day, the weather, the temperature indoors and outdoors, the exposure of the open windows, the humidity, the wind direction, "so that the readers are able to judge the care and precision with which I did my observations."

Although she did not use this language, she always had a control, writing that "to convince myself I always took care in every experiment to have as a point of comparison" some meat that was dry, or in ordinary water, of the exact same weight and under the exact same experimental conditions as the meat in the solution whose antiseptic vigor she was testing.[98] The book had ten foldout tables documenting her findings for the thirty-two classes of liquids, from those that preserved the meat for only a day, to those that kept it from spoilage for many months, to those that kept it intact indefinitely (figure 18). Her equipment was not fancy—jars, fabric and string to cover them, retorts for distillation, a double boiler, litmus paper—but she did have the ingredients for numerous solutions of salts, metals, and vegetable substances whose potency she needed to assess. She conversed and corresponded with doctors, parish priests, and directors of mineral spas in different regions, realizing, for example, that the Lisbon earthquake of 1755 would have disturbed the bedrock and wells from which these waters came and therefore altered them.[99]

Reading this book gives an idea of how thrilling d'Arconville's studies were for her. Far from being bored with the repetitiveness of these hundreds of experiments, she approached each opening of a jar with excitement, wondering what she would find. There was a palpable delight every time she returned to check on something, especially if she had been forced to be away for more than a day, absences to which she had to resign herself. Her method was to set up many

SUBSTANCES

QUI ONT RENDU LES ŒUFS, LE POISSON ET LA VIANDE INALTÉRABLES.

Toutes les Substances suivantes ont été dissoutes ou étendues dans deux onces d'eau commune, hors les Eaux Minérales, les Sucs de Plantes, & les Substances qui sont marquées avoir été employées à sec.

SELS Métalliques.	GOMMES & Résines.	LIQUEURS Vineuses.	EXTRAITS & Substances simples.	ACIDES.	ALKALIS Fixes.	ALKALIS Volatils.	TERRES.	SUCS.	SELS Neutres.	SELS à base terreuse.	EAUX.	SUBSTANCES Animales à sec.
Cristaux de lune, avec deux gros de bœuf.	Beaume du Pérou en poudre à sec, avec 2 gros de bœuf.	Vin d'Espagne blanc, avec 2 gros de bœuf.	Extrait de quinquina , avec deux gros de bœuf.	Vinaigre rouge, avec deux gros de bœuf.	Dissolution de sel de tartre, avec 2 gros de bœuf.	Sel volatil de corne de cerf à sec, avec 2 gros de bœuf.	Chaux vive à sec, avec 2 gros de bœuf.					
Turbith minéral. *idem.*	Camphre en morceaux à sec. *idem.*	Vin d'Arbois. *idem.*	Quinquina en poudre à sec. *id.*		Sel de tartre à sec , avec 2 gros de bœuf.	Sel volatil de corne de cerf à sec, avec un jaune d'œuf.	Chaux vive à sec, avec un blanc d'œuf.					
Nitre mercuriel. *id.*	Poix de Bourgogne broyée avec un peu d'esprit de vin. *id.*	Vin de Bordeaux. *id.*	Quinquina en poudre à sec, avec un jaune d'œuf frais.		Sel de Tartre à sec, avec un jaune d'œuf frais.	Sel volatil de corne de cerf à sec, avec un blanc d'œuf.						
Sublimé corrosif. *id.*	Storax calamite broyé avec un peu d'esprit de vin. *id.*		Quinquina en poudre à sec, avec un blanc d'œuf frais.		Sel de tartre à sec, avec un blanc d'œuf frais.							
Sel de mercure par le vinaigre. *id.*	Gomme ammoniaque en poudre à sec. *id.*		Sel essentiel de quinquina , avec 2 gros de bœuf.									
Dissolution de cuivre par l'acide nitreux. *id.*	Sarcocole en poudre à sec. *id.*		Sel essentiel de quinquina, *id.* pour rétablir.									
Dissolution de cuivre par l'acide du sel. *id.*	Gomme adragant en poudre à sec. *id.*		Gayac épuisé à sec , *idem.* pour rétablir.									
Sel de Mars par l'acide vitriolique. *id.*	Gomme arabique en poudre à sec. *id.*		Extrait d'opium. *id.*									
Dissolution de fer par l'acide nitreux. *id.*			Noir de gale en poudre *id.*									
Dissolution de fer par l'acide du sel. *id.*												
Sel de Saturne. *id.*												
Plomb corné. *id.*												
Sel de plomb nitreux. *id.*												
Vitriol d'argent. *id.*												
Vitriol bleu. *id.*												

Fig. 18. One of d'Arconville's many detailed fold-out tables from her *Essai* summarizing the preservative power of various substances. [History & Special Collections for the Sciences, UCLA Library Special Collections]

experiments at a time, and then wait and watch. Determined to see things through, to check regularly, she was nonetheless interrupted a lot, and frustrating as this must have been, she reported it matter-of-factly. We see frequent comments like "I went several days without being able to examine my jars," "different reasons prevented me from following this experiment any further," "I could not continue with this one." She needed to leave for Paris often, and at least once around the Christmas holidays when she was to be away from Crosne for a long time she carefully transported some jars to the city with her, making note of the changes in ambient conditions in case they affected the experiment's outcome.[100] Her Paris lab, for example, was one story up, the one on her country estate on the ground floor, the microclimates quite different, and such things could have an impact on the recorded results. Each series of putrefaction experiments, of which there were several hundred, was followed by an "Observation" with her interpretations.

Vivid descriptions abounded: foams making a mousse "like Champagne" around a piece of meat; thin and thick films forming of different opacities, sometimes interfering with what she wanted to observe but interesting in their own right and deserving of attention; specimens shrinking, swelling, hardening, or disintegrating; colors changing to hues both glorious and grotesque, greenish yellows and whitish reds; and particularly olfactory emanations. Her first chemistry teacher Rouelle was famous for teaching his students to activate their sensory perceptions, educating them in the "corporally disciplined analysis of smells, tastes, textures, and colors,"[101] and d'Arconville had certainly learned that lesson well, using her senses in a way that would have thrilled Mlle Ferrand and Condillac. She was attuned primarily to scents. They were the first things she noted: odors were occasionally "nulle," and others ranged the entire spectrum from "musky dank," "somewhat smokey," "lightly urine-like," "fetid" through "putrid" to "unbearable," at which point she peremptorily "threw the specimen out." Alain Corbin, in *The Foul and the Fragrant*, calls d'Arconville a "scientist ceaselessly alert to the smells of decomposition," an "incomparable observer of odors. Watching the incessant rhythm of putrefaction for months on end, she was overwhelmed by the variations in smell, bearing witness to immense mysteries. She found this variety more stimulating to the imagination than the change of color in decomposing substances or the hissings and bubblings of fermentation."[102] But while sniffing was paramount, looking, feeling, and listening were clearly important too. There is no record of her having tasted these samples.

D'Arconville did not hesitate to dispute Pringle's earlier work on the subject, giving his "claims"—not a neutral way to refer to his views—and then in many places presenting her objections to them, commenting, for example, "I never observed that fact," "he was mistaken," "he was wrong," "I believe I am certain he was wrong." His name occurs many times, both in the text and in her scholarly notes at the base of each page, and elsewhere she disputed his findings without naming him.[103] She had the courage of her convictions when it came to the details of these experiments.

Beyond this precise record keeping, however, we see something else going on. D'Arconville was searching for, and in fact approaching, an overarching cause that linked all of her disparate observations together. She had been gradually led to conclude that *exposure to air* was the reason for putrefaction. Having decided this on the basis of her many years of experimentation before publishing, she highlighted her idea regularly throughout her book, starting with the opening preface.[104]

She had not arrived at this judgment lightly but she did not adhere to it dogmatically. Her text is a revealing tribute to her earnestness as an experimentalist, as a scrupulously faithful, honest reporter of her results. She wanted to be practical of course, as in all her writings, but she stressed the importance of a guiding overall pattern as well, keenly eager to find or confirm some ordering principle, some "théorie certaine," mentioning this repeatedly as her fervent hope.[105] She sought a "general law of all putrefying bodies," an "invariable mechanism that the author of nature established in the universe" to explain the phenomenon she studied.[106] The word "law" is mentioned thirteen times, "laws" nine times.[107] She respected the attempts of previous writers but was ready to point out something that "throws doubt on what has been regarded until now as an invariable law of putrefaction." And there was something else "very contrary to the general laws of putrefaction, . . . but I still do not know the cause."[108] Earlier explanations seemed not to work, but it was hard to find a comprehensive new one to replace them. In seeking a cause—she used this word more than thirty times in her book—she believed she had come close, but she faithfully reported, for example, an experiment that "seems to destroy principles that I thought were established on very exact, very consistent observations." Such confessions were essential, because "truth is too dear to me to suppress anything that might unveil it."[109]

Yet d'Arconville kept returning to her focus on air. She believed air was enclosed in all organic matter, and that this internal "air, making an effort to break free, ruptures the tissues of the body, augments their volume, dilates them, gives access in this way to the exterior air, and consequently contributes to their corruption."[110] She said from the start that any liquid into which she submerged her different meats and eggs was "an obstacle to contact with outside air, without which no kind of fermentation can be executed," and in her search for the most powerful antiseptics she found that solutions did a better or worse job "shielding" the meat, "intercepting access to it." It seemed to her that without exposure to the destructive external air, "no kind of fermentation can be realized at all."[111]

D'Arconville was on the right track in so many ways, on the brink of a key discovery. But she did not consider what might be *in* the air. Familiar with the work of microscopists and the controversy over spontaneous generation, she sided with the nonbelievers, mentioning seventeen times that no insects had attacked her meat and therefore had not contaminated her experiments with their eggs, which ruled out any impurities of that kind. The implication here was that pollution in her experiments, had there been any, would have been caused by insect eggs rather than by anything spontaneously generated and magically arising from within.[112] She did observe "swimming molecules," some greasy, some farinaceous, some just described as white, but regarded them as inorganic and never seemed to suspect that these swirling particles were significant. Had she used a microscope to study them further, had she reflected that they could be alive and hungry, the germ theory of disease might have come along much earlier. Instead she stuck to her view that exposure to air in itself caused putrefaction in her specimens.

In the end, however, she capitulated quickly, suddenly, and unexpectedly, not to Pringle but to another British medical man, David Macbride, who put forth a theory of putrefaction relying on "fixed air." D'Arconville believed his idea clarified and unified everything. Her book was finished, she explained, and was being printed by the time she learned of Macbride's volume, in which he argued that the removal of so-called fixed air (carbon dioxide) caused putrefaction. He said antiseptics worked simply because they kept fixed air in. D'Arconville reported and described for several pages this alternative putrefaction theory and then commented: "We have seen on the contrary throughout the whole course of my work that I attribute the conservative power to the obstacles I place against contact with external air." But once she reminded her readers of her own theory, as she emphatically did here one last time, she conceded that Macbride's explanation seemed to her "simple, uniform," with obvious "advantages over mine and I yield to him without jealousy." Macbride's view that fixed air leaves organic substances resembled the prevailing idea that something called "phlogiston" leaves metals when they burn, only a "chaux" remaining behind. There seemed to her to be a logical sequence (*enchainement*), a symmetry, an analogy here between putrefaction and combustion that she had simply missed, in both cases something escaping to set one of nature's operations in motion. Bowing out graciously to Macbride, she wrote that the love of truth surpasses all self-love. "Those who rather than benefit from knowledge superior to theirs are wounded by it have no right to be enlightened."[113]

Was this a real surrender, and if so, why? Because d'Arconville knew that an all-embracing theory had eluded her. At the end of her *Essai* she wrote: "I am well aware of what is missing in this work. . . . Although there are multiple facts, there are not enough to lead to the discovery of nature's ways for operating putrefaction in bodies, or to delay its progress. . . . I only saw glimmers where I would have needed great clarity in my search for causes of these marvelous phenomena. The fear of going astray and misleading those I mean to instruct made me prefer to reenter obscurity, rather than let myself be led by these feeble glimmers which are sometimes only in the imagination. I take pride, then, that my readers will appreciate my restraint, and insufficient though this *Essai* may be, if I made few steps toward truth, they will at least not be able to blame me for substituting illusion."[114]

We should pause over d'Arconville's decision to "reenter obscurity." In a sense, as an anonymous author, she had been in obscurity all along, but what she meant was that she would no longer publish on this subject. Reflecting on it later, she said of her no-nonsense *Essai*, much less accessible to the general public than her other writings, that it "had no charm," that it "can only be read by those who want to work in this field, which has made huge progress since I pursued it."[115] Presenting three hundred experiments that required more than ten years to finish, it was made to be useful, not to please or entertain. After reading Macbride's text, hers seemed less relevant for the present or future, belonging instead with past efforts on the subject. She thus yielded and joined others whose work had been superseded.

However, "those who want to work in this field" did indeed pay attention to her work. Contemporary and later scientists took her book very seriously, and her research was appreciated as a pioneering contribution to antisepsis by sub-sequent scholars.[116] She had been correct, as far as she went, in suspecting out-side air to be the culprit, but she missed that the air transported living micro-scopic creatures which were themselves the real cause. This puzzle would not be figured out until a full century later with Pasteur's decisive refutation of spon-taneous generation, his definitive proof that extraneous microorganisms cause fermentation, and Lister's recognition that they also cause wounds to putrify. D'Arconville had been close to that insight too, even criticizing the famed Boer-haave—she saw this as a kind of "blasphemy" but did it anyway—for not real-izing that fermentation and putrefaction were related processes, precisely the insight that would later allow Lister to link the spoilage of wine to the rotting and infection of flesh.

Reviewers of d'Arconville's *Essai* refused to accept that Macbride's work was superior. Macquer's 1766 *Dictionnaire de chymie* called the study of putre-faction the most important undertaking in all of science, "the real key to the most hidden and essential secrets of the animal economy." While Pringle and Mac-bride had taken the first steps, he said, the new and improved findings were those of "a very able French physicist."[117] The *Journal Oeconomique* carried a detailed review by an unknown chemist who actually did research on putrefaction him-self and whose experiments, recently submitted to the École de médecine, cor-roborated d'Arconville's. He very much regretted the capitulation to Macbride, preferring the theory in the *Essai* and the "importance of the question treated in this work and especially the multitude of useful thoughts it contains," reit-erating that the author, assumed to be a man of course, should have stood his ground.[118] And there was an extremely enthusiastic review of the *Essai* in the *Journal des sçavans*. The "translator of Shaw" had, according to this reader, "perfected what M. Pringle only hinted at . . . [presenting] his reasons with a frankness and modesty that we would like to find in all writers. . . . The author who is unknown to us gives more than he promises, . . . experiments done with singular exactitude and precision. . . . One easily recognizes that this is the pro-duction of a distinguished physicist [*physicien*] and wishes that the illustrious Author, qualified to work on other equally useful subjects, will let the public know of his occupations. We have the right to expect it from a man less eager for the brilliant reputation of Savant than that of Useful Citizen."[119]

Despite such strong support for her science, d'Arconville was not per-suaded to go on. In fact she changed course very abruptly. She sold Crosne that same year, 1766, her beloved laboratory along with it. In her later memoirs she claimed that the sale was necessary so that she could gift to her two younger sons amounts equal to the substantial sum she had provided for the eldest, Thiroux de Crosne, at his marriage. But that wedding had occurred three years earlier, the next son Thiroux de Gervillier would not marry until 1767, and the third son Thiroux de Mondésir seems not to have married at all. So her memory was playing tricks on her. The sale might instead have triggered or been triggered by her choice to cease activity as an experimentalist; for at least fifteen years Crosne had been the main venue for her scientific research, and without her lab there was far less incentive to continue.

Ironically, Macbride acknowledged d'Arconville's putrefaction book in the second edition of his own *Experimental Essays on Medical and Philosophical Subjects* published in 1767, and very flatteringly. Of course, he could not have harbored even the remotest suspicion that the author who yielded to him so graciously was a woman. After listing many vegetable juices that seem to retard decay, he wrote: "Most of these, besides a great variety of other things, have been lately tried in France, with respect to their power as Antiseptics, and the result of the whole published under the title of *Essai pour servir à l'histoire de la putréfaction,* Paris 1766. It does not appear who the author is, but the experiments which are very numerous (not much short of 300) have been made with surprising patience and accuracy."[120] On the subject of scurvy in another chapter of his book Macbride lifted a lot from naval surgeon and hygienist James Lind, who had done painstaking work on that disease. It was said by many that Lind had good science but no influence, whereas Macbride had much influence but no science. In any case his explanation of putrefaction based on fixed air would soon fall apart, along with the whole discredited phlogiston theory, and both d'Arconville and Macquer would have to stand by and watch the old chemical edifice to which they were so accustomed collapse as Lavoisier's new views took precedence.

The year following the publication of her putrefaction book was one of the worst years of d'Arconville's life. Her husband's brother Thiroux d'Espersenne, who had never married and with whom she had an intensely close relationship, died suddenly in 1767.[121] The disappearance of this man, "my heart's dear love," made her sick with sorrow, causing a long-lasting depression. He was "the most intimate friend I had since my marriage, and who had a most tender attachment for me," her grief acute now "especially as he had bought a house where we were about to get together to spend the rest of our days. This plan had always been the object of his desires and of mine. I sought in vain to distract myself from a memory that my mind and heart were ceaselessly bent on." The two had been busy decorating his new home, close to where her beloved sister lived, d'Arconville designing and contributing all the woodwork and mirrors and outfitting a study for herself. "I was about to reunite with him and he was gone." She envisioned her future in his company, and now she could only hope that her heart would join his after her own death.[122]

It is unclear whether this living plan was to include her husband, in which case it was not even potentially scandalous, or whether she and Espersenne were planning to cohabit as a couple. If so, and it did sound rather like that was the case, whether they would have gone through with this liaison we will never know, but it was then, she said, that the "ink dried in her pen," that the physical act of writing became impossible for her because she would pause for too long, her bereavement and heartache knocking all other thoughts from her mind. On the advice of another brother-in-law, her sister's husband Denis-François Angran d'Alleray, of whom she was also very fond, she hired one Rossel, a friend of her sons from schooldays, as her secretary. He became her scribe and would take her dictation. She threw herself into some more translations of English literature and wrote a novel of her own, the 1767 *Mémoires de Mademoiselle de Valcourt*, whose tormented protagonist failed to devote herself to anything worthwhile but aspired to appear knowledgeable.[123] Perhaps she represented d'Arconville's nightmare, a window into her fears of what she herself might become, grieving, emotionally unmoored, and having given up her own steadying pursuit of scientific truth in the laboratory.

Before he died Espersenne bequeathed to d'Arconville his newly acquired mansion in the Marais and she now moved there, husband in tow. Built by Mansard a century earlier it was known as the Hôtel de Guénégaud. An art collector of fine taste, Espersenne also left her many other things, among them a magnificent statue by Falconet, *Pygmalion*, knowing that she already owned and loved the same sculptor's *La Mélancolie*. This new piece showed the naked beauty Galathée, her hand being kissed by Cupid, coming alive under the astonished gaze of the sculptor Pygmalion who had fashioned and fallen in love with her. But within a year Falconet, in Saint Petersburg at the time, wanted it back for no less a patron than Empress Catherine of Russia, and his good chum Diderot tried to secure it for him. D'Arconville wished to keep the statue. A fan of pagan antiquity, as seen in the allegorical images she chose to illustrate her works and the Greek and Roman myths she invoked in her writing, she "loved sculpture passionately" and specifically this statue which had immense sentimental value.[124] Others in her household, however, apparently deemed it shocking, probably her prudish husband who, if his library was an accurate indication, read religious books almost exclusively. The statue was concealed by a drape. Appalled that the exquisite masterpiece was shrouded from view, Diderot reported to the disappointed artist, "I saw, and saw again M and Mme d'Arconville. I solicited in writing and in person your *Pygmalion*. I am angry about it, my friend. There is nothing to do, and your animated statue will stay for a long time with these churchy rich people, covered by a satin gown which they take off from time to time for the curious."[125]

Yet it must also be said that d'Arconville was not inclined to oblige Diderot in any way. Since their first acquaintance studying with Rouelle at the Jardin du Roi they had had little contact, and her opinion of his works, as expressed in her memoirs, was dim. She loved the ancient philosophers and had some patience for Montesquieu and for Rousseau—although she called him "the one from Geneva," "not the great one" by whom she meant the poet and playwright Jean-Baptiste Rousseau—but most of the contemporary philosophes displeased her. She later blamed their books for plunging France into the hell of revolution, singling out Lalande, Helvétius, d'Alembert, Mably, Raynal, Condillac, Voltaire, Beaumarchais, and Diderot himself with his *Encyclopédie*—most, incidentally, players in this book—for particular scorn.[126] She was all for Enlightenment progress as long as it steered clear of political and social change and stayed focused on the scientific pursuit of understanding nature.

Did d'Arconville abandon science entirely after the *Essai?* Did she really "reenter obscurity"? The answer is complicated, for even though she had stopped her experiments and turned to composing histories, she remained connected to her chemistry circles. Letters from the 1770s show continued closeness with Macquer, whom she invited to musical soirées, consulted on stylistic matters in her historical writing, missed when his duties interfered with their visits, and confided in with blunt details about her various digestive ailments. Poulletier kept an eye on her, expressing concern when she became ill and telling Macquer to look after her in Paris when he himself was out of town. Guyton de Morveau, famed chemist from Dijon and a friend of Macquer's, wrote to him in a way that

made clear he was in on d'Arconville's scientific contributions despite her anonymity, and that he knew she was "no stranger to the translator of Shaw."

They were all following the emerging chemistry of gases that Lavoisier
was spearheading and sending books back and forth, including the new edition
of Macquer's *Dictionniare de chymie* from 1778 and its English translation with
updates on the studies by Joseph Black—discovering what would be called carbon dioxide—and Henry Cavendish—discovering what would be called hy-
drogen. At one point in 1778 Macquer forwarded to d'Arconville a letter from
Lavoisier, who had just given the name oxygen to the new gas discovered by
Joseph Priestley. The mineralogist Antoine-Grimoald Monnet, inspector of
mines, sent regards to d'Arconville through Macquer and inquired whether she
was still working on gums and resins or now only on history.[127] By then she had
composed two lengthy, time-consuming biographies, *Vie du Cardinal Ossat* in
1771 and *Vie de Marie de Médicis* in 1774, this last with three volumes of some
six hundred pages each, so the question was understandable. How could she
find a moment for anything else? But the answer was that she had both interest
and energy left over for science, and she may indeed have been experimenting
on rubber—"gums and resins"—with Macquer in his labs during this period, as
Monnet implied.

And in 1775, a year after publishing her long biography of Marie de Medici,
the terse conclusion of which was "how dangerous and imprudent it is to want to
rule others when one is incapable of governing oneself,"[128] d'Arconville brought
out a seven-volume *Mélanges de littérature, de morale, et de physique,* which the
editor, none other than her secretary Rossel, explained was a collection of some
of the author's previous work, revised, corrected, and considerably augmented,
with the addition of at least one volume's worth of new material.[129] It was, like
all of her work, anonymous, with added comments and updates. In her reprint
of *Pensées et réflexions morales sur diverses sujets* from 1760, for example, there
were now many new passages, in particular on science, which she opined "does
not so much consist in knowing a lot, but in knowing well."[130] The third volume
of her *Mélanges* was entirely devoted to her scientific writings, her *Discours
préliminaire* to the Shaw, and the prefaces to her own *Essai* and to the Monro
translation, now assembled as "Discourses on Different Subjects in Natural Philosophy [*physique*]" and called First Discourse on Chemistry, Second Discourse
on Putrefaction, and Third Discourse on Osteology.

There was, in other words, no attempt by d'Arconville to forget or tuck away her scientific efforts after 1766; on the contrary, she offered them to the public again and even included some of her previously unpublished work, translations of fifteen articles from the *Philosophical Transaction of the Royal Society*, ten on anatomy, four on medicine and plagues, and one on botany which she said she worked on for over fourteen years.[131] More specifically, these concerned the salivary glands, fractures and surgical techniques, the colon, bones and their fibrous membranes, tumors, overripe fetuses, muscles, epidemics, maladies of livestock, and the pores in leaves. This was clear evidence of an interest in histology, physiology, and pathology that continued long after her 1766 putrefaction essay. She also included her translations of some other medical writings by Monro, originally bound as back matter with his *Anatomy of the Human Bones*, such as his "Account of the Reciprocal Motions of the Heart," and his "Description of the Human Lacteal Sac and Duct."[132] In 1778 when the second edition of Macquer's *Dictionnaire de chymie* appeared it was impossible for him to contain his enthusiasm for d'Arconville's science any longer, and with or without her permission he distinctly named her in the index of his book, though still not in the body of the text.[133] Perhaps she had finally acquiesced to letting her scientific identity leak out?

Yet this was an exceedingly trying time for both of them scientifically, and chiefly for Macquer who was much in the public eye. The phlogiston theory, and along with it Macbride's associated fixed air explanation of putrefaction, was being threatened, especially since the appearance of Lavoisier's 1777 paper on combustion. Much as Macquer tried to dismiss it, he knew he could not ignore the new developments, specifically the discovery of different gases in air, and maintain his credibility. On 4 January 1778 he wrote to Guyton de Morveau: "For some time now M Lavoisier has been frightening me by a great discovery that he was reserving in petto (privately) and which was quite simply going to destroy completely the theory of phlogiston or compound fire. His confident air was making me deathly afraid. What would have happened to our old chemistry if an entirely different structure had had to be built? As for myself, I confess, I would have given up the fight. M Lavoisier has just described his discovery: I confess that I now feel much less weight on my stomach."[134]

But Guyton himself was slowly being won over by Lavoisier's new ideas, as were Fourcroy, Monge, and Berthollet, so Macquer had felt constrained to say in the new edition of his *Dictionnaire* in 1778: "M. Lavoisier . . . seems to have had a rather strong temptation to . . . destroy the entire phlogiston doctrine. . . . However, this good physicist has resisted the temptation, at least until now, and has refrained from making a clear-cut decision on this delicate point. His prudence is all the more commendable because it is the distinctive characteristic of all those who truly possess the spirit of chemistry."[135]

In fact Lavoisier had already made the discoveries that would spell the death of the old chemistry, defining oxygen as an element and combustion and calcination as chemical reactions.[136] He went on to show that burning was completely explicable without phlogiston, and in the next decade he would do his ground-breaking work on respiration, solidifying his "doctrine pneumatique," establishing the "école des chimistes français," reforming nomenclature (1787), refuting point for point the phlogiston theory (1788), creating the *Annales de chimie* (1789), and finally in that same year publishing his definitive *Elements of Chemistry*. Although Macquer and d'Arconville had both known this sea change was coming, he did not live to see it, and she was distracted at that point by the political turmoil of the approaching revolution which would take her eldest son, her sister's husband, and Lavoisier himself to the guillotine before it had run its course.

Meanwhile d'Arconville had finally turned away from science. In 1783, she wrote her last book, a third history, this time *Histoire de Francois II, Roi de France et de Navarre*. A characteristically scholarly two-volume analysis of the child-king, including translations of some Italian manuscripts, it was thorough and scrupulous like her two earlier histories; using manuscript sources made available to her by learned friends and never before mined in this professional way, the resulting tomes were wonderfully detailed as all the reviews asserted.[137] The librarian J.-B.-M. Gence, one of many who granted her unlimited archival entry, was in awe of her scholarship.[138] Not surprisingly, these historical works were undertaken with the same determination to do original research that had shaped her science. She did some translations as well, among them a favorite history of England from a Jacobite perspective, showing sympathies she shared with Mlle Ferrand.[139]

D'Arconville had a sophisticated view of the historian's task, the need to ferret out bias, as explained in the preface to her 1774 *Vie de Marie de Médicis*. She meant to write a biography, not an "indigestible compilation," and gave profuse thanks to those who granted her access to manuscripts, which she called the only reliable building blocks of history. Her list of sources for the Medici biography filled eight pages. A critical mind, she warned, was needed to detect untrustworthy renditions and recognize subjective distortions, every author unavoidably swayed by the realities of his position. Some accounts were blind, some unjust, some whitewashed, some slanderous, some fearful, some crassly ambitious, and through it all the researcher had to pursue the truth.[140] Truth was a goal d'Arconville never abandoned. The sources that interested her most were not military histories, always partisan and which bored her to tears, but instead the minute details to be found in eye-witness accounts that revealed the private life and personality of important historical actors and "showed mankind in all its nudity."[141] In her works of history, as in her science, d'Arconville was an uncompromising seeker of facts.

After 1783 she never published again. Macquer died in 1784, and perhaps she thought printing too risky without the undying support that she had always enjoyed from him in editorial as well as scientific matters. Poulletier died in 1788. Her husband expired on 28 March 1789, four months before the Bastille fell. Her three sons lived on but only the eldest, Thiroux de Crosne, seemed good company for her. Tolerant and progressive, he had early helped to rehabilitate Calas and later, as intendant of Rouen, instituted training techniques and kits for the revival of apparent drowning victims. Among provincial administrators he stood out as a particularly enlightened one who, for example, treated the traveling midwife Mme du Coudray exceptionally well when she brought her obstetrical teaching to his region, seemingly unthreatened by this formidably strong woman.[142] He had become the chief of police in Paris in 1785, and in this capacity he ordered and supervised the transfer of corpses from the festering city cemeteries to the catacombs, making the air in the teeming capital more breathable. Were these activities inspired by his very unconventional mother? Had she made him secure, encouraged in him a kind of daring, caring, and independence similar to her own?

Soon, and for five infernal years, the Revolution took over d'Arconville's life. Her two other sons had early joined the military, but they now emigrated, and the youngest, who signed a document saying his mother had urged him to flee France, was in her opinion responsible for her seizure and detention. She had been alone in Meudon, terrified, when the Bastille was stormed, then under house arrest in Paris, then incarcerated at the notoriously cruel Saint-Lazare jail, then moved to Picpus where she could at least rejoin her son Crosne and her sister's husband d'Alleray, who was accused of having sent money to his daughters abroad. To this he bravely admitted for he did not see providing for and nourishing one's children as a crime. In d'Arconville's memoirs she recounted the humiliation of first being guarded in her home by "Cerberus," how most of her servants became popular "patriots" who wouldn't even mail her letters for her, and then how she rejoiced when reunited with family members, albeit in prison. They ate communally in a big room, "malnourished," with "cold food . . . [and] very bad meat." Now and then a faithful cook from her former household brought the family patés and fatted chickens to cheer them up, but they were denied any light, and it was freezing.[143] She, her son, and brother-in-law read in the papers of the accelerating daily executions. Malesherbes, some of whose manuscripts she was able to preserve, was the first of their close circle to be decapitated, along with his family, on 23 April 1794.[144]

Less than a week later, on 28 April, Crosne and d'Alleray were themselves sent to the guillotine, at which point d'Arconville, who no longer wished to live, assumed that would be her imminent fate too, thinking "that such a death was less painful and quicker than an inflammation of one's entrails. The only idea that revolted me extremely was that I would have to be driven in a tumbril before the eyes of a populace mostly paid to applaud at these massacres."[145] Shocked to be released—it was the sudden fall of Robespierre that saved her— she found that the disasters did not stop back out in the world, and she grew increasingly infirm. The death of her beloved sister in 1802 was followed by the return of her youngest son Thiroux de Mondésir, who came back from Belgium penniless and in rags with his domestic, forcing her to lodge, feed, and clothe him although she herself had lost her fortune with the fall in value of paper currency, the ill-fated *assignat*. Here is her sign-off to her autobiographical sketch "Sur Moi" in her memoirs: "The extraordinary pains that I suffer . . . only increase every day, and I doubt any being in the world is as miserable as I am. It is in this cruel situation that I conclude what my few friends still alive might be interested to read."[146]

In fact she did have friends, both old and new. Predictably, she had continued to choose learned men. The erudite Gence was one, her relative Bodard, physician and botanist, was another, and her relationship from before the Revolution with the chemist Antoine François Fourcroy endured. Pascal François Joseph Gosselin, geographer, antiquarian, and numismatist, to whom in her testament she would bequeath all of her books and who ended up also in possession of her manuscript memoirs for a time, was very important to her in the last ten years of her life. Prudent, loyal, and frank, he lived on the rue Vielle du Temple very near her and gave her the courage to go on after the trauma of prison and the loss of her loved ones.[147]

And working on her memoirs sustained her. This manuscript disappeared early in the nineteenth century and only resurfaced, unaccountably, in Mauritius at the beginning of the twenty-first. Dictating both her recollections and many new thoughts was the project that kept her going until the very end, for as she said, "One dies once one no longer has plans."[148] They came about because "one of my grandnieces, seeing me sad and melancholy especially since two years before when my weakening eyesight made reading impossible" suggested that she reminisce in this way. "I had to pull everything out of my eighty-two-year-old head," she reported, in order to get her mind off her troubles.[149]

Yet she was often writing about precisely those troubles. She was also very ill during the last years of her life, not only losing her vision but systemically sick on and off as she had been ever since the early bout with smallpox, her friends referring frequently in letters to her fragility, although she herself blamed her extra weight for everything. She thanked a doctor Jeanroy for so attentively keeping her alive and may have sought some new galvanic treatments.[150] She continued her critique of idle women, now mainly those who, knowing nothing, gave medical advice that interfered with doctor-patient relations. There were no bounds to the pretensions of such meddlesome and ignorant old ladies.[151] She railed against charlatans, complaining that the very act of writing about the travesties of quacks like Mesmer and Delon [sic] soiled her pen.[152] After the death of her sister, who was three and a half years younger and was, as she said, not supposed to go first, she wrote that her heart was irreparably torn and pleaded with God to abridge her days or to help her use them to purify herself so she could join her dear ones in eternity.[153] But grief, decrepitude, and mostly fear of death also caused her to be skeptical about God and the hereafter, and even to wonder "if eternity exists." She flirted briefly with the views of Spinoza and Montesquieu, and with atheism, doubt rearing its head now and again, although she eventually decided that there was no danger in accepting the dogmas of religion and living a life according to those principles, and that belief in the divine plan was, after all, still the best bet.[154] In her eighties and going both blind and deaf, she was ready to die and leave behind the suffering that nothing and nobody could alleviate.

What mark did she mean to leave? Despite d'Arconville's lifelong determination to remain publicly invisible—she stated in her memoirs, "I was true to my vow on this subject, having reflected that there was always much to lose for a woman to declare herself an author and very little to gain"—she was nonetheless outed, after her death of course but during her lifetime too, in hints and even direct naming.[155] While Joseph de la Porte listed all the works of a certain Mme d'*** up to 1769, saying that as she preferred to enlighten the public than to be known by it he would respect her privacy, Diderot announced in a 1771 review that although her *Vie du Cardinal Ossat* was written in a style that some said "wore a beard," he knew it to be by "la presidente d'Arconville." A few years later Portal, in his *Histoire de l'anatomie,* named her as the translator of the *Traité d'ostéologie.* In 1778 her devoted friend Macquer himself, as we saw, perhaps tired of continuing the decades-long hide-and-seek and eager for her to get well-deserved credit, discussed her chemical books in the second edition of his widely read *Dictionnare de chymie,* this time disclosing her full name in his new index. Fourcroy in several works between 1786 and 1805, including his volumes on chemistry in the *Encyclopédie méthodique,* praised her "large number of intellectual publications" and fine contributions to experimental science, and the English physician John Berkenhout listed her by name along with Macquer and Lavoisier in a "Chronological Chart of Eminent Chemists" in 1788. Agronomist and physiocrat Louis-Paul Abeille, calling d'Arconville a "virtuoso," was reluctant to divulge her secret but did in 1798, and Irishman William Higgins identified her in his *Essay on Bleaching* the next year. In 1804, just a year before she died, Fortunée Briquet's *Dictionnarie historique . . . des Françaises* listed each of her books, commented specifically on their being worthy of the esteem they received, and exposed her authorship.[156] This female compiler of a group biography felt strongly that women should openly claim credit for their accomplishments. There is no telling how d'Arconville reacted to her anonymity being breached in these various ways.

Posthumous accolades were naturally much freer with details of her life. Grandnephew Bodard, reconnecting with her shortly before her death, wrote later about their meeting. He found her still lively and full of "fire" despite the grave illness of her last years. They were in fact related, but he valued their deep, sincere, inalterable friendship more than their blood ties. Realizing that in rendering this homage he was "perhaps indiscreetly lifting the veil with which she always wanted to cover herself," Bodard nonetheless risked "throwing some flowers on her tomb" and then, for himself and other fans who had outlived her, assuaged his sadness by enumerating all of her accomplishments.[157] Gence in 1829 and 1830 would reveal still more about her life, and Hippolyte de La Porte in 1835 provided the fullest version. He knew her relatives, not just Bodard but also Mme Joubert of the Poulletier family, and had obviously seen her manuscript memoirs before they disappeared.[158] D'Arconville expected them to be read eventually, frequently addressing potential readers with commentary, updates on her deteriorating condition, apologies for repeating herself, but also pride in her continuing mental acuity even as her body gave out.

She died in her home at 15 rue du Chaume (now the rue des Archives) on 24 December 1805, at the age of eighty-five. Her testament mentions the nearby Notre Dame des Blancs-Manteaux, the neighborhood district church, as the place where she requested masses be said and as the "resting place for my soul." One of her descendants whom I met, a courtly and kind gentleman, wanted to reassure me that she was always "très correcte dans sa famille," and her biographers also insisted that she never neglected her domestic and social duties. She reported that she was obliged to give balls until she was sixty-six years old, so until the year 1786.[159] But she said repeatedly in her memoirs that the only good thing about wealth was being able to help the indigent. She genuinely admired the stoics although, far more emotional than they, she was never able to approach their level of equanimity. She thought, wrote, and labored purposefully throughout her life, with no wish for attention or glory.

Reiterating that her most profound joy had been her life in science, d'Arconville remarked that her favorite field, chemistry, was now so radically transformed she scarcely recognized it but understood that was all part of the onward march of ideas. For her nothing compared to the pleasure of studying nature.[160] When not with other savants of her choosing in real life, d'Arconville had always conversed in her head with the writers of the texts she read, and with those whose views she discussed in her copious footnotes, convening them in her imagination and then listening, agreeing, and arguing with them in the "virtual salon" of her many scholarly annotations.[161] This was the company she always tried to keep, those were the conversations she wished to have, that was what heartened her. Only by constant inquiry could she have held her own with the savants of the past and of her own day, which she brilliantly did. There was no ready model for the kind of probing she undertook and her way of learning by devouring, digesting, amending, and then communicating knowledge over her whole lifetime. As Bodard said of his brave, self-fashioned aunt, "Having only herself to guide her studies, she could justifiably say that she was her own creation."[162]

Dear Geneviève,

I have been walking by your haunts for decades without even knowing it, because the Archives Nationales where I do much of my research are right in the middle of the neighborhood you inhabited your
whole life: rue Pecquay where you lived when first married, various
addresses on what is now the rue des Archives, including the stunning
Hôtel de Guénégaud mansion that is today the Fondation de la Chasse.
You died on what was then called the rue du Chaume, were buried at
the church of Notre Dame des Blancs-Manteaux. All of these locations are within a few short blocks of each other, narrow streets that
I have rushed through on the way to get a seat in the *salle de lecture*
for work or ambled around to meet friends at favorite cafés or visit
nearby museums. This was your beat long before it was mine, rue des
Haudriettes, rue des Quatre-Fils, rue des Francs-Bourgeois, passage
Sainte Avoie, and rue du Temple, although some streets had different
names in your day. Reading documents in the Archives, working my
way back to your century, here in your zone I feel I am getting close.

You explain that you asked to marry at fourteen, desperate to
escape the boredom of the paternal home. But the devil you know
sometimes trumps the devil you don't, and marriage seems quickly
to have lost whatever charm it had for you in your loveless union.
So you turned to "nourishing your head," first becoming a polyglot.
Learning other languages exercises the mind, enhances memory,
concentration, creativity, and flexibility. A cognitive workout, it
improves overall brain function and even delays decline in old age,
keeps senility at bay. Your head stayed active and engaged to the very
end as you marveled repeatedly in your memoirs. Perhaps that is not
so mysterious after all.

Before going any further I must thank you for those memoirs,
your final say. Accessing the interior life of the preceding five women
in this story has been very difficult, but however unreliable some of
your reminiscences might be, I have a better sense of what made you
tick than I do any of the others. Those twelve volumes you dictated

are a window into your world. For two centuries they were lost, sur-
facing only a couple of decades ago and quickly authenticated. Un-
believable, really, that you recounted details from your whole life,
conversations you couldn't possibly remember with such accuracy
although of course you recalled the gist of them. And reciting entire
scenes from plays and novels—they must have become part of your
very being. We historians seek movements of the soul as we struggle
to understand our subjects in depth. Your memoirs have provided us
with a feast.

You should know that two recent books have been devoted to
you, and some of your writings have been translated into English, just
as you translated English works into French. Also, two paintings of
you, the one at fourteen by Coypel (done before your February 1735
wedding, it is called Fille Darlus), and another by Roslin—here you
are all gussied up and look much older than your thirty years at the
time, heavily rouged to cover the ravages of smallpox—hang today
for tourists to see in the Chateau de Cheverny, home to descendants
of the Vibraye family into which your sister's daughter married.
There is also a picture of your wealthy tax farmer father by the
painter Hyacinthe Rigaud. Depicted casually with vest undone as
if to show his humble beginnings—he would say proudly that his
grandfather was a ploughman—his disdain for formality and pretension suggests, as you reported, that he was a kind parent to his two
motherless girls. Crowds at Cheverny are told that he was "gentle,
honest, polite, and charitable." The guides do not say anywhere near
enough about you and the woman of science you became.

The nightmare of the Revolution. How did you hang on? Keeping busy after your release indeed appears to have been your salva-
tion. Working on those memoirs . . .

One of your husband's brothers, Thiroux de Montregard, was
an outrageous womanizer, which must have greatly offended your
sensibilities. You never once mention him. It was the other brother,
Thiroux d'Espersenne, whom you loved. Whom you really loved.

Turgot's copy of your *Essai* on putrefaction was bequeathed
to Fourcroy's student Vauquelin, who would go on to discover the

elements chromium and beryllium. Interesting to think of that book being passed along, chemist to chemist.

Did you meet Elisabeth before she died? You certainly had things in common. A member of her close circle, La Curne de Palaye, was (along with his twin brother) a cousin of yours, and you wrote warmly of these two eccentric gentlemen. Elisabeth studied the senses, which might have inspired your sensory experimental style, implementing in your chemical research what she advocated theoretically. And both of you were cerebral, scholarly, intense. You shared an appreciation of Pygmalion, she the myth that perhaps inspired her awakening statue, you a sculptural representation. And both of you were impatient with feminine frivolity, in her case the Princesse de Talmont, in your case most women you met, but at least you blamed society for depriving them of education and opportunity. And neither of you were shy. Elisabeth flat out told Condillac when he was wrong, and you fearlessly refuted claims by famous men of science. You even made terrible fun of Condillac himself for starting his *Traité des sensations* with "I forgot," mocking the rhetorical device of his opening "Avis" at least twice in your memoirs, so you clearly knew this work well, with its lengthy heartfelt homage to Mlle Ferrand.

And did you see her portrait "meditating on Newton" at the Salon exhibition of 1753? You were already working on chemistry and anatomy and beginning your scientific translations that appeared full-blown in 1759. Maybe Marie-Marguerite was next to you, two beholders drawing sustenance from that picture for the work you yourselves were embarking on. Reine could have been standing there too, perhaps inspired to do the pendulum calculations she would print two years later. You and she could have met through your male colleagues, who knew each other well, Macquer even suggesting that Lalande marry his daughter. Macquer socialized with Commerson during the years 1764–66 when he lived in Paris, so you probably met Jeanne too. Madeleine Françoise you surely knew through your dear friend Bernard de Jussieu and from all the time you spent in the Jardin. And so the web of possible connections intertwines.

Two chemists you may well have inspired: Marie Anne Paulze, the lively thirteen-year-old who married Lavoisier in 1771, an even

younger child bride than you had been. You certainly knew her husband. Did you commiserate with her after his execution in 1794?
What about Guyton de Morveau's lively colleague and later wife,
the Dijonaise-turned-Parisienne Claudine Picardet, like you a fine
translator and experimentalist? But they each worked in partnership
with a husband. Unlike all of you, the soloists.

So let's stick with the six of you in this book. You each figured
out ways to work on your science independently, doing extraordinary
things on your own, not for spouses or brothers or fathers but for
yourselves. Still, I like to imagine what it would have been like had
you known and inspired one another in a mutually supportive group
for women only, how you might have cross-fertilized, pooled your
skills: Elisabeth's rigorous mathematical mind, Reine's understanding
of celestial mechanics, Jeanne's vigorous observational acuity, Madeleine Françoise's keen ability to represent nature, Marie-Marguerite's
knowledge of how the body works, and your discerning expertise in
the lab—what a society that could have been.

Short of it, I have gathered you together in my mind.

Epilogue

IN 1787, TWO YEARS BEFORE THE REVOLUTION, THE CHEMIST A. F. Fourcroy wrote of women that although their education was lacking, "It is luckily recognized for the common good that they are capable of the efforts necessary to successfully cultivate the sciences. Today we no longer doubt their zeal and their facility."[1] This was overly optimistic, for increasing numbers of men *did* doubt women's abilities and did not see their involvement with science as offering anything for the "common good." The situation was to get only more difficult for women who had scientific aspirations like the ones presented in this book.

Fourcroy was writing at the end of the Enlightenment, a period of scientific and institutional malleability that favored or at least permitted the contributions of outsiders, including women. They could show their worth, be acknowledged for it, and receive recognition, even gratitude. The scientific establishment was still in a state of formation, of flux. There were opportunities for un-networked women—as no official networks of or for them existed—to make meaningful contacts with men who helped and then relied on them as essential equals in joint endeavors, or praised them for their solo work. Women were not admitted to the existing brotherhoods and did their important work outside of the hallowed halls reserved for men.[2] But the boundaries were porous, ideas flowed more freely in and out, and women participated at the forefront of exciting developments in their day, with cutting-edge contributions to, for example, what contemporaries called *cometomania* and *botanophilia*.

And their work mattered. Without Ferrand, Condillac would not have explored cognition before the acquisition of language. Without Lepaute, the accurate prediction of the return of Halley's Comet would not have been possible. Without Barret, Commerson would have collected and catalogued far less of the rich flora in the distant lands of South America and the Pacific and Indian

Oceans. Without Basseporte, the botanical documentation in the *vélins* collection of the Jardin du Roi would have been far poorer. Without Biheron, a real driver of innovative pedagogy, anatomy teaching would not have been available year-round and women would have learned much less about their bodies. Finally, without d'Arconville, scientists might not have begun to consider, as early as they did, that putrefaction could be caused by something airborne. As their admiring colleagues repeatedly said, these women knew they were needed, redoubling their efforts in the face of challenges.

Such challenges there surely were. Ferrand was an invalid, which did not stop her from bringing acute thinkers into her home, keeping abreast of mathematical research, and maneuvering behind the scenes of the Académie. Lepaute had a highly skilled but irascible husband whose work she supported even though this was not reciprocated. Somehow maintaining her self-esteem, she bookended her period of activity by winning academic membership toward the beginning and requesting a government pension for her work in science toward the end. Barret did not allow her lowly social status, or being an orphan, or being a female for that matter, to stop her from doing the exploration and science she wanted to do, choosing to assume a male disguise and persona to achieve her ends. Basseporte, whose widowed mother was in financial straits, did portraiture to provide for the two of them but then, once an adult herself, vowed to find a regular salary through science and secured a royally pensioned post. Biheron conquered the isolation involved in constant dissections by going public with her museum and her touring. Finally, d'Arconville dealt with the demands of her lofty rank and station—others may have envied her but she found such social obligations onerous—and despite lifelong depression managed to stay active and engaged. For all six women, the pursuit of science gave their lives meaning.

They successfully raced the clock, aware of the ravages of time that threatened to upend their work. Ferrand knew her health was precarious, but she continued to host savants at her home and collaborate with her philosophe until the very end. Lepaute stayed ahead of the celestial phenomena that wait for no mortal, successfully predicting comets, transits of Venus, and a solar eclipse. Barret had to grab whatever new vegetation she could find during fleeting landfalls so that Bougainville's ships would not leave her behind. Basseporte captured, in stunning images, countless fragile plants before they wilted and died, Biheron deftly took casts from each cadaver before decay rendered its organs useless, and d'Arconville caught and recorded every evolving stage of her numerous rotting specimens in a laboratory juggling act. That they were swift in these specific ways relates to their more general ability to tactically—in Certeau's sense—seize openings and take advantage of chances that presented themselves, creating hospitable spaces and rhythms in which to do their science.

Their male contemporaries honored them enthusiastically. During their lifetimes they were defended vociferously, as in Condillac's correspondence insisting that Ferrand's ideas and formulations were original and preceded similar ones by others, Lalande's sharp criticism of Clairaut for failing to give Lepaute due credit, Commerson's irate letter to his brother-in-law in which he upheld Barret's indispensability and his effusive dedication of a plant to her, whose physical stamina and intelligence he compared to both Diana and Minerva; Buffon's and Jussieu's protection of Basseporte's right to her coveted place even in old age, and Diderot's furious missive to his brother about Biheron's uniqueness as an expert anatomist and teacher. Only d'Arconville was left out of such encomiums because of her insistence on remaining anonymous and her admirers being sworn to secrecy. After the women's deaths those with whom they had associated most closely and who were best equipped to evaluate their science sang their praises in impassioned mourning: Condillac's fifteen-page homage to Ferrand in the "Dessein de l'ouvrage" of his *Traité des sensations;* Lalande's several lengthy eulogies for Lepaute in newspapers and in his astronomical works; geographer Mentelle's long obituary of Basseporte in the *Nécrologe des hommes célèbres;* testimonials to Biheron's unique, path-breaking contributions in reports by several revolutionary *officiers de santé;* and tributes to d'Arconville by her grandnephew Bodard and many others who had been forced to remain silent while she was alive.

Given the vibrancy of these women's lives, we might well wonder how they could have disappeared from the historical record and why it is therefore necessary to write them back into it. Later writers minimized their contributions in ways that smacked of misogyny. Ferrand was said to have been no more than a sounding board for Condillac, merely the object of his gallantry, and even the reason that the *Traité* was flawed. Lepaute was just a tool, a mere servile calculator with no original ideas, Barret nothing but a sturdy "beast of burden"—of course Commerson did not help here by calling her that himself, no doubt to head off any suspicion on the ship that she was a female. Basseporte was appraised as artistically inferior to the men who preceded and succeeded her as *dessinateur du roi,* Biheron's models were deemed primitive compared to those made later by men, and d'Arconville's work was dismissed as naïve and old-fashioned because of the chemical revolution that occurred soon after her period of activity (and that, it must be said, caused the same obsolescence in the work of every male chemist before Lavoisier).

The Enlightenment was a phenomenon in many respects, not least because it thrived under an absolute monarchy that did all in its power to censor and squelch it. But there were fissures that the philosophes exploited, weaknesses in the authority of church and state, moments of elasticity that allowed challenges, if only temporarily, to despotism on all levels, possibilities within the repressive system. The women in this book took advantage of similar plasticity in the male monopoly on science to realize their ambitions. In some respects it was a golden moment, before the Revolution sent women back into their homes, disenfranchised and out of sight. The artist Vigée LeBrun said of the Old Regime, "Women reigned supreme then; the Revolution dethroned them."[3]

Women's ambitions began to subside, their vision narrowing, their claims becoming more modest, many frowning upon those who dared to penetrate the traditionally male realm of science. We see this already at the start of the Revolution, in the *Petition of Women of the Third Estate to the King 1 January 1789:* "If we are left at least with the needle and the spindle, we promise never to handle the compass or the square. . . . Sciences? . . . They serve only to inspire us with a stupid pride, lead us to pedantry, go against the wishes of nature, make of us mixed beings who are rarely faithful wives and still more rarely good mothers of families."[4]

Soon the scientific establishment itself tightened. There was far less fluidity in the next two centuries as professionalization and regulations set in, with a new insistence on credentials, certifications, high-level training, formal qualifications, and advanced degrees. Opportunities for women closed up, they lost whatever access they had had. Marginalized even more, definitively excluded from the traditional universities and other such institutions, women's ideas were appropriated if they were valuable and dismissed or reviled if they were anything less, exactly as d'Arconville had predicted. Stricter, standardized, formalized science became a profession, insisting on expertise in various approved and specialized fields. Power dynamics kicked in along with the definition of the "scientist" as masculine. In 1834 William Whewell first coined the term to describe polymath Mary Sommerville, actually praising her book *The Connexion of the Physical Sciences,* but because initially tainted by its association with a woman, the word "scientist" took decades to be accepted.[5] Once it was, however, it became strictly gendered and connoted exclusively men of science, elite, rational, self-controlled.[6] Not only were women barred from official scientific venues, but beginning with the Industrial Revolution and its sexual division of labor, when the spheres of public and private were rent asunder and became polarized, patriarchal medical experts conspired to declare women sick, weak, hysterical by nature, and fit only to stay inside.[7] This was a toxic mix for women of science, the rigidity of exclusion persisting and strengthening. They could not call themselves scientists without incurring sexist mockery.

Although the nineteenth and most of the twentieth century denied recognition to women in science, things began to look up slightly in the late 1900s. Some Nobel prizes were even awarded to female researchers whose subject matter and style align with the women in my story. Barbara McClintock won in 1983 for work on the maize, or corn, plant. Her approach would have pleased botanists Barret and Basseporte, for she had a close connection with her plants, a "feeling for the organism" as her biographer put it. Devoted entirely to her research, she had worked for years on her own, never married, concentrating with total, patient absorption, "ensouling" the plants she observed with a passion that kept her going.[8] In 2004 Nobel laureate Linda Buck was honored for her work on odorant receptors and overall elucidation of the olfactory system, which would have interested Ferrand trying to parse out what each sense contributes to knowledge, d'Arconville who used her sense of smell to such positive effect, and Biheron who studied the structure and function of every bodily organ. Tu Youyou in 2015 was recognized for her work on the sweet wormwood herb as a cure for malaria, inspired by her earlier studies of traditional Chinese medicine's approach to intermittent fevers. This winner of a plant-based prize had no post-graduate degree, had done no study or research abroad, and belonged to no scientific academy. She desired no credit, she said, just wanted to help the world. Five more awards were given to females in the last few years. Despite such bright spots, however, since the inception of the Nobel Prize in 1901, only 3 percent of its scientific awards have gone to women.[9]

Turning back for one last look at France, the Eiffel Tower, built between 1887 and 1889, sports the names of seventy-two of France's illustrious scientists, engineers, and mathematicians engraved on all four sides under its lowest balcony. Not a single woman is represented, not even the brilliant mathematician Sophie Germain whose work on elasticity in the early nineteenth century was a crucial contributor to the construction of this architectural wonder. The Académie des Sciences, now part of the Institut de France, famously refused membership to Marie Curie—who won the Nobel Prize twice, in 1903 and 1911—and did not elect a woman member until 1979.

But the Fondation L'Oréal, in partnership with UNESCO, created a prestigious and lucrative Women in Science Award, with the first laureates chosen in 1998. It offers fellowships for female scientists-in-training as well. Other foundations, organizations, and government agencies have committed themselves to similar missions. The Association Femmes et Sciences was founded in 2000 to elevate scientific education for girls and discover stimulating openings for women, and in 2001 the French National Center for Scientific Research, the CNRS, established a program to measure, transparently, the progress of women at various levels of scientific activity and to publish recruitment and promotion information. It initiated an action plan that used statistics to monitor gender disparities and created more opportunities for women of science to rise through the ranks. But there is attrition, and the higher the professional status, the fewer women there are to be found at those levels. In 2015, 65.1 percent of technicians but only 18.8 percent of research directors were women. They are being lost along the career ladder.[10] Clearly, this is a steep uphill climb and there is still a very long way to go. But as Marie Curie is said to have said, "We must have perseverance and above all confidence in ourselves. We must believe that we are gifted for something, and that this thing, at whatever cost, must be attained."

Notes

ABBREVIATIONS

AAS	Archives de l'Académie des Sciences
AN	Archives Nationales
AN MC ET	Archives Nationales Minutier Central Etude (for notarial documents)
BCMNHN	Bibliothèque Centrale du Muséum National d'Histoire Naturelle
BnF	Bibliothèque national de France
BN MS	Bibliothèque national manuscripts

INTRODUCTION

1. See, for example, Dena Goodman's *The Republic of Letters: A Cultural History of the French Enlightenment* (Ithaca: Cornell University Press, 1994), Carla Hesse, *The Other Enlightenment: How French Women Became Modern* (Princeton: Princeton University Press, 2003), and Olivier Blanc, *Olympe de Gouges: Des droits de la femme à la guillotine* (Paris: Tallandier, 2014).

2. Emilie Du Châtelet, "Translator's Preface" to Mandeville's *Fable of the Bees*, in *Selected Philosophical and Scientific Writings*, ed. Judith P. Zinsser (Chicago: University of Chicago Press, 2009), 48–49.

3. Mary Wollstonecraft, *Vindication of the Rights of Women* (New York: A. J. Matsell, 1833), 36, 210.

4. Paula Findlen, Lucia Dacome, Rebecca Messbarger, and Emily Winterburn are among the fine scholars who write on these women.

5. See one of the earliest discussions of her work by a historian of science, Carolyn Merchant's publication as Carolyn Iltis, "Madame Du Châtelet's Metaphysics and Mechanics," *Studies in History and Philosophy of Science* 8, no. 1 (1977): 29–48. Numerous scholars have pursued various aspects of her work since then, as in Mary Terrall's "Emilie Du Châtelet and the Gendering of Science," *History of Science* 33 (1995): 283–310. For a fine comprehensive treatment, see Judith Zinsser, *La Dame d'Esprit: A Biography of the Marquise Du Châtelet* (New York: Viking, 2006). Recent research is exploring the influence Du Châtelet had on articles in the flagship *Encyclopédie*.

6. For a sample, see the variety of contributions to Judith Zinsser and Julie Chandler Hayes, eds., *Emilie Du Châtelet: Rewriting Enlightenment Philosophy and Science* (Oxford: Voltaire Foundation, 2006).

7. This in a March 1737 letter to Crown Prince Frederick of Prussia, found in all editions of Voltaire's collected works.

8. For the relationship between the concepts of natural philosophy and science and the eventual widespread acceptance of the latter, see Judith P. Zinsser, ed., *Men, Women and the Birthing of Modern Science* (De Kalb: Northern Illinois University Press, 2005), "Introduction" and the articles in section 2, "Shifting Language, Shifting Roles."

9. "Natural History," in *The Encyclopedia of Diderot & d'Alembert Collaborative Translation Project,* trans. Marc Olivier and Valerie Mariana (Ann Arbor: Michigan Publishing, University of Michigan Library, 2015), http://hdl.handle.net/2027/spo .did2222.0000.189 (accessed 1 March 2019). Originally published as "Histoire naturelle," *Encyclopédie ou Dictionnaire raisonné des sciences, des arts et des métiers,* VIII (Paris, 1765), 225–230.

10. *Leçons de Chymie, propres à perfectionner la physique, le commerce, et les arts. Par M Pierre Shaw, premier médecin du Roi d'Angleterre, Traduites de l'Anglais* (Paris: Herissant, 1759), xciii.

11. Nina Rattner Gelbart, "Introduction" to the Hargreaves translation of Bernard Le Bovier de Fontenelle, *Conversations on the Plurality of Worlds* (Berkeley: University of California Press, 1990).

12. Coste explained that his translation of the *Opticks* was inspired by the very learned and "enlightened" Princess of Wales, Caroline d'Ansbach. *Traité d'Optique,* "Preface du Traducteur." She had received a solid and liberal education at the hands of her guardian, Queen Sophie Charlotte of Prussia.

13. Isaac Newton, *Opticks* (New York: Dover, 1952), 405, 339.

14. See J. E. Force and S. Hutton, eds., *Newton and Newtonianism: New Studies* (Netherlands: Springer, 2004), and especially in that volume Sarah Hutton, "Women, Science and Newtonianism: Emile Du Châtelet versus Francesco Algarotti," 183–203.

15. Newton, *Opticks,* Query #30, with more mentions of putrefaction specifically in Query #31, and d'Arconville's anonymous *Essai pour servir à l'histoire de la putréfaction* (Paris: Didot le Jeune, 1766), xi.

16. There is a recent article on Ferrand's portrait by La Tour, references to Lepaute as part of more general treatments, writings on Barret which often embellish her story beyond recognition, an article on Basseporte's flower paintings but not her key involvement with scientists, some good discussions of Biheron but usually in connection with the Italian waxes of Bologna and Florence, and two books on Thiroux d'Arconville which treat also, and in one case mostly, her nonscientific writings on history, morals, and her fiction. These works will be referred to in detail in the appropriate chapters.

17. Carolyn G. Heilbrun, *Writing a Woman's Life* (New York: W. W. Norton, 1988).

18. See Louis Crompton, "The Myth of Lesbian Impunity: Capital Laws from 1270 to 1791," *Journal of Homosexuality* 6, nos. 1/2 (Fall/Winter 1980–1981): 11–25. See also Elizabeth Susan Wahl, *Invisible Relations: Representations of Female Intimacy in the Age*

of Enlightenment (Stanford: Stanford University Press, 1999). And for the contemporary view, see Antoine-Gaspard Boucher d'Argis, "Sodomy," in *The Encyclopedia of Diderot & d'Alembert Collaborative Translation Project*, trans. Bryant T. Ragan, Jr. (Ann Arbor: Michigan Publishing, University of Michigan Library, 2003), http://hdl.handle.net/2027/spo.did2222.0000.037 (accessed 19 June 2014). Originally published as "Sodomie," *Encyclopédie ou Dictionnaire raisonné des sciences, des arts et des métiers,* XV (Paris, 1765), 266. See also "Lesbian," in *The Encyclopedia of Diderot & d'Alembert Collaborative Translation Project*, trans. Bryant T. Ragan, Jr. (Ann Arbor: Michigan Publishing, University of Michigan Library, 2003), http://hdl.handle.net/2027/spo.did2222.0000.036 (accessed 15 March 2013). Originally published as "Tribade" [author unknown], *Encyclopédie ou Dictionnaire raisonné des sciences, des arts et des métiers,* XVI (Paris, 1765), 617.

19. Karen Offen, *European Feminisms, 1700–1950: A Political History* (Stanford: Stanford University Press, 2000).

20. On the gendering of science as masculine, see *Osiris* 30 (2015), an issue devoted to "Scientific Masculinities" edited by Erika Lorraine Milam and Robert A. Nye.

21. Dianne Millen, "Some Methodological and Epistemological Issues Raised by Doing Feminist Research on Non-Feminist Women," *Sociological Research Online* 2, no. 3 (1997), http://www.socresonline.org.uk/2/3/3.html.

22. See Michel de Certeau, *The Practice of Everyday Life* (Berkeley: University of California Press, 1984).

23. Jansenists were oppositional, nonconformists, often freemasons, many of them patriotic *parlementaires* who stressed that the king had duties to his people. They were anti-Jesuit, enlightened, and *frondeur* (rebellious) in numerous ways, some even anti-royalist and republican. They were "political heretics more than theological ones" as Dale K. Van Kley writes. See his many works on this subject, in particular, "The Jansenist Constitutional Legacy in the French Revolution, 1750–1789," *Historical Reflections/Réflexions Historiques* 13, nos. 2/3 (summer/fall 1986): 393–453.

24. See Donald L. Opitz, Staffan Bergwik, Brigitte Van Tiggelen, eds., *Domesticity in the Making of Modern Science* (Cham, Switzerland: Springer, 2016), Introduction. Patricia Fara, in her *Pandora's Breeches: Women, Science and Power in the Enlightenment* (London: Pimlico, 2004), writes of women working in pairs with husbands, fathers, and brothers, but that model does not fit the women discussed here. For more on spaces of investigation, see David N. Livingstone, *Putting Science in Its Place: Geographies of Scientific Knowledge* (Chicago: University of Chicago Press, 2003), 40–45, 48–51, 55, 62.

25. Judy Long, *Telling Women's Lives: Subject/Narrator/Reader/Text* (New York: New York University Press, 1999), 7, 105–107.

26. Germaine Greer, *The Obstacle Race* (London: Secker and Warburg, 1979). A reviewer of "The Newtonian Moment" exhibit in New York City called the few women portrayed as scientists "colorful, fascinating *creatures*" (emphasis mine), while admitting

they were some of the "strongest, smartest women in history." See "The Ways of Genius," *New York Review of Books*, 2 December 2004.

27. Jenny Uglow, "Friends United," *Guardian Review* 29 (April 2005).

28. Alison Booth, *How to Make It as a Woman* (Chicago: University of Chicago Press, 2004), 10. Emphasis mine.

29. As Elaine Tyler May wrote, "adding women and stirring provided a hint of spice, but the flavor of history remained pretty much the same." "Redrawing the Map of History," *Women's Review of Books* (February 2000): 26.

30. Nina Rattner Gelbart, "Adjusting the Lens: Locating Early Modern Women of Science," *Early Modern Women* 11, no. 1 (Fall 2016): 116–127.

31. Sarah Benharrech, "Botanical Palimpsests or Erasure of Women in Science: The Case of Mme Dugages de Pommereul," *Harvard Papers on Botany* 23, no. 1 (2018): 89–108.

32. Steven Shapin, *Never Pure: Historical Studies of Science as if It Was Produced by People with Bodies, Situated in Time, Space, Culture and Society, and Struggling for Credibility and Authority* (Baltimore: Johns Hopkins University Press, 2010).

33. Kathleen Barry, "The New Historical Synthesis: Women's Biography," *Journal of Women's History* 1, no. 3 (Winter 1990): 75–105, especially 76, 101–102.

34. Barry, "New Historical Synthesis," 84.

35. Nina Rattner Gelbart, *The King's Midwife: A History and Mystery of Mme du Coudray* (Berkeley: University of California Press, 1998).

36. See, for example, d'Arconville's *Traité d'ostéologie* (Paris: Guillaume Cavelier, 1759), xiv, xxvii and xxviii.

37. Nina Rattner Gelbart, *Feminine and Opposition Journalism in Old Regime France: Le Journal des Dames* (Berkeley: University of California Press, 1987). See especially chapter 3 on Mme de Beaumer, and one of the epigraphs in the present book.

38. Margaret Rossiter, "The ~~Matthew~~ Mathilda Effect in Science," *Social Studies of Science* 23, no. 2 (May 1993): 325–341; Naomi Oreskes, "Objectivity or Heroism? On the Invisibility of Women in Science," *Osiris* 11 (1996): 87–113; Hilary Rose, "Hand, Brain, and Heart: A Feminist Epistemology for the Natural Sciences," *Signs* 9, no. 1 (1983): 73–90.

39. Phyllis Rose, *Writing of Women: Essays in a Renaissance* (Middletown, Conn.: Wesleyan University Press, 1985); Carolyn G. Heilbrun, *Writing a Woman's Life* (New York: W. W. Norton, 1988); Paula Backsheider, *Reflections on Biography* (Oxford: Oxford University Press, 2001); Janet Beizer, *Thinking Through the Mother: Reimagining Women's Biography* (Ithaca: Cornell University Press, 2009); Jenny Uglow, "Writing Group Biography," *Guardian Review* 29 (April 2005); Judy Long, *Telling Women's Lives* (New York: New York University Press, 1999); and Amy Richlin, *Arguments with Silence* (Ann Arbor: University of Michigan Press, 2014). See also a summary of some of these positions in Diane Mehta, "The New Candor," *Paris Review* (10 April 2014).

40. See Rosenstone's numerous innovative suggestions in the journal he founded, *Rethinking History: The Journal of Theory and Practice,* and inventive articles by Söder-

qvist, Dorinda Outram, Roy Porter, and others in Michael Shortland and Richard Yeo, eds., *Telling Lives in Science: Essays on Scientific Biography* (Cambridge: Cambridge University Press, 1996).

41. As recent evidence of the resurgence of life writing as history of science, see the Focus on Scientific Biography in *ISIS* 97, no. 2 (June 2006): 302–329; Paola Govoni and Zelda Franceschi, eds., *Writing about Lives in Science: (Auto)Biography, Gender, and Genre* (Göttingen: V&R Unipress, 2014); and the panel led by Ted Porter at the 2019 History of Science Society meeting, "What a Life Means: The Uses of Biography in the History of Science."

42. Ruth-Ellen B. Joeres and Elizabeth Mittman, eds., *The Politics of the Essay* (Bloomington: Indiana University Press, 1983).

43. See, for example, Susan Mann, *The Talented Women of the Zhang Family* (Berkeley: University of California Press, 2007); Jill Lepore, *Book of Ages: The Life and Opinions of Jane Franklin* (New York: Vintage, 2014); and Lisa Brooks, *Our Beloved Kin: A New History of King Philip's War* (New Haven: Yale University Press, 2018).

44. See William J. Bouwsma's eponymous *A Usable Past* (Berkeley: University of California Press, 1990), in which he harkens back to Nietzsche's view of the study of history as something of a "public utility."

45. For other fields, see, for example, Dale Spender, *Women of Ideas and What Men Have Done to Them* (New York: Harper Collins, 1991); Joanna Russ, *How to Suppress Women's Writing* (Austin: University of Texas Press, 1983); Rosalind Miles, *Who Cooked the Last Supper?* (New York: Broadway Books, 2001); and *The Guerilla Girls Bedside Companion to the History of Western Art* (London: Penguin Books, 1998).

46. See, for example, Eileen Pollack's "Why Are There Still So Few Women in Science?," *New York Times Magazine*, 3 October 2013. She then wrote a book, *The Only Woman in the Room: Why Science Is Still a Boys' Club* (Boston: Beacon Press, 2015). See also Londa Schiebinger's Gendered Innovations project at Stanford.

CHAPTER ONE. MATHEMATICIAN AND PHILOSOPHER

1. *Journal Historique par E. J. F. Barbier,* or *Journal de Barbier,* quatrième série (Paris: Charpentier, 1885), May 1750, 438.

2. La Tour's telescopes were to be taken especially good care of, as he stipulated in his testament. There he bequeathed the best one to the acclaimed astronomer Bailly, and digressed to explain how looking through them helped him appreciate the antiquity of the universe and all its suns of different sizes and all the globes going around those suns, "including our little earth." See *Correspondance inédite de Maurice Quentin de La Tour,* ed. Jules Guiffrey and Maurice Tourneux (Paris: Charavay, 1885), 36–37. See also Jean-Dominique Augarde, "The Scientific Cabinet of Comte d'Ons-en-Bray and a Clock by Domenico Cucci," *Cleveland Studies in the History of Art* 8 (2003): 80–95, 80.

3. Elie-Catherine Fréron and Joseph de la Porte, *Lettres sur quelques écrits de ce temps,* XI (1753), 190.

4. See Elisabeth Badinter, *Les Passions intellectuelles* (Paris: Fayard, 2002), II, 107. She believes this was a kind of rebuke.

5. Meister still refers to her that way decades later. See Grimm, Diderot, Raynal, Meister et al., *Correspondance Littéraire*, ed. Tourneux (Paris: Garnier frères, 1877–82), XII (November 1779), 343.

6. Jean Levesque de Burigny, *Lettre de M de Burigny a M l'abbé Mercier . . .* (Paris: chez Valade, 1780), 26–27.

7. See Olivier Courcelle's excellent Clairaut website, Chronologie de la vie de Clairaut (1713–1765), www.clairaut.com.

8. Jacques Derrida, in books like *Politiques de l'amitié* (1994) and *The Work of Mourning* (2001), characterizes friendship as fidelity, finitude, and loss.

9. See, for example, comments by Georges Le Roy in his edition of Condillac's *Oeuvres philosophiques* (Paris: PUF, 1947–51). But in fact other scholars—Laurence L. Bongie, Olivier Courcelle, and Neil Jeffares—have brought to light some bits and pieces of her life, as will be discussed below.

10. Rudolf Wolf, "Auszug aus Johann II Bernoullis Reisjournal vom Jahre 1733," *Mitteilungen der naturforschenden Gesellschaft in Bern* (1851), 96–104, 103.

11. Troubling about this story are the facts that Pierre Rémond de Montmort died of smallpox in 1719, and that the Bernoullis' brother Nicolaus II is thought to have come to France only in 1725, and he died in 1726. Perhaps Nicolaus II made an earlier undocumented trip to France? Much more likely, the traveling brothers heard of the precocious Ferrand from their older cousin Nicolaus I who lived with them, and who would have met the young girl during many-month visits to Montmort's estate during their work together on game theory in 1709 and 1712–13. Ferrand lived nearby and later sold some fiefs to Rémond's son. See Baron J. de Baye, "Notes sur le Chateau de Montmort," *Revue de Champagne et de Brie* 16, 8e année (1884): 21–35, 24. But that Bernoulli mentioned meeting her then suggests that she was not just a neighborly visitor but rather involved somehow in the math the men were doing.

12. Montmort had met Newton twice and had sent him bottles of fine champagne. On Pierre Rémond de Montmort (1695–1719), see the many references in Richard S. Westfall, *Never at Rest: A Biography of Isaac Newton* (Cambridge, Cambridge University Press, 1980); J. B. Shank, *The Newton Wars and the Beginning of the French Enlightenment* (Chicago: University of Chicago Press, 2008); Lenore Feigenbaum, "The Fragmentation of the European Mathematical Community," in P. M. Harman et al., eds., *The Investigation of Difficult Things: Essays on Newton and the History of the Exact Sciences in Honor of D. T. Whiteside* (Cambridge: Cambridge University Press, 1992), chapter 15; Henry Guerlac, *Newton on the Continent* (Ithaca: Cornell University Press, 1981).

13. See Pierre Speziali, "Une correspondance inédite entre Clairaut et Cramer," *Revue d'histoire des sciences et de leurs applications* 8, no. 3 (1955): 193–237, 198. Cramer was living on the rue Guénégaud not far from Ferrand, so they might well have seen each other then.

14. See Mary Terrall, *The Man Who Flattened the Earth* (Chicago: University of Chicago Press, 2014), chapter 3, "Mathematics and Mechanics in the Paris Academy of Sciences." Terrall speaks of Newton becoming something of a fad among women, as does J. B. Shank in *The Newton Wars*, p. 11, but Ferrand was the genuine article.

15. See, most recently, the fine biographies of Du Châtelet by Judith P. Zinsser and her excellent chapter "The Many Representations of the Marquise Du Châtelet" in the volume she edited, *Men, Women and the Birthing of Modern Science* (De Kalb: Northern Illinois University Press, 2009), 48–67.

16. See *Mercure de France* (March 1736): 600, and AN MC ET XLVI/330, 30 December 1751, for the death dates of her sons.

17. *Journal de Bruxelles* (28 January 1738). There are many *mémoires* concerning the Vassé family. See at the Bibliothèque nationale de France, for example, 4-FM-12180, and 4-FM-12181 (1) and (2), protesting the "libels" written about the sister-in-law but detailing the complaints against her, including that she was a "furie capable de mettre toute la terre en combustion."

18. See her protection of la dame de Planström de Pelletot, http://www.clairaut .com/ncoDecembrecf1762p01pf.html. Maupertuis was involved in this effort also.

19. See Neil Jeffares's finding that her only sibling, a brother, died young, making her the sole heir of her father's domain. http://www.pastellists.com/Genealogies /Ferrand.pdf.

20. Camille Piton, *Marly-le-Roi, son histoire* (Paris: A. Joanin, 1904), 34, 245, 304, 362, 433. See the plan of Vassé's estate and some mentions of the king needing to pave the route in front of her house, granting her rights to water, to ice, etc. She frequented the lofty and the lowly, attended the wedding of two of her domestics, 356. See also AN, O¹ 96, fol. 141, 20 April 1752, in which Vassé secured water rights for a surgeon lodged in her *domaine*, perhaps brought there to treat the ailing Ferrand in her last months?

21. See AN MC ET LVIII/369, 20 and 30 January 1751, and XLVI/328, 21 April 1751.

22. Andrew Lang, *Companions of Pickle, Being a Sequel to "Pickle the Spy"* (London: Longmans, Green, 1898), 96. See also the article on Quentin de La Tour's Ferrand portrait by Neil Jeffares in the online version of his *Dictionary of Pastellists before 1800*, http://www.pastellists.com/Essays/LaTour_Ferrand.pdf, 9, and Frank J. McLynn, *Bonnie Prince Charlie* (London: Pimlico, 2003).

23. Alan Bray, *The Friend* (Chicago: University of Chicago Press, 2006), 6.

24. Lionel Gossman, *Medievalism and the Ideologies of the Enlightenment: The World and Work of La Curne de Sainte-Palaye* (Baltimore: Johns Hopkins Press, 1968), 36, 41, 54, 69.

25. See Kent Wright, *A Classical Republican in Eighteenth Century France: The Political Thought of Mably* (Stanford: Stanford University Press, 1997), 16, 21, and passim. See also Laurent-Pierre Bérenger, *Esprit de Mably et Condillac relativement à la morale et la politique* (Grenoble: Le Jay, fils, 1789), I, 11–13.

26. *Oeuvres de Voltaire*, ed. Beuchot, LXII (Paris: Lefevre, 1832), 123.

27. Clairaut's mathematician friend Charles Bossut believed that dissipation destroyed his constitution: "He was focused on dining and evenings, coupled with a lively taste for women, and seeking to take his pleasures into his day-to-day work, he lost rest, health, and finally life at the age of fifty-two." See *W. W. Rouse Ball, A Short Account of the History of Mathematics*, 4th ed. (London: Macmillan, 1908), 374.

28. See my discussions of how the Jansenism of the *parlementaires* was often an opposition strategy in Nina Rattner Gelbart, *Feminine and Opposition Journalism in Old Regime France: Le Journal des Dames* (Berkeley: University of California Press, 1987). See also Arnaud Orain, "The Second Jansenism and the Rise of French Eighteenth-Century Political Economy," *History of Political Economy* 46, no. 3 (2014): 463–490.

29. J. Bauquier, *Les provençalistes du dix-huitième siècle . . . lettres inédites* (Paris: Maisonneuve, 1880), 39 note 1, where La Curne's eulogist says he exchanged roses for pine needles, gold for lead, by turning his back on the classics of antiquity to study medieval chronicles.

30. In addition to Gossman's *Medievalism*, see also Geoffrey J. Wilson, *A Medievalist in the 18th Century: Le Grand d'Aussy* (Netherlands: Springer, 1975), 69–70, 279. Although La Curne's main writings appeared after Ferrand's death, he had already written a few short pieces.

31. Rex A Barrell, *Bolingbroke and France* (Lanham, Md.: University Press of America, 1988), 16.

32. On Pouilly's introduction of Newton's ideas into France, see Jean-Vincent Genet, *Une famille Rémoise au 18e siècle* (Reims: Imprimerie Coopérative, 1881), especially 27.

33. See Chaussinand-Nogaret, "Une élite insulaire au service de l'Europe : Les Jacobites au 18e siècle," *Annales* 28e année, no. 5 (September–October 1973): 1097–1122.

34. Lévesque de Pouilly, *Théorie des sentiments agréables* (Geneva: Barrillot et fils, 1747).

35. Online University of Michigan, *The Encyclopedia of Diderot & d'Alembert Collaborative Translation Project*, "Zèle (de religion)."

36. Testament de Mme de Vassé, AN MC ET XCII/715, 30 May 1768.

37. *Encyclopédie*, XV (1765), 24. Some have said he was referring here to Locke, but it was in fact the "modern" younger Frenchman.

38. John Nichols, *Literary Anecdotes of the Eighteenth Century*, 9 vols. (London: Nichols, Son, and Bentley, 1812–15), IX, 401–402.

39. *Correspondance générale d'Helvétius* (Oxford: Voltaire Foundation, 1981–2005), 5 vols., I, 291.

40. *Correspondance d'Helvétius*, IV, 292–295; I, 281–284, 294–297; I, 349.

41. There is a portrait of Graffigny commonly attributed to La Tour and a strong feeling among scholars of both the writer and the artist that such a portrait was done,

but it is still debated. See G. J. Mallinson, *Françoise de Graffigny, femme de lettres* (Oxford: Voltaire Foundation, 2004), 196.

42. See Michael Sonenscher, *Sans-Culottes: An Eighteenth-Century Emblem in the French Revolution* (Princeton: Princeton University Press, 2008), 378.

43. *Oeuvres de Turgot* (Paris: Belin, 1808), II, "Seconde Discours sur les progrès successifs de l'esprit humain," 52–92, especially 63.

44. R. Taton, "Esquisse d'une bibliographie de l'œuvre de Clairaut," *Revue d'histoire des sciences et de leurs applications* 6, no. 2 (1953): 161–168, 166, mention of November 1759 letter to Cramer. Condillac first appears on the list of censors in the *Almanach Royal* of 1752.

45. Wright, *A Classical Republican*. See also Wright's review on H-France of Sonenscher's *Sans-Culottes*.

46. *The Confessions of Jean-Jacques Rousseau*, trans. J. M. Cohen (New York: Penguin, 1953), 324–325.

47. Condillac, *Lettres inédites à Gabriel Cramer*, ed. Georges Le Roy (Paris: PUF, 1953), 12. See Jean le Rond d'Alembert, *Preliminary Discourse to the Encyclopedia of Diderot*, trans. Richard N. Schwab with the collaboration of Walter E. Rex (Chicago: University of Chicago Press, 1995), ix–lii.

48. Jean Sgard, ed., *Corpus Condillac, 1714–1780* (Geneva: Slatkine, 1981), 40, 113–115, 58–60.

49. Condillac, *Essay on the Origin of Human Knowledge*, trans. and ed. Hans Aarsleff (Cambridge: Cambridge University Press, 2001), 57–58.

50. Ibid., 47–49, 100.

51. Ibid., 11, 12, 41–44.

52. Jørn Schøsler, "La réception de Condillac dans les périodiques français du XVIIIe siècle," *Revue Romane* 27, no. 1 (1992): 121–139, especially 121–126.

53. Condillac, *Essay*, trans. Aarsleff, 171–172, 74, 199. He did praise Newton for his use of mathematical signs in communicating his ideas, 187.

54. Condillac refers to these "conversations" in the *Traité des sensations* (Londres et se vend à. Paris: Chez de Bure, 1754), "Dessein de cet ouvrage," 12, and in numerous letters.

55. Condillac, *Lettres inédites*, ed. Le Roy. I give page numbers to the Le Roy book, but I follow here the chronological reordering proposed by Piero Petacco of the letters edited by Le Roy. See Petacco, "Note sul carteggio Condillac-Cramer," *Belfagor* 26 (1971): 83–95. Le Roy places the "Mémoire" last in this series of exchanges, Petacco places it first. I have also reordered slightly on the basis of Jean Sgard's *Corpus Condillac,* putting Petacco's letter #5 as #2 but otherwise respecting his order.

56. Condillac, *Lettres inédites,* ed. Le Roy, 89.

57. Ibid., 95–96.

58. Ibid., 105–106.

59. Ibid., 107–108.

60. Ibid., 109.

61. Ibid., 108.

62. Ibid., 35–36. Petacco numbers this letter the fifth in the sequence, but Jean Sgard believes it is the second, and I agree. See *Corpus Condillac*, 129.

63. Condillac, *Lettres inédites*, ed. Le Roy, 43. Letter of 6 July 1747. Much later, Condillac will make reference to another work of Clairaut's, his *Éléments de l'algèbre* (1746).

64. Ibid., 44.

65. Condillac, *Essay*, 191–192.

66. Ibid., 61–62.

67. *Dissertation qui a remporté le prix propose par l'Académie royale des Sciences et Belles Lettres sur le system des monades, avec les pièces qui ont concouru* (Berlin: s.n., 1748).

68. Another translation is "How fine it is to be willing to admit in respect of what you do not know, that you do not know, instead of causing disgust with that chatter of yours which must leave you dissatisfied too." See Laurence Bongie's "Introduction" to *Les Monades*, in *Studies on Voltaire and the Eighteenth Century* (henceforth *SVEC*) 187 (1980): 50, note 2.

69. See Jean-Pierre Deschepper's review of Bongie's presentation of Condillac's *Les Monades*, *Revue philosophique de Louvain*, 81, no. 51 (1983): 508.

70. See Condillac's *Traité des animaux* (1755), in the chapter called "How man acquires the knowledge of God." The first note reads, "This chapter is almost entirely taken from a Dissertation I did several years ago which is printed in a collection of the Berlin Academy on which I did not put my name."

71. Speziali, "Une correspondance," 223.

72. Bibliothèque de Genève, MS fr, 657b, fol. 43: letter of May 1748. Hundreds of thousands of fighters were killed in this war, and even more civilians. The Treaty of Aix-la-Chapelle would not be signed until October, but the last holdout, Maastrich, surrendered 7 May just before Cramer's letter.

73. Bibliothèque de Genève, MS Suppl. 384, fols. 212–213: letter of 1 September 1748.

74. Condillac, *Lettres inédites*, ed. Le Roy, 52.

75. *Traité des systèmes, où l'on en démêle les inconvénients et les avantages.* Par l'Auteur de *l'Essai sur les origines des Connaissance Humaines* (La Haye: Neaulme, 1749), 437.

76. Ibid., 433, 440.

77. Ibid., 364, 376–378.

78. Ibid., 393–395.

79. Ibid., 396–397.

80. Jørn Schøsler, "La réception de Condillac dans les périodiques français," 125–129. *Systèmes* was reviewed in the *Mercure*, the *Journal de Trévoux*, the *Journal Helvétique*, the *Cinq Années Littéraires*, the *Mercure Suisse*, and the *Bibliothèque raisonnée*. All of these journals cautioned, however, that the author not slide into materialism. And Grimm was glad to see Condillac attack Leibnitz.

81. This La Tour portrait, incidentally, was one that the prince prized and wanted handled with care. See Andrew Lang, *Pickle the Spy, or the Incognito of Prince Charles* (London: Longmans, Green, 1897), 136.

82. Lang, *Pickle the Spy*, 79, note 1, and letters of 30 June and 23 July 1749, 82–83.

83. Condillac, *Lettres inédites*, ed. Le Roy, 53–54. Raynal, Baculard d'Arnaud, and others agreed in their misogynistic reviews. Only Fontenelle, the censor for the play, complimented both it and its author, an "amazone du parnasse" in her own right.

84. Colm Kiernan, "Science and the Enlightenment in Eighteenth Century France," *SVEC* 59 (1968): 131.

85. Diderot, *Lettre sur les aveugles à l'usage de ceux qui voient* (1749) was indeed filled with the highest praise for young Condillac as a philosopher on a par with Locke and Molyneux, and as a stylist, and more specifically for his *Essai* and his very new *Traité des systèmes*, which Diderot said spelled doom for all the "systématiques." See 97–100, 142–145, 150–151, 163, 181–185.

86. Diderot was in Vincennes from 24 July to 3 November 1749. Clairaut visited him on 1 August, the conciliatory letters were sent 10 August. See Courcelle's Clairaut website.

87. Condillac, *Lettres inédites*, ed. Le Roy, 59.

88. Ibid., 76–77. Here Le Roy dates the letter September, but that is not correct.

89. See Aarsleff's translation of the *Essai* where Condillac uses the word "sentiment" twenty-six times and the translator renders it variously as opinion, view, and thought.

90. Achille Le Sueur, *Maupertuis et ses correspondants* (Montreuil-sur-mer: Notre-Dame des Prés, 1896), 386–387.

91. Sgard, ed., *Corpus Condillac*, 132, full text on 147.

92. McLynn, *Bonnie Prince Charlie*, chapter 27, note 5.

93. Grimm, et al., *Correspondance Littéraire*, ed. Tourneux, XII (November 1779), 343.

94. See Speziali, "Une correspondance," letters of 5 August, 14 September, and 18 November 1750.

95. François Weil, "La correspondance Buffon-Cramer," *Revue d'histoire des sciences et de leurs applications* 14, no. 2 (1961): 97–136. Letters of 19 January and 4 August 1750.

96. Letter from Clairaut to Cramer, 2 February 1748, http://www.clairaut.com /n2fevrier1750p01pf.html.

97. Speziali, "Une correspondance," 229.

98. We know his address from the *Almanach Royal* of 1752, p. 364, which now listed him in his new capacity as royal censor. This is also the address he gives for himself in a letter to Maupertuis of 12 August 1750.

99. Laurence L. Bongie, "A New Condillac Letter and the Genesis of the *Traité des sensations*," *Journal of the History of Philosophy* 16, no. 1 (January 1978): 94, note 36.

100. Ibid., 93.

101. Le Sueur, *Maupertuis,* 390–391.

102. Speziali, "Une correspondance," 231, letter of 14 September 1750.

103. 25 November 1750, in Bibliothèque Centrale du Muséum National d'Histoire Naturelle (henceforth BCMNHN), MS KYPH 293. This MS is a collection of letters from various people to Cramer. The originals are in the British Museum.

104. Lang, *Pickle the Spy,* 97, letter of May 1750.

105. Lang, *Pickle the Spy,* 120. The author in question was Joseph-Louis Vincens de Mauleon de Causans who insisted he had squared the circle and was now writing philosophy in the form of a *Spectacle de l'Homme.*

106. Lang, *Pickle the Spy.* These letters, from 19 November 1750, 5 March 1751, 19 October 1751, and 21 October 1751, are on 115–116 and 120–122. For the entire correspondence between Ferrand, Vassé, and the prince, see 82–83, 97, 112–116, 120–122, 136, and 143–144.

107. Clairaut seems to have played some role in Malesherbes acquiring this post. See his letter to Cramer in November 1750 congratulating himself for accomplishing this, in Speziali, "Une correspondance," 233.

108. Le Sueur, *Maupertuis,* 391–394, letter of 25 June 1752, from Paris.

109. Testament de Mlle Ferrand, 8 February 1752, AN MC ET XCII/575.

110. Christian Albertan and Anne-Marie Chouillet, "Autographes et documents," in *Recherches sur Diderot et sur l'Encyclopédie* 20, no. 20 (1996): 174–175. D'Alembert's letter was written 27 December 1751, so although Ferrand would live another nine months, she was clearly on death's door several times before.

111. *Annonces, Affiches et avis divers* (1752): 568. See also Courcelle's Clairaut website.

112. AN MC ET XCII/578, Inventaire après le décès de Mlle Elisabeth Ferrand, 11 September 1752. I thank Kate Norberg for her help parsing this long document.

113. "Avis important au lecteur," in the original 1754 edition, is in I, iii–vi.

114. Isaac Newton, "Scholium" in the *Principia.* Condillac, "Avis important au lecteur," I, v–vi.

115. In volume I, 155–156 of the 1754 original, he explains one of the topics on which he and Ferrand did not see eye to eye, so to speak. It concerned the combination of taste, smell, and hearing, and the statue's understanding of which thoughts came from each of these senses. She thought he was too optimistic in his imaginings about the statue's powers of understanding at this point. Because of her hesitance at this juncture, he put in this note of caution.

116. This long homage to Ferrand, in volume 1 of the original edition of 1754, fills pages 1–16.

117. See Bongie, "Introduction" to Condillac's *Les Monades.*

118. See Julia Douthwaite, *The Wild Girl, Natural Man and the Monster* (Chicago: University of Chicago Press, 2002), 245, 76–77, 80, for some interesting thoughts on this.

119. *Traité des sensations*, 1754, II, 286–300. Condillac mentions Ferrand right at the start of the "Response," on 286.

120. See Condillac's letter to Formey of 22 February 1756, describing this unpleasant scene that occurred when the *Traité* was published, as quoted in Bongie, "New Condillac Letter," 86, note 11.

121. Buffon, *Histoire Naturelle*, III (1749), 364–370.

122. See note in *Traité*, I, 341, regarding Buffon's views of the hand in his *Histoire Naturelle*, III, 359.

123. BCMNHN, MS KYPH 293. Letter of 22 April 1747 where Helvétius invites Cramer to dinner at his hotel d'Anjou (rue Croix des Petits Champs) with Buffon.

124. Grimm et al., *Correspondance Littéraire*, III (November 1755), 111–113. Even Laurence Bongie, a present-day admirer of Condillac, calls his works "massively bolted together." See Bongie's "Diderot's femme savante," *SVEC* 166 (1977): 156.

125. See Keith Baker, "Un 'éloge' officieux de Condorcet: sa notice historique et critique sur Condillac," *Revue de synthèse* 88, nos. 47–48 (1967): 227–251.

126. J. L. Carr, "Pygmalion and the Philosophes: The Animated Statue in 18th century France," *Journal of the Warburg and Courtauld Institutes* 23 (July–December 1960): 239–255.

127. Deslandes's *Recueil de différents traités de physique et d'histoire naturelle*, first printed in 1736, had several subsequent editions. He had been to England, had even met Newton, and his knowledge of engineering had gotten him admitted to the Académie des Sciences as an "apprentice geometer."

128. Boureau-Deslandes, *Pygmalion ou le statue animé* (London: Samuel Harding, 1741), 47–48.

129. Anne Deneys-Tunney, "Le roman de la matière dans Pigmalion," in Béatrice Fink, et al., *Être matérialiste à l'âge des Lumières* (Paris: PUF, 1999), and Sébastien Drouin, "Allégorisme et matérialisme dans Pigmalion," *Oxford University Studies in the Enlightenment* (formerly *SVEC*) 7 (2003): 383–393.

130. See, for example, Yves Citton, "Fragile euphorie: La statue de Condillac et les impasses de l'individu," *SVEC* 323 (1994): 279–321.

131. See interesting discussions of this in Elizabeth Goldsmith, *Writing the Female Voice* (Boston: Northeastern University Press, 1989), 179ff., and Sylvie Romanowski, *Through Strangers' Eyes: Fictional Foreigners in Old Regime France* (West Lafayette, Ind.: Purdue University Press, 2005), 135–154.

132. See the discussion of Janet Altmans's analysis in John C. O'Neal, *Authority of Experience: Sensationist Theory in the French Enlightenment* (University Park: Penn State Press, 2010), chapter 5, "An Exemplary yet Divergent Text: Graffigny's *Lettres d'une Péruvienne*," 125–146, 131 note 10. Foucault, in *Birth of the Clinic*, pointed to two characteristic preoccupations of the Enlightenment, the "foreign spectator in an unknown country, and the man born blind restored to light." See O'Neal, *Authority*, 134 note 18.

133. O'Neal, *Authority*, 132.

134. Formey, *Bibliothèque Impartiale* 11 (May–June 1755): 342–359. See also Formey, *Souvenirs d'un citoyen* (Berlin: François de la Garde, 1789), II, 290–295.

135. *Journal de Trévoux* (March 1755): 641–669. See also Jeffrey Schwegman, "Etienne Bonnot de Condillac and the Practice of Enlightenment Philosophy" (PhD diss., Princeton University, 2008), which shows that this journal praised Condillac's earlier and later works too.

136. Grimm et al., *Correspondance Littéraire*, ed. Tourneux, II (1877); Raynal, "Nouvelles Littéraires," December 1754, 204, and Grimm, "Correspondance Littéraire," December 1754, 438–444.

137. Schwegman, "Etienne Bonnot de Condillac," 275–277.

138. Condillac, "Extrait raisonné du *Traité des sensations*," published originally with the *Traité des animaux* in 1755 but later augmented. These quotations are from the version in the posthumous edition of *Oeuvres de Condillac* (Paris: Ch. Houel, 1798).

139. Lang, *Companions of Pickle*, 64.

140. S. Eltis and W. Eltis, "Introduction" to Condillac's 1776 *Commerce and Government Considered in Their Mutual Relationship* (Indianapolis: Liberty Fund, 1997), says Vassé died in Condillac's home 2 June 1768. And that is the street address listed for him in the *Almanach Royal*. So the brothers at this time were living together.

141. Testament de Mme de Vassé, AN MC ET XCII/ 715, 30 May 1768.

142. Jean-Luc Malvache, "Correspondance inédite de Mably à Fellenberg, 1763–1778," in *Francia* 19, no. 2 (1992): 47–93. Mably describes how Vassé endured many months of treatments, "des nuits agitées, et pendant le jour des angoisses d'estomac. . . . Vous sentez combien les amis de Mde de Vassé doivent être inquiets, car elle n'est pas faites pour inspirer des attachements superficiels" (letter 13, début 1768). Mably describes more fully the constant vomiting and agitations (letter 14, March 1768) and then the end; he had "l'âme déchiré en voyant qu'on est séparé pour toujours d'une amie avec qui on était accoutumé à passer les moments les plus doux de la vie" (letter 15, summer 1768).

143. *Nécrologe des hommes célèbres de France par une société de gens de lettres* 16 (1781): 3–47, "Éloge de Condillac" par M. de Sivry.

144. See Baker, "Un 'éloge' officieux de Condorcet."

145. And Condillac did continue to liken mathematical rigor to precise language and clear thinking. See Robin Rider, "Measure of Ideas, Rule of Language: Mathematics and Language in the 18th Century," in Tore Frängsmyr et al., eds., *The Quantifying Spirit in the 18th Century* (Berkeley: University of California, 1990), 113–140, especially 115–120. This later emphasis on linguistic clarity was to influence Lavoisier's chemical nomenclature profoundly.

146. Quoted in James Roy Newman, ed., *The World of Mathematics* (Mineola, N.Y.: Dover, 2000), I, 268.

147. Is she clothed in a dressing gown, a bed jacket, and sleeping cap? Or is this a society outfit for receiving? See, for example, Christine Debrie and Xavier Salmon,

Maurice Quentin de La Tour, Prince des Pastellistes (Paris: Somogy, 2000); Elise Goodman, *The Portraits of Mme de Pompadour: Celebrating the Femme Savante* (Berkeley: University of California Press, 2000); and Neil Jeffares's entry on his website, http://www.pastellists.com/Essays/LaTour_Ferrand.pdf.

148. Mary Sheriff, "Disciplinary Problems in the History of Art, or What to Do with Rococo Queens," in *The Interdisciplinary Century,* ed. Julia Douthwaite and Mary Vidal, *Oxford University Studies in the Enlightenment* (formerly *SVEC*) 4 (2005): 79–101, especially 89, 93, 100–101.

149. She cannot be reading Colin Maclaurin's volume, translated in 1749 by Lavirotte as *Exposition des découvertes philosophiques de M le Chevalier Newton,* as John Heilbron suggests because Maclaurin's recto page, in addition to saying DE M NEWTON, also gives the book and chapter numbers (e.g., LIV. I CHAP. III) and so has a quite different format. The 1745 edition of Voltaire's *Éléments de la philosophie de Newton* has the spelling NEUTON instead. Coste's translation of the *Opticks, Traité d'Optique,* looks completely different. Newton's originals also look different; the *Opticks* has nothing but page numbers at the top of each page, and the *Principia* does not have the author's name at the top. Du Châtelet's translation of the *Principia,* though completed before her death in 1749, had not yet been published.

150. Kathryn Norberg of UCLA, a historian and expert in fashion and furniture in this period, has suggested this interpretation, which I find the most persuasive, in a private communication.

151. For different views, see Patricia Fara, *Newton, the Making of Genius* (New York: Columbia University Press, 2002), 136–137; Anthony Grafton, "The Ways of Genius," *New York Review of Books,* 2 December 2004; Adrian Bury, *Maurice-Quentin de La Tour, The Greatest Pastel Portraitist* (London: Charles Skilton, 1971), #46 (this is not a page number); and those cited in above.

152. See Deloynes, *Collection de pièces sur les beaux-arts (1673–1808) dite Collection Deloynes* (Paris: Bibliothèque nationale, 1980), V, catalogue #58, Abbé Garrigues de Froment, *Sentiments d'un amateur sur l'exposition des tableaux du Louvre et la critique qui en a été faite.* Deuxième lettre, 4 September 1753. This critic and several others (see #56, for example) said these portraits were coarser than usual and should not be looked at up close. But then Cochin insisted they were "également faits d'une manière ferme et hardie," #61.

153. Robin Nicholson, *Bonnie Prince Charlie and the Making of a Myth: A Study in Portraiture* (Lewisburg, Pa.: Bucknell University Press, 2002), 84–85.

154. See his letter of 13 July 1752 in Charles Desmaze, *Le Reliquaire de M. Q. de La Tour, peintre du roi Louis XV, sa corrrespondance et son œuvre* (Paris: Leroux, 1874), 20–21, where he speaks of having a crisis, and so much trouble with his portraits: "je me trouve dans un abattement, un anéantissement qui me fait craindre . . . je ne sait que devenir." Was this just his response to pressure? To Ferrand's being so ill that he had to really rush?

155. Gossman, *Medievalism*, 133.

156. Jean-Bernard Leblanc, *Observations sur les ouvrages de MM de l'Académie de peinture et de sculpture* (Paris: [s.n.], 1753), 37ff.

CHAPTER TWO. ASTRONOMER AND "LEARNED CALCULATOR"

1. Bachaumont, *Mémoires secrets* (London: John Adamson, 1783–89), 36 tomes in 18 volumes, XVI, 18–19 (9 October 1780). There are numerous editions of Bachaumont, even several published by Adamson, in which volume and page numbers differ, but the date of the entry remains the same in all of them.

2. Jean Haechler includes her in his group of headstrong subjects in *Les Insoumises: 18 portraits de femmes exceptionelles* (Paris: Nouveau Monde, 2007), 141–150.

3. Joseph-Jérôme Lalande, *Bibliographie astronomique avec l'histoire de l'astronomie depuis 1781 jusqu'à 1802* (Paris: De l'Imprimerie de la République, An XI. = 1803), 676.

4. Lalande and the members of the Béziers academy referred to the high quality of these now-lost astronomical "mémoires."

5. Elisabeth Badinter, *Les Passions intellectuelles* (Paris: Fayard, 2002), II, 260. This and other cited letters are from a private archive belonging to M. Jean-Denis Bergasse of Béziers, to which Badinter was allowed access.

6. For examples of Lepaute's fluent, comfortable handwriting, see p. 70 of Elisabeth Badinter, "Un couple d'astronomes: Jérôme Lalande et Reine Lepaute," *Bulletin de la Société archéologique, scientifique et littéraire de Béziers*, dixième série, I (2004–2005): 70–76.

7. Lalande, *Bibliographie astronomique*. Pages 677–678 refer to these letters from Clairaut but the originals seem lost.

8. See the interesting documents unearthed by Alain Demouzon, a descendant of the Lepaute family, on his genealogical website called "Au fil du temps." See "De la bruyère dans l'étable." Last revised 2012. http://alain-demouzon.fr/Alain_DEMOUZON _%E2%80%93_site_officiel/Au_fil_du_temps.html.

9. Etable continued to define himself in relation to the Duchesse de Berry long after her death, as seen in Demouzon's documents.

10. Lalande, *Biblio. astro.*, 676.

11. A Lepaute clock is still there on the façade, but as those made by Jean-André have the name in script, this with its bold capital letters is clearly a more modern incarnation. It does, however, say H.ger DU ROI at the bottom.

12. J. Monnier et al., *Philibert Commerson, le Découvreur du Bougainvillier* (Châtillon-sur-Chalaronne: Association Saint-Guignefort, 1993), 113.

13. *Réflexions de M. LeRoy l'aîné, fils, sur un Écrit intitulé "Copie d'une Lettre."* For examples of more such accusations and the retorts, see the pamphlet war between LeRoi and Lepaute at the BnF, cotes V-8736 through V-8738. For some examples of defenses by Lalande, see *Mercure de France* (August 1751): 156–173, and *Mercure* (July 1755):

183–192. For a response to one of Beaumarchais's attacks, see *Réponse du Sieur Le Paute, . . . à une lettre du Sieur Caron fils* (Paris: veuve David, 1753).

14. *Copie d'une letter écrite à M. le Duc de ***, par le Sieur Lepaute, Horlogier du Roy* [sic] *au Palais de Luxembourg: servant à la justification dudit Sieur Lepaute, contre différentes imputations du Sieur le Roi, fils ainé du Sieur Jullien le Roi, Horlogier du Roy.* This nine-page pamphlet is dated January 1752.

15. *Réflexions de M. LeRoy l'ainé, fils.*

16. For these deliberations, see AAS, Dossier Beaumarchais and Dossier Lepaute. See also the *Pochette de séance* of 11 June 1755. Elizabeth Sarah Kite, *Beaumarchais and the War of American Independence* (Boston: R. G. Badger, 1918), I, has many interesting quotes from Caron's feud with Lepaute, although she provides no footnotes. Haechler, in *Les Insoumises,* also gives a spirited account of the conflict, 142–144.

17. Lalande, *Biblio. astro.,* 677.

18. This *Réplique du sieur Lepaute . . .* is bound in BnF, cote V-8738, cited in note 13.

19. Lalande, *Supplément pour le Dictionnaire des Athées,* 11, in Sylvain Marechal, *Dictionnaire des Athées,* 2nd ed., expanded (Brussels: chez l'éditeur, 1833).

20. See, in Wikimedia Commons, https://commons.wikimedia.org/wiki/File: Hotel_de_Cluny_XIX_century.jpg, a nice engraving of this marine observatory, now demolished. The similar one on the top of the Luxembourg Palace is also gone.

21. Constance de Salm, *Éloge Historique de M de la Lande* (Paris: Sajou, 1810), 29. See also Helène Monod-Cassidy, "Un astronome-philosophe, Jérôme de Lalande," *SVEC* 56 (1967): 907–930.

22. Ken Alder, *The Measure of All Things* (New York: Simon and Schuster, 2002), 79, 309–310.

23. Quoted in *Rencontre de l'Ain,* 13e année, no. 50 (October 1982). In AAS, Dossier Lalande.

24. Jean-Pierre Luminet has written an entire novel, *Le Rende₂-vous de Vénus* (Paris: J. C. Lattès, 1999) on this "romance," which, though based on some interesting research, has not been welcomed by Lalande scholars. But Elisabeth Badinter also took the couple seriously in her article, "Un couple d'astronomes."

25. Unlike Diderot, Lalande was disdainful of artisans, although he was obviously not free to say this about M. Lepaute. For more on his true feelings, see F. Gordon and P. N. Furbank, *Marie Madeleine Jodin, 1741–1790; Actress, Philosophe and Feminist* (London: Routledge, 2001), 3–4.

26. AAS, Dossier Lepaute.

27. Demouzon, "Au fil du temps."

28. Jean-André Lepaute, *Traité d'horlogerie* (Paris: J. Chardon, 1755), 33, 138, 205, for example. There would be two more editions of the *Traité,* in 1760 and 1767.

29. Badinter, "Un couple d'astronomes."

30. Letter from Lalande to Bonnet of 11 February 1760, quoted in Badinter, *Passions,* II, 256 note 5.

31. Lalande, *Biblio. astro.,* 679. The bride was Jacques Chardon's daughter, Marie Thérèse Victoire. Lalande said this was arranged by Mme Lepaute.

32. Lepaute, *Traité,* vi.

33. Ibid., planches XVI and XV.

34. Ibid., 192.

35. Ibid., 282. The three-page table, "calculée par Madame Lepaute," is on xx–xxii. "Table de la longueur que doit avoir un Pendule simple pour faire en une heure un nombre de vibrations quelconque, depuis 1 jusqu'à 18000."

36. Tardy, *Dictionnaire des Horlogers français* (Paris: chez les libraires associés, 1971–1972), II, 378–384. There is a list of some clocks with their prices in Jean-André Lepaute's *Descriptions de plusieurs ouvrages d'horlogerie* (Paris, 1764). See also Galignani's *New Paris Guide* (Paris: A. and W. Galignani, 1830), which lists numerous Lepaute clocks in the Palais Bourbon, the Elysée Bourbon, and the Palace of Saint Cloud, 157, 201, 204, 219, 224–225, 672–673.

37. Gabriel-Joseph Lepaute, *Notice sur la Famille Lepaute* (Paris: Paul Dupont, 1869), 13.

38. See Paola Bertucci, *Artisanal Enlightenment: Science and the Mechanical Arts in the Old Regime* (New Haven: Yale University Press, 2017), 153, 160.

39. See, for example, Londa Schiebinger's "Maria Winkelmann at the Berlin Academy: A Turning Point for Women in Science," *ISIS* 78, no. 2 (June 1987): 174–200. This extremely talented woman made a name for herself in astronomy only as long as her husband Gottfried Kirch was alive, after which she was mocked and excluded.

40. See, for example, the *Journal de Trévoux* (November 1757): 2850–2863.

41. Quoted in Mark Littman, *Planets Beyond: Discovering the Outer Solar System* (New York: Dover, 2004), 29.

42. Lalande, *Biblio. astro.,* 677–678. Earlier, 466, Lalande had also referred to working on the comet calculations "with such assiduity that I became sick." Jacques Babinet, in *Études et lecture sur les sciences d'observation et leur applications* (Paris: Mallet-Bachielier, 1858), V, 82, called this Lalande's "arithmetic fever."

43. See Craig B. Waff, "The First International Halley Watch: Guiding the Worldwide Search for Comet Halley, 1744–1759," in *Standing on the Shoulders of Giants: A Longer View of Newton and Halley,* ed. Norman J. W. Thrower (Berkeley: University of California Press, 1990), 373–411, 409 note 68.

44. Waff, "First International," 373–411.

45. Curtis Wilson, "Clairaut's Calculation of the Eighteenth-Century Return of Halley's Comet," *Journal of the History of Astronomy* 24 (1993): 1–15.

46. Quoted in Camille Flammarion, *Popular Astronomy: A General Description of the Heavens* (London: Chatto and Windus, 1894), 486 note 1.

47. *Tables astronomiques de M. Halley . . . Et l'histoire de la comète de 1759* de M. Lalande (Paris: Durand, 1759), 110.

48. Ferner wrote in his journal an account of this visit in Paris. See http://www .clairaut.com/n1avril1761p03pf.html, on Olivier Courcelle's excellent Clairaut website, www.clairaut.com.

49. Jean-Marie Homet, *Le Retour de la Comète* (Paris: Imago, 1985).

50. C. Wolf, *Histoire de l'Observatoire de Paris de sa fondation à 1793* (Paris: Gauthier-Villars, 1902).

51. Harry Woolf, *Transits of Venus: A Study of Eighteenth-Century Science* (Princeton: Princeton University Press, 1959), chart on 139.

52. See *Histoire de l'Académie des Sciences* (1757): 77–99 for Lalande's upcoming role in the 1761 transit, and *Mémoires de l'Académie royal des sciences* (1769): 417–425, for the transit of that year.

53. This letter is quoted in full in Gabriel-Joseph Lepaute, *Notice sur la Famille Lepaute*, 37ff.

54. See the entries on Gouilly on Courcelle's website, at http://www.clairaut.com /ncocjuincf1757p02pf.html.

55. Jeanne Peiffer, "Gendered Working Realities in Enlightenment Mathematics," CNRS (Centre national de la recherche scientifique), EWM (European Women in Mathematics), 2011.

56. Luminet, *Le Rendez-vous*, chapter 6, "Le Retour de la Comète."

57. *Oeuvres Complètes de Diderot. . .* , ed. Jules Assezat and Maurice Tourneux (Paris: Garnier, 1875), VI, 473–476, "Notice sur Clairaut."

58. See, for example, Bachaumont, *Mémoires secrets* (London: John Adamson, 1777–83), III, 204 (19 April 1767).

59. Olivier Courcelle, "La mathématicienne la moins connue du monde (II)"— Images des Mathématiques, CNRS, 2011. See Courcelle's website, www.clairaut.com, for the entries on Gouilly.

60. Lalande, *Biblio. astro.*, 679.

61. Badinter, *Les Passions intellectuelles*, cites these letters from the Bibliothèque publique et universitaire de Genève, II, 258.

62. C. Wolf, *Histoire de l'Observatoire de Paris*, chapter 14.

63. We know Lalande's addresses from the yearly *Almanach Royal* which listed them for all members of the Académie. Addresses for the Lepautes can be found in various works on ornamental clocks and their mechanisms. The two sets of addresses match up.

64. See, for example, *Connaissance des temps* (henceforth *CDT*) of 1753, before their tenure, and that of 1777, after their departure.

65. *CDT* (1760): 212ff.

66. *CDT* (1763): 213.

67. On Harrison's invention, see Dava Sobel, *Longitude* (New York: Walker, 1995).

68. Lalande, "Voyage en Angleterre," Bibliothèque Mazarine, Ms 4345. See, for example, 115.

69. Badinter, "Un couple d'astronomes," 73.

70. "Passage de l'ombre de la lune au travers de l'Europe dans l'Éclipse de Soleil centrale et annulaire qui s'observera le 1e avril depuis le Cap St Vincent extremité méridionale de l'Espagne, jusqu'au Cap Wardhus dans la Mer Glaciale, calculé par Mme le Paute de l'Académie de Béziers, gravé par Mme Lattré, gravé [this was the ornamental cartouche] par Mme Tardieu. A Paris chez Lattré Graveur, rue St Jacques au coin de celle de la Parcheminerie, à la Ville de Bordeaux, avec privilège du Roi [1764] . . . 47x51 cm." Several copies of this work are now digitized on the BnF's website Gallica. The color of the original, however, is very faint.

71. "Explication de la Carte qui représente le Passage de l'Ombre de la lune au travers de l'Europe. . . . Presentée au Roi le douze Août 1762. Par Madame le Paute de l'Ac. Roy des Sci de Béziers." 8 pages. This tiny pamphlet is bound with others, cote VP-4712, at BnF. This was entirely her work, but a longer 1764 version carried ads in the back for Lalande's publications on the past (1761) and future (1769) transits of Venus.

72. See Mary Terrall, *The Man who Flattened the Earth* (Chicago: University of Chicago Press, 2002).

73. These passages and those in the following paragraphs are from the 1764 edition of the "Explication de la Carte."

74. Lalande, *Biblio. astro.*, 678. See also Guy Boistel, "Nicole-Reine Lepaute et l'Hortensia," *Cahiers Clairaut* 108 (Winter 2004): 13–17.

75. This figure was altered by a few minutes over the two years between the explanatory booklet of 1762 and the actual eclipse, showing that as the event approached Lepaute continued to update and refine her calculations.

76. Bibliothèque de l'Observatoire de Paris, Papiers Delisle, MS A3–6, piece 29.10 I (on microfilm role 269–270 [A3:6–8]), "Figures des 12 phases principales de la grande Éclipse de Soleil qui s'observera le 1 avril 1764. Calculées pour Paris par Madame Lepaute de l'Académie Royale des Sciences de Béziers. A.P.D.R. Chez Lattré rue St Jacques près la fontaine St Severin, Permis de Graver et distribuer ce 22 mars 1764, De Sartine." This image has recently appeared on Gallica.

77. This eclipse map is reproduced in Peter Barber, ed., *The Map Book* (New York: Walker, 2005), 220–221, where the name of Mme Tardieu can be more easily seen under the cartouche.

78. *Journal de Trévoux* (June 1762): 1534. The article spans pages 1529–1534.

79. Montucla and Lalande, *Histoire des mathématiques* (1802), IV, 98. There are four pages devoted to Lepaute in this book, a long rave of her eclipse maps of 1764, especially the one that was by three women, and a discussion of her invaluable work on the comet calculations. This includes the revelation that it was raining so hard in Paris that the 1764 eclipse, after all that, was not visible there.

80. *New-York Mercury* 665 (14 May 1764): 1. Headline "Utrecht, February 23" in News/Opinion section.

81. See Geoff Armitage, *The Shadow of the Moon: British Solar Eclipse Mapping in the 18th century* (Tring, Herts, Eng.: Map Collector Publications, 1997), for an image of Desnos's chart.

82. *Journal de Trévoux* (February 1764): 147–148. Quoted in Mary Sponberg Pedley, *The Commerce of Cartography: Making and Marketing Maps in Eighteenth-Century France and England* (Chicago: University of Chicago Press, 2005), 213 and 272 notes 35–37. As Pedley explains, a former employee of Lattré, the geographer Rizzi-Zannoni, now worked for Desnos, and had probably stolen the copper plate with Lepaute's chart, which the police confiscated, although Desnos protested that Rizzi-Zannoni "had no need for the wings of others to lift himself to the stellar regions."

83. *Journal des sçavans* (May 1766): 306 and (October 1766): 644–657, especially 653 and 657 where the "Académicienne de Béziers" is defended. The date of the letter is 24 May 1766. Its author, one Trébuchet, had an ongoing feud with the astronomer Delisle in the newspapers earlier on, and seemed to enjoy such controversies. See Woolf, *Transits of Venus*, 54–58.

84. Jean Etienne Montucla, and now Lalande, *Histoire des mathématiques* (Paris: Henri Agasse, 1802), IV, 97. Montucla, who had started this many-volume work, died in 1779, so subsequent volumes were by Lalande.

85. *CDT* (1764): 204–206.

86. *CDT* (1765): 222–251.

87. Paola Bertucci, "The In/visible Woman: Mariangela Ardinghelli and the Circulation of Knowledge between Paris and Naples in the Eighteenth Century," *ISIS* 104 (2013): 226–249.

88. *CDT* (1766): 232–236.

89. *Journal des sçavans* (May 1766): 291–295. In 1854 Michaud's *Biographie universelle* says Lepaute's map is beautiful, well executed, and that all who own an original one value it highly. See vol. 24, 218 note 2.

90. As quoted in Woolf, *Transits of Venus*, 64.

91. Lalande, *Biblio. astro.*, 679.

92. Ibid., 678. Dagelet was born in 1751 so was seventeen, not fifteen, when he came to Paris in 1768.

93. See, for example, her signature in "Un couple d'astronomes," 70.

94. Lalande, *Biblio. astro.*, 680.

95. See Meghan Roberts, "Learned and Loving: Representing Women Astronomers in Enlightenment France," *Journal of Women's History* 29, no. 1 (Spring 2017): 14–37.

96. Johann III Bernoulli on his 1768–69 trip to France, in "Lettres astronomiques" in *Oeuvres de Bernoulli* (Berlin: Chez l'auteur, 1771), letter 12.

97. Bernoulli, "Lettres," letter 13. In a later list Bernoulli compiled of astronomers

still living and active in 1776, he would list "Mme le Paute" (under the letter P for Paute), 23, but otherwise he does not seem to mention her.

98. James Bruce, et al., *Travels to Discover the Source of the Nile in the Years 1768 . . . 1773* (Edinburgh: G. G. J. and J. Robinson, 1790), I, lx.

99. See the website called *Auprès de nos racines—Blog de généologie*, "Lepaute-Dagelet: un astronome au service de Lapérouse," https://www.aupresdenosracines .com/lepaute-dagelet-un-astronome-au-service-de-laperouse.

100. Barber, *The Map Book*, incorrectly attributes this to Mme Lepaute, 220. The British Library integrated catalogue cites it correctly as Dagelet's. See "Figure de l'Eclipse de Soleil du 24 juin 1778 ou l'on voit les phases de cette éclipse pour tous les pays de la Terre, calculée par M le Paute Dagelet" (Paris: Lattré, 1778).

101. Lapérouse named it Isle Dagelet, and this name continued to be used until after World War II when the volcanic island, east of South Korea, became Ullûngdo, today a popular tourist attraction.

102. This information is from a notebook left by Bonaparte's classmate, Alexandre des Mazis, found in the appendix to Paul Bartel, *La Jeunesse inédite de Napoleon* (Paris: Amiot-Dumont, 1954).

103. AAS, Dossier Dagelet, Joseph Lepaute.

104. Bachaumont, *Mémoires secrets* (1777–1783), VI, 369–370, 372–375, 378–379 (6, 9, 13, 14, and 27 May).

105. *Réflexions sur les comètes qui peuvent approcher de la terre* (Paris: chez Gibert, 1773).

106. "Avertissement" in *Réflexions sur les comètes*.

107. *Réflexions sur les comètes*, 19–20, 31–32.

108. [Voltaire], *Lettre sur la prétendue comète*, dated 17 May, circulated as a separate pamphlet with the imprint Lausanne on that day and was later printed in the 1 June 1773 issue of the *Journal Encyclopédique*.

109. Guillaume Bigourdan, *L'Astronomie à Béziers: L'Observatoire; La Querelle Cassini-Lalande* (Paris: Imprimerie nationale, 1927).

110. *CDT* (1775): 335–337.

111. AN, O¹ 123, fols. 152–153, 27 September 1776.

112. Lalande, *Biblio. astro.*, 539.

113. *Éphémérides des mouvemens célestes pour le méridien de Paris*, 8, iv.

114. See his long manuscript on this subject, at the Sorbonne réserves, Fonds Victor Cousin, MS 99.

115. *Dictionnaire historique, littéraire et bibliographique des Françaises, et des étrangères naturalisées en France . . . par Mme Fortunée B. Briquet* (Paris: Treuttel et Würtz, 1804), 207.

116. See Peiffer's "Gendered Working Realities."

117. Lalande, *Biblio. astro.*, 677.

118. As quoted in David Smith, "Nouveau Regards . . . Mme Du Châtelet et Saint

Lambert," in Terry Pratt and David McCallam, eds., *The Enterprise of Enlightenment: A Tribute to David Williams* (Oxford: Peter Lang, 2004), 339.

119. Grimm et al., *Correspondance Littéraire*, XI (March 1777): 436–437.

120. In *Souvenirs de la marquise de Crequy, 1710–1800* (Paris: Fournier, 1834), especially 108–119, but "la divine Emilie" is referred to throughout with unremitting sarcasm.

121. Voltaire thanked Du Bocage for her kindness when Du Châtelet died, in a letter of 12 October 1749.

122. Badinter, *Les Passions intellectuelles*, II, 320.

123. Louis-Sébastien Mercier, *Tableau de Paris*, nouvelle édition corrigée et augmentée (Amsterdam: s.n., 1782), IV, 25–26.

124. Louise Elisabeth Félicité Pourra de la Madeleine, Mme Du Pierry (née en 1746 à la Ferté-Bernard). See Jérôme Lalande, *Lalandiana: Lettres à Mme Du Pierry et au juge Honoré Flaugergues*, ed. Simone Dumont and Jean Claude Pecker (Paris: J. Vrin, 2007), I, 38, including the gushing letter written by Lalande from England on 29 July 1788. But Lalande had been in love with her since April 1779.

125. See Isabelle Lémonon, "Gender and Space in Enlightenment Science: Mme Dupiéry's Scientific Work and Network," in *Domesticity and the Making of Modern Science*, ed. Donald L. Opitz et al. (Cham, Switzerland: Springer, 2016), 41–60.

126. It was said by many that Euler lost his vision for similar reasons, although others attributed it to botched cataract operations. See Babinet, *Etudes et lectures*, V, 82.

127. Lalande, *Biblio. astro.*, 679.

128. Simone Dumont, *Un Astronome des Lumières: Jérôme Lalande* (Paris: Vuibert, 2007), 148.

129. Archives de Paris, *État civil reconstitué, Paris*, 5 Mi 1 1127, Note de décès.

130. Archives de la Seine, fiches de décès, bobine 5 m 12/544. He died in paroisse Saint Germain l'Auxerrois and was buried in that church.

131. *Journal de Paris* (29 December 1788): 1556–1557, "Nécrologie."

132. Elizabeth Conner, "Mme Lepaute, an 18th Century Computer," *American Astronomical Society of the Pacific*, leaflet #189 (November 1944).

133. Lalande, *Biblio. astro.*, 680.

134. Ibid., 679–680.

135. See Catherine Voiriot, "Guillaume Voiriot, portraitiste de l'Académie royale de Peinture et de Sculpture," *Bulletin de la Société de l'histoire de l'art français*, année 2004 (2005): 111–157, 124–129. Her other article on this is forthcoming.

136. Fragonard too did a nice painting of Lalande. These were the exceptions.

137. I devoted a chapter to La Louptière in my book *Feminine and Opposition Journalism in Old Regime France: Le Journal des Dames* (Berkeley: University of California Press, 1987), 67–94.

138. Lalande, *Biblio. astro.*, 680. The flower's name was changed by A. L. de

Jussieu to Hortensia, which led some to think that that was one of Lepaute's names which it was not. In France it is still called the Hortensia.

139. Lalande, *Biblio. astro.*, 681.

140. Badinter, "Un couple d'astronomes," 75.

141. Lalande, *Biblio. astro.*, 681.

142. This never stopped. Michaud's 1819 *Biographie universelle* commented, "Mme Lepaute, douée de tous les avantages extérieurs, portait dans la société cette politesse et cette fleur d'esprit que semblent exclure les études." The 1849 *Biographie universelle* spoke of Lepaute as "un modèle de dévouement conjugal," and a 1898 issue of *La Femme* reassured that "la famille n'est pas en péril parce que les filles s'adonnent aux mêmes études que les garçons et osent aspirer à des carrières libérales et scientifiques. . . . Mme Lepaute nous en donne une très noble preuve" (29–30). A recent article titled "The Comet Calculator" shows a cheery illustration of Lepaute's familial loyalty, again upholding this view of easy harmony between her duties. The plot of course was thicker. https://cosmosmagazine.com/mathematics/the-comet-calculator-nicole-reine-lepaute.

143. It was then called rue du Dauphin, but its name changed to rue Saint Roch.

CHAPTER THREE. BOTANY IN THE FIELD AND IN THE GARDEN

1. Commerson letter to Bernard de Jussieu 6 February 1770, AAS, Dossier Bernard de Jussieu.

2. On Bernard's relentless search for laws, see the *Éloge* of him by Condorcet in AAS, Dossier Bernard de Jussieu. This is also discussed in Roger L. Williams, *Botanophilia in Eighteenth-Century France: The Spirit of the Enlightenment* (Dordrecht: Springer, 2001). Bernard's system was not actually published until after his death, by his nephew A. L. de Jussieu in his 1789 *Genera Plantarum*.

3. Natania Meeker and Antónia Szabari, "Inhabiting Flower Worlds: The Botanical Art of Madeleine Françoise Basseporte," *Arts et Savoirs*, 6 | 2016, mis en ligne le 29 August 2016, https://doi.org/10.4000/aes.757.

4. There is a full, fascinating eulogy of Basseporte in the *Nécrologe des hommes célèbres de France par une société de gens de lettres*, XVI (Paris: Moutard, 1781), 159–187. As this periodical was rare and hard to find, the obituary was reprinted in the *Revue universelle des arts* 13 (1861): 139–147. The *Revue* is online and easily accessible, so I will henceforth refer to the page numbers in that reprint of Basseporte's eulogy. Rousseau is quoted in *Nécrologe*, 170, in the reprint in *Revue universelle*, 142. The poet Mlle Adèle Sauvan, later Madame Legouvé, wrote a poem with a similar sentiment regarding Basseporte: "Un jour, un même jour, de la rose nouvelle / Unit la tombe et le berceau / Mais prêtez-lui votre pinceau / Et vous la rendrez immortelle."

5. Both of them were to become very attached to her when she returned from her round-the-world travels. I have reconstructed this account of Barret's childhood from a genealogy by Sophie Miquel, private communication, December 2018.

6. *Déclaration de grossesse* 22 August 1764 in Digoin, seventeen miles from Toulon

sur Arroux—Archives Départementale de Saône-et-Loire, étude du Notaire Labeloyne, 3E 22802.

7. F. B. de Montessus, *Martyrologie et Biographie de Commerson* (Chalon-sur-Saône: L. Marceau, 1889), 5–6.

8. Ibid., 212. This letter is dated 9 November 1765.

9. Henriette Dussourd, *Jeanne Baret, 1740–1816: Première femme autour du monde* (Moulins: Pottier, 1987), 24.

10. See the little pamphlet *Testament singulier de M. Commerson, docteur en médicine, médecin botaniste et naturaliste du Roi, fait le 14 et 15e décembre 1766* (Paris: s.n., 1774).

11. Paul-Antoine Cap, *Philibert Commerson, naturaliste-voyageur* (Paris: Victor Masson, 1861), 84. Letter to brother-in-law Beau, 20 October 1766.

12. This flower had already been named the Rose of Japan and had acquired several other appellations, so Commerson's name for it did not stick. But Jean-Marie Pelt, in *Canelle et Le Panda: Les naturaliste explorateurs autour du monde* (Paris: Fayard, 1999) writes that Commerson thought he had discovered it and found varieties in 1771 on Bourbon and in February 1773 on Isle de France one month before dying.

13. *L'Herbier du monde,* ed. Philippe Morat et al. (Paris: Les Arènes/L'iconoclaste, 2004), 78.

14. Pierre-Antoine Véron was Lalande's student who came along on Bougainville's voyage but died; his flower, *Veronia Tristifloria,* was star-shaped but only lasted a few hours and had the "perfume of tears." Commerson named plants for many other male associates, and for his wife, and of course for Barret, about which more later. For some of these detailed dedications, see BCMNHN, MS 198 (YL 51). They are each a full page of tiny writing, in Latin.

15. Lalande, "Éloge de Commerson," *Journal de Physique* 5, pt. 1 (February 1775): 89–120. See 118.

16. Commerson's *Inventaire après décès* (death inventory) shows this third-floor apartment (second floor French style) looked out on the courtyard, was filled to the brim with the couple's natural history collections, and consisted of a room used as a kitchen and probably also dining room, a bedroom, and a study. See AN MC ET LXXXIV/537.

17. André Role, "Vie aventureuse d'un savant, Philibert Commerson, martyr de la botanique, 1727–1773: bi-centenaire de la mort de Philibert Commerson" (Réunion: Académie de la Réunion, 1973).

18. Samuel Pasfield Oliver, *The Life of Philibert Commerson, D.M., Naturalist Du Roi: An Old-World Story of French Travel and Science in the Days of Linnæus* (London: J. Murray, 1909), 53. This letter was dated 11 April 1758.

19. Cap, *Philibert Commerson,* 158–159.

20. Louise Audelin, "Les Jussieu; Une dynastie de botanistes au 18e siècle" (thesis, École des Chartes, March 1987), 330–352. Jacques Barbeu Dubourg's *Le Botaniste Français* (Paris: Lacombe, 1767), I, 215 note 1, recorded the success of Jussieu's snake bite remedy on one such occasion in July 1748.

21. *Almanach Royal* (1736): 326. Here she is shown as holding Aubriet's *survivance*. This changed at his death when she assumed the full position herself.

22. *Nécrologe,* 170–171, note. In *Revue Universelles des Arts* 13 (1861): 142.

23. See, for example, the *Journal du Citoyen* (1754): 173.

24. Lalande, *Éloge de Commerson,* especially 105, 111.

25. Cap, *Philibert Commerson,* 84. Letter to brother-in-law Beau, 20 October 1766.

26. Louis Antoine de Bougainville, *Voyage autour de Monde par le frégat du Roi la Boudeuse et la flûte Étoile en 1766, 1767, 1768 et 1769* (Paris: Saillant et Nyon, 1771), 253–254.

27. Nicole Crestey, "L'Affaire Jeanne Barret," speech delivered 29 April 2008, Réunion.

28. See Cap, *Philibert Commerson,* 87, 91–94.

29. The manuscript version of this testament was left with their trusted friend Vachier and deposited with his notary Regnault in Paris on 30 August 1773, as soon as he received the news from Barret that Commerson had died. The significance of her nickname Bonnefoi is not known, but Commerson would refer to it later when he named a plant for her.

30. Leon Bultingaire, "Les Peintres du Jardin du Roy au XVIIIe siècle," *Archives du Muséum,* 6e série, 3: 29–32 on Madeleine Basseporte.

31. The geographer Edmé Mentelle wrote her *éloge* in the *Nécrologe.*

32. Henri Bourin, *Paul-Ponce-Antoine Robert (de Sery)* (Paris: Alphonse Picard, 1907). See page 9 for the location of the house they shared.

33. Neil Jeffares, *Dictionary of Pastellists before 1800* online, http://www.pastellists .com/Articles/Basseporte.pdf.

34. *Nécrologe,* 162. *Revue Universelle,* 140. See also *Journal de Rosalba Carriera pendant son séjour à Paris en 1720–1721,* trans. Alfred Sensier (Paris: Techener, 1865).

35. *Oeuvres Complètes de Madame la Comtesse de Genlis* (Brussels: De Mat, 1828), XV, 46–47.

36. Abbé l'Attaignant, *Chansons et autres poésies posthumes* (Paris: Duchesne, 1780), 199–200.

37. Honoré Gabriel de Riquetti Comte de Mirabeau, *Catalogue des livres de la bibliothèque de feu M. Mirabeau l'ainé* (Paris: Rozet et Belin, 1791), 176.

38. *Recueil de dessins de fleurs,* BnF Département des Estampes, RESERVE JD-33-PET FOL. See especially number 7. One hundred and seventeen informal flower paintings done by Basseporte or those in her "circle" are at the Morgan Library in New York.

39. See *Almanach Royal* (1736): 325–326. The royal acts of her *survivance* and then regular appointment are in the Maison du Roi series at the AN, O^1 36, fols. 162/163, O^1 79, fols. 162/163, and O^1 85, fols. 252/253, as cited in Aline Hamonou-Mahieu, *Claude Aubriet: Artiste naturaliste des Lumières* (Paris: MNHN, 2010).

40. M. Flourens, "The Jussieus and the Natural Method," *Annual Report of the Board of Regents of the Smithsonian* (1867): 259–260.

41. See the mentions of Basseporte in Albert van Spaandonck, *Le Comte Gerard Van Spaendonck, le dernier peintre du Roi* (Brussles: n.s., 1984), I, 35–41, 46–49, 58–60, 78, 80, 91. This work is a fictionalized but interesting biography of the man who took over Basseporte's position after she died. The author is probably a descendant with access to family papers. His name differs by just one letter.

42. Alfred Lacroix, *Notice historique sur les Cinq de Jussieu* (Paris: Gauthier-Villars, 1936), 30–31. See also M de Caze, "Sur une correspondance inédite entre Linné et Bernard de Jussieu," *Précis analytique des travaux de l'Académie Impérial de Rouen* (1857): 58–64, 61. There are other letters from Linnaeus translated from Latin into French at the AAS in the Dossier Bernard de Jussieu.

43. Jeffares says it is the profile of Jussieu studying a leaf with a magnifying glass, but others attribute to Basseporte a portrait of Jussieu facing fully forward.

44. Letter of 23 April 1749, available at The Linnaean Correspondence, http:// linnaeus.c18.net/. L1029.

45. Jules Guiffrey, *Scellés et inventaires d'artistes, partie, 1741–1770* (Paris: Charavay, 1886), 32. Aubriet lived in relative luxury at the Jardin, had a second apartment, and owned valuable possessions.

46. See the review of the 1781 issue of the *Nécrologe* in the *Mercure de France* (1781): 226. And *Revue Universelle*, 143–145. Basseporte's assistance to Bougainville's brother can be figured out from clues in the notice although he is not named. He was a classicist with a sea captain brother, and the mentions of his teachers and eulogists who *are* named reveal his identity.

47. "Notice sur la vie et les oeuvres de Edmé Mentelle," in *Oeuvres complètes de Mme la princesse Constance de Salm*, tome IV in vol. 2 (Paris: Firmin Didot, 1842), 181–228, especially 192–198, 220. For more on Mentelle, see David N. Livingstone, ed., *Geography and Revolution* (Chicago: University of Chicago Press, 2005).

48. Edmond Pilon, "Autour de Buffon: Mme Daubenton et sa famille," *Mercure de France* 44 (1911): 30–60. Pilon calls Marguerite Daubenton a bluestocking, 35.

49. Roger L. Williams, "Malesherbes: Botanist, Arborist, Agronome," *Journal of the Historical Society* 7, no. 2 (June 2007): 265–284. See also *Duhamel du Monceau, 1700–2000. Académie d'Orléans. Actes du Colloque du 12 mai 2000.*

50. Bruno Dinechin, *Duhamel du Monceau, un savant exemplaire au siècle des lumières* (Paris: CME, 1999), 317–328. This work of Malesherbes's was published posthumously in two volumes with an introduction by the physiocrat Louis Paul Abeille, *Observations de Lamoignon de Malesherbes sur l'Histoire Général et Particulier de Buffon et Daubenton* (Paris: Charles Pogens, 1798).

51. Buffon, *Histoire Naturelle*, III (1749), facing 228. This was not a botanical drawing but one from a wax model of the inside of a human head. It was in this anatomical direction that Basseporte would push her prize pupil, Biheron.

52. *Nécrologe,* 165–167, *Revue Universelle,* 141, note.

53. BCMNHN, MS 440 shows he was sketching this out in 1757 and then in 1759 he began implementing it in the Trianon (BCMNHN, MS 1169, VIII "Ordres des plantes etablis par M B de J dans le Jardin de Trianon").

54. See the preface to Dalibard's *Florae parisiensis prodromus, ou Catalogue des plantes qui naissent dans les environs de Paris* (Paris: Durand, 1749).

55. *Nécrologe,* 169 note. *Revue universelle,* 142 note 1. That she was hired to give flower painting lessons to the royal children was rumored, although I have found no evidence of this in the records.

56. *Correspondance inédite de Linné avec Claude Richard et Antoine Richard,* trans. and annotated by A. Landrin (Versailles: Imp d'Auguste Montalant, 1863), 42–43.

57. AN, O^1 1293, #317.

58. BCMNHN, MS 3515 and 3516.

59. She did considerable work for members of the Académie des Sciences, for which some tidy sums were paid, but they mostly went to Aubriet as long as he was alive. Some examples can be found in the AAS, Fonds Lavoisier, chemises 1089, 1097, 1098, and 1107.

60. See the interesting discussion of this in Lorraine Daston and Peter Galison, *Objectivity* (New York: Zone Books, 2007), 86. See also 20, 22, 35, 39, 42, 58–60, 67–68, 70–74, 82, 97.

61. Michel Adanson, *Famille des Plantes* (Paris: Vincent, 1763), I-Partie, 168 of the alphabetical "Table Raisonnée." She is also mentioned on cxlii and 12 of his *Table Chronologique.*

62. See Lissa L. Roberts, Simon Schaffer, and Peter Dear, eds., *The Mindful Hand: Inquiry and Invention from the Late Renaissance to Early Industrialization* (Chicago: University of Chicago Press, 2008).

63. Wilfred Blunt and William T. Stearns, *The Art of Botanical Illustration* (London: Colling, 1950), 4.

64. Bultingaire, "Peintres du Jardin du Roy," 32.

65. See, for example, BCMNHN, MS 92, part 2, fols. 452–456.

66. See BCMNHN, MS 92, "Recueil des dessins de plantes par Aubriet et Mlle Basseporte," sketches of *vélins* "réalisés pour illustrer des travaux des Jussieu."

67. See, for example, Hamonou-Mahieu, *Aubriet,* 82–83.

68. BCMNHN, old fichier "Biographies," "Dossier documentaire biog. Basseporte." Basseporte's *vélins* are scattered throughout the many portfolios of the *Collection des Vélins du Muséum national d'Histoire naturelle.* Because some are unsigned, it is unknown who created them, but as many as 340 may be hers.

69. In BCMNHN, MS 92 we can still find many of Basseporte's (undated) preliminary pencil sketches (with some color added) (268). In her case the number of the corresponding painting on vellum is given too. She did some of the quinine tree for La

Condamine's *mémoire* to the Académie in 1738 (286). Many are attributed to her in a later curator's notes, e.g., that of the Magnolia Grandiflora (307). She did zizania and put her age of 79 on it (101), as she did with ixia (131)—these are also *vélins*. This MS 92 has sketches on paper (407ff.). Many were done "pour M. Jussieu" and show her practice/process.

70. Antoine Laurent de Jussieu, *Annales du Muséum national d'histoire naturelle*, 5e notice, VI (1805): 18.

71. AAS, Dossier Bernard de Jussieu. This *éloge* is a pamphlet, but it was also printed elsewhere as part of the Académie des Sciences *mémoires*.

72. This memo, written in 1788 after Buffon died to inform the Jardin's new intendant about its workings, is quoted in Charles Coulton Gillispie, *Science and Polity in France at the End of the Old Regime* (Princeton: Princeton University Press, 1980), 183–184.

73. The original of this letter from Commerson to Bernard de Jussieu, 6 February 1770, is in the AAS, Dossier Commerson. But it and many others are also now reproduced in the excellent website, Pierre Poivre et Compagnie, put together by Jean-Paul Morel, http://www.pierre-poivre.fr/.

74. Cap, *Philibert Commerson*, 94–95. Letter to Beau dated 22 December 1766.

75. Cap, 81. Letter of January 1767 (must be at the end of the month, though no exact date specified), to old friend Bernard.

76. Cap, 102.

77. *Bougainville et ses compagnons autour du monde: 1766–1769, journaux de navigation*, compiled and with commentary by Etienne Taillemite (Paris: Imprimerie nationale, 1977), II, 267–268.

78. Louis Antoine de Bougainville, *Voyage autour du Monde par le frégat du Roi la Boudeuse et la flûte Étoile en 1766, 1767, 1768 et 1769* (Paris: Saillant et Nyon, 1771), 253–254.

79. Taillemite, *Bougainville et ses compagnons*, II, 408.

80. Ibid., 267–268.

81. Ibid., 237–241.

82. He voluntarily joined the expedition, a "noble savage" who was taken back to Paris for study by La Condamine, Jacob Rodrigues Péreire who was an expert on deaf mutes, and the English physician Pringle visiting with Franklin from London. But that is another story.

83. See Bougainville's entry of 29 May, Taillemite, *Bougainville et ses compagnons*, I, 349–350.

84. François Vivez, *Autour du monde sous les ordres de Bougainville: Carnet de voyage* (France: CLAAE, 2018), especially "Suite de l'histoire particulière." Vivez was an alternate spelling of Vivès.

85. Jeannine Monnier et al., *Philibert Commerson: Le Découvreur du Bougainvillier*

(Châtillon-sur-Chalaronne: Association Saint-Guignefort, 1993), 100. BCMNHN, MS 887 III, 42 (Y.L. 72), "Perruche à Baret, ou de Bourou." From the description it was probably what is now called the tricolor Moluccan king parrot.

86. See, for example, Cap, *Philibert Commerson*, 105–107, a letter of 30 November 1768.

87. Taillemite, *Bougainville et ses compagnons*, I, 349–350.

88. This is in Bougainville's onboard journal. See Taillemite, I, 443.

89. Marguerite Duval, *The King's Garden* (Charlottesville: University of Virginia Press, 1982), 83–85.

90. See Louis Malleret, *Pierre Poivre* (Paris: Ec. Fr. d'Extrême-Orient, 1974), 350, 612, 643ff.

91. Lalande, *Éloge de Commerson*.

92. Dussourd, for example, says she stayed by Commerson's side there, and on his trips to Madagascar and Bourbon, along with other domestics (67).

93. See, for example, BCMNHN, MS 337, for some beautiful black-and-white drawings from Buenos Aires, Montevideo, the Straits of Magellan, and Isle de France. See also Cap, *Philibert Commerson*, 99, letter of 28 May 1767 where Commerson writes that necessity makes him an artist, perhaps because there was more to draw than Barret could handle alone.

94. Dussourd, *Jeanne Baret*, 70. See Commerson's letter to Beau of 25 February 1769, in Montessus, *Martyrologie*, 115.

95. Madeleine Ly-Tio-Fan, *Pierre Sonnerat, 1748–1814: An Account of His Life and Work* (Cassis (Mauritius): Imprimerie et papeterie commerciale, 1976), 59.

96. Cap, *Philibert Commerson*, 143. Letter of 29 September 1770 to Cossigny. A letter with no date on page 152 indicated that Mme Poivre was at the intendance and didn't like animals.

97. Jacques-Bernardin-Henri de Saint Pierre, *Harmonies de la Nature*, 3 vols. (Paris: Louis Aimé-Martin, 1815), II, 193–195.

98. Jean-Paul Morel, "Philibert Commerson à Madagascar et à Bourbon," http://www.webcitation.org/6iBjoHcCe, 5. Also Cap, *Philibert Commerson*, 117. The date on this letter is 18 April 1771, when he was already on the Isle Bourbon but reflecting back on the big island. Commerson's writings on quadrupeds also sound Darwinian.

99. BCMNHN, MS 887, letter of 17 February 1771.

100. Jean-Baptiste Lislet-Geoffroy's account. See Poivre website, http://www.pierre-poivre.fr/doc-71-11-mois-b.pdf.

101. Letter to Jussieu 6 February 1770, http://www.pierre-poivre.fr/doc-70-2-6b.pdf.

102. Malleret, *Pierre Poivre*, 350, 410–413, 433, 440, 506.

103. Much work remains to be done on this Tahitian native and his treatment in the metropole, but sources are scant.

104. BN MS N.A.F. 5071, fols. 92–106, *Lettre de Mr. Beauvais, artiste vétérinaire et directeur de haras entretenu par le Roy aux Isles de France et de Bourbon*, 24 August 1784.

105. Dussourd, *Jeanne Baret*, 75. Gilles Pacaud of Autun who did research on Mauritius told me that this house accommodated Commerson's collections amply, and that its contents were inventoried by an official, but I have as yet not seen the evidence.

106. Cap, *Philibert Commerson*, 115. Letter of 16 October 1772.

107. Poivre website, http://www.pierre-poivre.fr/doc-72–10–27.pdf. The original manuscript is in the Bibliothèque municipale de Nantes, MS 2423.

108. See Bézac's report on Poivre website, http://www.pierre-poivre.fr/doc-73-3 -14.pdf.

109. Dussourd, *Jeanne Baret*, 75. Beau did eventually manage to get a 1,000 livres pension for Archambault, some of which he may have used himself.

110. Cap, *Philibert Commerson*, 129. Also http://www.pierre-poivre.fr/doc-70 -an-d.pdf.

111. Cap, *Philibert Commerson*, 154. This letter has no date.

112. The herbier of Montpellier has a particularly beautiful specimen of *Baretia*.

113. My translation of the French translation of Commerson's original Latin, found in Jean-Marie Pelt, *La Canelle et Le Panda: Les naturaliste explorateurs autour du monde* (Paris: Fayard, 1999), 156. Commerson's Latin manuscript is in BCMNHN, MS 198 (YL 51), a big folio with many different plants, including the *Baretia*, described.

114. Carole Christinat, "Une femme globe-trotter avec Bougainville: Jeanne Barret, 1740–1807," *Revue française d'histoire d'outre-mer* 83, no. 310 (1er trimestre 1996): 83–95, 91.

115. See *Herbier du Monde*, 88–89. And Monnier, *Philibert Commerson*, 99. The name did not stick, however. The plant was later called Quivisia, and today is known as Turrea, possibly obtusifolia, which has changing leaves, even changing height, beautiful white flowers, and a green pumpkin-like fruit with bright red seeds inside.

116. Montessus writes this in his *Martyrologe*, 253; as do A. Role's *Vie aventureuse d'un savant;* Oliver's biography, 206; Roger Espitalier-Noel's entry in *Dictionnaire de Biographie Mauricienne;* John Dunmore's *Monsieur Baret: First Woman Round the World, 1766–1768* (New York: Heritage, 2002); Dussourd, and almost all others.

117. Oliver, *Life of Philibert Commerson*, 201–202.

118. *Herbier du Monde*, 82, speaks specifically of the care she took gathering, drying, and arranging samples. So do the articles by Sophie Miquel cited below.

119. Oliver, *Life of Philibert Commerson*, 30.

120. Monnier, *Philibert Commerson*, 163.

121. Letter of 15 March 1773 on Poivre website, http://www.pierre-poivre.fr/doc -73-3-15.pdf.

122. AN MC ET LXXXIV/537. Barret's letter, to which this is Vachier's reply, is missing.

123. See Poivre website, http://www.pierre-poivre.fr/doc-73–11–9.pdf. See also, in the relevant online documents from the Archives d'Outre-mer, with the reference ANOM Col E 89, Mihiel's letter of 15 November 1773, view 574–575. On Jossigny's enmity toward Commerson, see the letters in Poivre website, pierre-poivre.fr, on that subject. See also Ly-Tio-Fan, *Pierre Sonnerat,* 65–66.

124. Archives numérique de l'Ile Maurice, Z2B1. The fine is dated 22 December 1773. See also Dunmore, *Monsieur Baret,* 18.

125. Mauritius Archives, Registre de Port Louis, 1774, KA61-Vue 023. Dunmore, *Monsieur Baret,* 194 note 108 cites the non-online version as K.A. 61/D, f.21, p. 75. There were five witnesses to the marriage, all of whom did sign their names. They were probably soldier friends of Dubernat's.

126. Roger Bour, "Paul Philippe Sanguin de Jossigny (1750–1827), artiste de Philibert Commerson. Les dessins de reptiles de Madagascar, de Rodrigues et des Seychelles," *Zoosystema* 37, no. 3 (September 2015): 415–448. http://dx.doi.org /10.5252/z2015n3a1, p. 422.

127. AN MC ET LXXXIV/537.

128. In his testament Commerson had indicated he would like Adanson or his old friend Gerard to take this role, but they were not in Paris at the time, and Barret was still in Isle de France, so these two other botanists were chosen instead.

129. BCMNHN, MS 3515 and 3516.

130. Dalibard wrote *Parisiensis prodromus, ou catalogue des plantes* in 1749, and Dubourg wrote *Le Botaniste Français* in 1767. These men were also the first translators of Franklin's works into French.

131. Basseporte's niece Aillaud refers to this meeting. It was probably after 1776 when Franklin came to Paris to settle in for many years. See letter from Aillaud to Franklin, 26 August 1780, on the easily searchable Founders Online website, founders .archives.gov.

132. Dubourg, *Botaniste Français,* I, preface. See *Correspondance Complète de Rousseau,* 1965 edition, vol. 35, letter #6367, 309–310; and vol. 37, letter #6589.

133. Gavin de Beer, "Jean-Jacques Rousseau, Botanist," *Annals of Science* 10, no. 3 (September 1954): 189–223, 195.

134. Ibid., 207, 209.

135. Alexandra Cook, "Jean-Jacques Rousseau and Exotic Botany," *Eighteenth Century Life* 26, no. 3 (2002): 181–201.

136. Condorcet, *Éloge de Bernard de Jussieu,* AAS, Dossier Bernard de Jussieu, 18.

137. Musée du Louvre, Département de Peintures, Documentation, *Dossier Biographique, Reboul-Vien.*

138. Anna Raitieres, "Lettres à Buffon dans les 'Registres de l'Ancien Régime, 1739–1788,'" *Histoire et Nature* 17–18 (1980–1981): 85–148. My quote is from letter #51 on 108. Raitieres cites this as AN, O^1 411, fol. 517.

139. Muséum d'Histoire Naturelle, laboratoire de Phanérogamie, Fonds Jussieu,

Dossier Basseporte, côte Per-K-130(1) "Correspondance des botanists." This letter was recently moved to the main museum library.

140. Bougainville, *Voyage autour du Monde*.

141. Denis Diderot, *Supplément au Voyage de Bougainville*, http://www.gutenberg .org/cache/epub/6501/pg6501-images.html, chapter 2. "Frêles machines" is a term for women that Diderot used repeatedly, in *Le Rêve de D'Alembert*, for example, and in his writings for Catherine of Russia (Tourneux ed., *Diderot et Catherine*, II, 385).

142. BCMNHN, MS 3515. A. L. de Jussieu also wrote directly to Malesherbes that they were "bon patriots," "bon citoyens" who cared deeply about the liberties of the people. See Audelin, "Les Jussieu," 153–156.

143. Musée du Louvre, Département de Peintures, Documentation, *Dossier Biographique, Basseporte*, which contains a part called *Dossier Bourin*. On an envelope it says these notes were taken by a Mme Briere-Misn. See also *Revue Universelle*, 142.

144. Bachaumont, *Mémoires secrets* (edition of 1780),VII, 159 (15 April 1774).

145. Lalande, *Éloge de Commerson*.

146. BCMNHN, MS 3515 and 3516.

147. Laurent's letters are in Yvonne Letouzey, *Le Jardin des Plantes à la croisée des chemins avec André Thouin, 1747–1824* (Paris: Edition du Muséum Nationale d'Histoire Naturelle, 1989), 130–131.

148. Sophie Miquel, "Les testaments de Jeanne Barret, première femme à faire le tour de la terre, et de son époux perigordin Jean Dubernat," *Bulletin de la Société Historique et Archéologique du Périgord* 144 (2017): 771–782. On page 774 she mentions that Dubernat signed a document at his brother François's wedding in Sainte-Foy-la-Grande in late 1775.

149. Letter of 6 February 1770 from Commerson to Bernard de Jussieu, www .pierre-poivre.fr/doc-70-2-6b.pdf.

150. Montessus, *Martyrologie*, 178.

151. Last known letter of Commerson's, to Lemonnier, 27 October 1772, http:// www.pierre-poivre.fr/doc-72-10-27.pdf.

152. Poivre website, http://www.pierre-poivre.fr/doc-74-an-d.pdf.

153. Private communication from Sophie Miquel on 15 October 2018.

154. AN Col E 89, dossier Philibert Commerson.

155. Montessus, *Martyrologie*, 196 note 1.

156. BCMNHN, Fonds Thouin non-classé, Bôite 1. The date on this letter seems to read 13 December 1776, but Malesherbes was no longer minister then, having been dismissed with Turgot on 12 May of that year. So it must be 1775.

157. AN AJ/15/512, fol. 494, 11 January 1776.

158. Oliver, *Life of Philibert Commerson*, 229. Beau actually did not manage to get his funds until 1781. See digitized Archives d'Outre-mer, with the reference ANOM Col E 89, views 605–625.

159. Montessus, *Martyrologie*, 200.

160. AN MC ET LXXXIV/537, fols. 30 and 31. Barret's signature and that of her husband Dubernat appear on these last two pages.

161. AN MC ET XVIII, fol. 811, 7 March 1776. This original will was amended on 18 May 1777 to say that if both her executors, Dubourg who was her doctor and Mme Dubourg, were not alive, that "my friend Biheron" should take over those duties. These executors were to be paid 550 livres. The will was amended again 29 June 1780 asking that scellés be placed on her things after her death.

162. See letter from Laurent of Brest to Thouin regarding Basseporte as part of the household, 26 February 1777, in Letouzey, *Le Jardin des Plantes*, 130. See also 27. To this sister, Marie-Jeanne Thouin, Basseporte bequeathed her copy of Buffon's *Histoire Naturelle*. Mlle Thouin was in turn a tutor to the Duchesse de Chartres, future mother of Louis-Philippe.

163. A. L. de Jussieu's letters attest to this. Bernard would die 6 November 1777.

164. Cap, *Philibert Commerson*, 27.

165. Rochon, *Voyage to Madagascar and the East Indies* (London: Robinson, 1792), 160–163.

166. Lucille Allorge, *La fabuleuse odyssée des plantes* (Paris: J. C. Lattès, 2003), 510–511.

167. See AN AJ/15/512, fol. 503, where Lamarck, in January and April 1793, unabashedly blends Commerson's collection with all the others.

168. Baron Georges Cuvier et Magdeleine de Saint-Agy, *Histoire des sciences naturelles*, V (1845), 95–97. Georges Cuvier had died in 1832 so this was published posthumously. Ironically, he was himself one of the offenders, claiming some of Commerson's work as his own.

169. Roger L. Williams, *French Botany in the Enlightenment: The Ill-fated Voyages of La Pérouse and His Rescuers* (Netherlands: Springer, 2003), 11.

170. As examples, see BCMNHN, *Collection des Vélins*, tome 46, #33, and tome 28, #54.

171. She was still listed in the *Almanach Royal* (1780): 517 with nobody named *en survivance*. Even Buffon had a designated successor by then, publicly named on page 516, but Van Spaendonck's name did not appear until the 1781 volume, after Basseporte's death.

172. AN AJ/15/510, #334. See also #331 for Basseporte's various pensions.

173. After Basseporte died Van Spaendonck paid Desève yearly, but in 1794 the matter arose again as the Revolution gave his job a different title. The judgment was that because Van Spaendonck's new position paid even better, and because in a government where truth and justice were the order of the day, ingratitude a crime of which a real republican could not make himself guilty, he should continue to pay. And yet he wriggled out of it. See *Les Tribunaux civil de Paris pendant la Révolution* (Paris: Quantin, 1905), I, 781–782, and II (pt. 1), 34–35, 70–71.

174. See all of these letters, arranged by date and name of sender, in the digitized Franklin papers to be found at the website Founders Online, founders.archives.gov.

175. Abbé Riballier and Charlotte Catherine Cosson, *De l'Education physique et morale des femmes, avec une notice alphabétique de celles qui se sont distinguées dans les différentes carrières des sciences et des Beaux-arts, ou par des talens et des actions mémorables* (Paris: Etienne, 1779), 143–144. This was copied verbatim into the *Dictionnaire Portatif des femmes célèbres* (1788), I, supplément, 755. The earlier 1769 edition of the *Dictionnaire Portatif* had no mention of Basseporte (or Biheron).

176. *Nécrologe*, 185, *Revue universelle*, 146.

177. BCMNHN, MS 3517, unnumbered folio.

178. BCMNHN, MS 1978, fol. 1467.

179. AN MC ET XVIII, fol. 811.

180. AN MC ET XVIII, fol. 812.

181. *Ouevres complètes de . . . Constance de Salm* (1842), 198.

182. See, for example, his letter of 31 May 1766. BCMNHN, MS 3515.

183. *Revue universelle*, 142.

184. These details are based on documents available online from the Archives Départementales de la Dordogne and from a genealogy given to me by Sophie Miquel. For preliminary information, see her "Jeanne Barret (1740–1807) en Périgord, après son tour du monde," *J. Bot. Soc. Bot. France: Le Journal de Botanique* 77 (March 2017): 49–55.

185. Registre des naissances, 17 July 1740, AD Saône-et-Loire, La Comelle 4^E 142/2 for Barret's birth, and AD Dordogne, 5^E 365 for her death. See Miquel, "Jeanne Barret."

186. Williams, *French Botany*, 13–14.

187. AN, Marine, C^7 17, Dossier "Baré."

188. AAS, Dossier Castries. This is a cross-reference from Dossier Bougainville. Castries was secretary of the navy and colonies from 1780 to 1787 and a highly distinguished military general.

189. Benoît Carré, "Femmes, pensions et autres grâces royales à la cour de Versailles au XVIIIe siècle," in Caroline zum Kolk and Kathleen Wilson-Chevalier, eds., *Femmes à la Cour de France* (Lille: Presses Universitaires de Septentrion, 2018), 168, 170, 175, 178–179, and 181.

190. Miquel, "Jeanne Barret," 51.

191. The original letters regarding the pursuit of her pension are at the AAS, reclassified now as MS 1J32, and there are copies in Dossier Commerson. Scrambled out of order, they also bear dates from the revolutionary calendar. I have changed them into "old style" Gregorian dates so the chronological sequence of events can be more easily followed.

192. AAS, MS 1J32.

193. Miquel, "Jeanne Barret," 49–55. Contrary to what many others have written, she definitely did not leave her earthly goods to Commerson's son Archambault with whom she had surely lost contact more than thirty years after his father's death.

194. Miquel, "Les testaments de Jeanne Barret," 771–782.

195. Archives départementales de Dordogne, série 5^E, 365. See also Jeannine Gerbe "La famille bressane de Philibert Commerson: essai généalogique," *REGAIN: Recherches et études généalogiques de l'Ain* (1994), 41–47.

196. Montessus, *Martyrologie*, 25.

197. Miquel, "Les testaments," 780, 777.

198. Taillemite, *Bougainville et ses compagnons*, II, 408.

199. Barret is the subject not only of scholarship, but of graphic novels, plays, songs, and numerous fictionalized versions of her adventure.

CHAPTER FOUR. ANATOMIST AND INVENTOR

1. *Gazette d'Epidaure* 3 (6 February 1762): 88, "Talens récompensés." Her model of the male organ, "une verge soigneusement confectionée en cire," was listed in Biheron's collection by Vicq d'Azyr, for example, whose inventory will be discussed below. The *Gazette d'Epidaure* was a medical journal edited by her friend, doctor, and mentor Jacques Barbeu Dubourg whom we have already met in connection with Basseporte.

2. Grimm et al., *Correspondance Littéraire*, ed. Tourneux, IX (Paris: Garnier, 1879), 276. While there have been no comprehensive studies of Biheron, aspects of her life and work are discussed in: Michel Lemire, *Artistes et Mortels* (Paris: Chabaud, 1990), 80–85; Georges Boulinier, "Une femme anatomiste au siècle des Lumieres: Marie-Marguerite Biheron, 1719–1795," *Histoire des Sciences Médicales* 3, no. 4 (2001): 411–423; Adeline Gargam, "Marie-Marguerite Biheron et son cabinet d'anatomie: une femme de science et une pédagogue," in I. Brouard-Arends and M. Plagnol-Diéval, eds., *Femmes éducatrices au siècle des Lumières* (Rennes: Presses universitaires de Rennes, 2007), 147–156, and comparative articles by Margaret Carlyle and Lucia Dacome.

3. Leopold Auenbrugger's method of percussion of the chest, also published in 1761, decades before Laennec's invention of the stethoscope, revealed much about the heart and lungs without opening the thorax.

4. J.-J. Sue, *Abrégé de l'anatomie du corps de l'homme* (Paris: Simon fils, 1748), 2–3.

5. *Mémoires inédites de Madame la Contesse de Genlis* (Paris: Ladvocat, 1825), I, 308–309.

6. See Andrew Cunningham, *The Anatomist Anatomis'd: An Experimental Discipline in Enlightenment Europe* (Farnham: Ashgate, 2010).

7. Jean-Jacques Rousseau, *Reveries of a Solitary Walker,* trans. Charles E. Butterworth (Indianapolis: Hackett, 1992), 97.

8. Denis Diderot, *Correspondance générale*, texte établi par J. Assézat et M. Tourneux (Paris: Garnier), XX, 61–65, Diderot to Betzky, 15 June 1774; Diderot, *Mémoires pour Catherine II*, ed. Paul Vernière (Paris: Garnier, 1966), 88; Mentelle's Basseporte

eulogy in *Nécrologe des hommes célèbres*, 1781, reprinted in *Revue universelle des arts* (1861): 145.

9. *Mémoires inédites de . . . Genlis*, I, 308–310.

10. See Grandjean de Fouchy, "Éloge de Morand," in *Histoire de l'Académie Royal des Sciences* (henceforth *HARS*) *pour l'année 1773* (Paris, 1777), 99–117.

11. The chemist Macquer, whom we will meet again in chapter 5, had a decades-long collaboration with the learned apothecary Baumé. They could not have given their highly successful course without each other.

12. Ads for Maille were ubiquitous. See also Morag Martin, "Il n'y à que Maille qui m'aille: Advertisements and the Development of Consumerism in Eighteenth Century France," *Proceeding of the WSFH* 23 (1996): 114–121. I have relied on eighteenth-century descriptions of apothecary shops and on various websites.

13. Basseporte's image was modeled on Zumbo's head of 1701 in the Cabinet du Roi. See Takuya Kobayashi, "Ecrits sur la botanique de Jean-Jacques Rousseau: Edition Critique" (thesis, Université de Neuchatel, 2012), on the books Rousseau consulted, annotated, and took with him on his rambles.

14. There is evidence of their closeness in their respective testaments, and they are always mentioned in the same breath in the letters of A. L. de Jussieu to his mother. Jussieu does mention a Mlle Lainé as another "très bonne amie" of Biheron, so the plot thickens, but Lainé seems absent from Biheron's scientific or political life and is not named in her testament.

15. Today this is 3 rue de l'Estrapade, marked by a historical plaque (about Diderot, of course, not Biheron). Grimm and many others confirm that they both lived in the same house. The upholsterer Fleury had his shop on the ground floor, and Dubourg's *Carte Chronographique* was distributed "chez Fleury" by Biheron. She must have had the back apartment facing the court, where she installed a glass enclosure for her ongoing dissections.

16. According to Jean La Tynna's *Dictionnaire topographique, étymologique et historique des rues de Paris,* the strappado was used until the construction of the new Sainte Geneviève church (today the Pantheon) in 1757. The *Encyclopédie* defines it matter-of-factly in the present tense, with no comment on its being an earlier or frowned-upon practice. Finding this quarter sinister, Balzac would later use it as the setting for the dismal rooming house of his novel *Le Père Goriot*.

17. Biheron inherited these houses in 1752 when her mother died. See Boulinier, "Une femme anatomiste," 411–423. See also Archives de Paris, DC6 258, fol. 151: Vente de quatre maisons, enregistrée le 30 mars 1780.

18. Jean Mayer, *Diderot, homme de science* (Rennes: Imprimerie Bretonne, 1959), 30–31.

19. See Diderot, *Éléments de physiologie*, édition critique . . . . Introduction by Jean Mayer (Mâcon: Protat, frères, 1964), ix. Also Andrew H. Clark, *Diderot's Part* (Aldershot: Ashgate, 2008), 39.

20. *Première lettre d'un citoyen zélé qui n'est ni chirurgien ni médecin: à M.D.M. [. . .] où l'on propose un moyen d'apaiser les troubles qui divisent depuis si longtemps la médecine et la chirurgie.* (Paris: sn, 1748).

21. Morand, "Éloge de Cheselden," *Mémoires de l'Académie de chirurgie* 2 (1757): 107–108.

22. "Cadavre" in *Encyclopédie,* II (1751), 510–511.

23. Mme Vandeul, "Vie de Diderot," in Diderot, *Oeuvres choisies* (Paris: Garnier frères, 1934), 36.

24. Diderot complained to Sophie Volland regarding the errands Grimm was making him do when Grimm himself was not in town. *Oeuvres complètes,* ed. Assezat and Tourneau, XIX, 296.

25. All letters from, to, or about Franklin in this chapter come from the National Archive website, Founders Online, which includes the American Philosophical Society Franklin Papers, and much more. See letter of 31 January 1768.

26. *Encyclopédie,* III (1753), 400–401. Stephen Ferguson's "The 1753 *Carte Chrono-graphique* de Jacques Barbeu-Dubourg," *Princeton University Library Chronicle* (1991), notes on p. 206 that according to Vicq d'Azyr, Biheron taught Diderot and Dubourg and that she sold the explanatory publication that accompanied Dubourg's chart. Vicq's "Éloge de Barbeu Dubourg" does indeed say this chart was distributed "A Paris, chez Mlle Biheron" in *Oeuvres de Vicq d'Azyr: Éloges Historiques,* ed. Jacq. L. Moreau (Paris, 1805), 2: 185 note 1. An extant *Carte Chronographique* of Dubourg's is at the Princeton University Library.

27. "Eh bien, marchand de hasard, avez-vous assez d'esprit pour nous faire con-cevoir que le hasard en ait tant." L.-G. Michaud, *Biographie universelle, ancienne et moderne* (Paris: C. Delagrave, 1870–75), IV, 315.

28. This is the title of an essay on the website of the Wildgoose Memorial Library (wildgoose@janewildgoose.co.uk), where Jane Wildgoose presents also such intriguing topics as The Business of the Flesh; Who Really Owns Our Bodies; Curiouser and Curio User; The Hybrid Nature of Specimens; Displaying the Dead, etc.

29. Abbé Riballier and Charlotte Catherine Cosson de la Cressonière, *De l'Educa-tion physique et morale des femmes, avec une notice alphabétique de celles qui se sont dis-tinguées dans les différentes carrières des sciences et des Beaux-arts, ou par des talens et des actions mémorables* (1779), 152–154. Her name in this article is spelled Bieron. The long entry on Biheron was copied verbatim in de la Croix and de la Porte: *Dictionnaire his-torique portatif des femmes célèbres contenant l'histoire des femmes savantes, des actrices & généralement des dames qui se sont rendues fameuses dans tous les siècles* (Paris: Belin, 1788), II, 763–765. The earlier 1769 version of de la Croix's *Dictionnaire* had made no mention of Biheron.

30. *Mémoires inédites de . . . Genlis.* For the meaning of "boudoir" in this context, see Valérie Lastinger, "The Laboratory, the Boudoir and the Kitchen: Medicine, Home and Domesticity," in Anne Kathleen Doig and Felicia B. Sturzer, eds., *Women, Gender*

and *Disease in Eighteenth Century England and France* (Newcastle upon Tyne, Eng.: Cambridge Scholars Publishing, 2014), 119–148, 123.

31. L.-M. Prudhomme, *Biographie universelle et historique des femmes* (Paris: Lebigre, 1830), 363. See also the Wildgoose essay "Zones of Morbidity" on the enormous number of corpses needed for modeling, wildgoose@janewildgoose.co.uk.

32. For a moving description of dissection, which I have very closely relied on here, see Christine Montross, *Body of Work: Meditations on Mortality from the Human Anatomy Lab* (London: Penguin, 2007).

33. Diderot, *Mémoires pour Catherine II*, 87.

34. Regis Olry, "Wax, Wood, Ivory, Cardboard, Bronze, Fabric, Plaster, Rubber, Plastic . . .," *Journal for the International Society of Plastination* 15, no. 1 (2000): 30–35.

35. See the thoughts of Veronique Roca and Jeanne Peiffer in Thérèse Chotteau, et al., *Rencontres entre artistes et mathematiciennes; Toutes un peu les autres* (Paris: L'Harmattan, 2001), "Corps moulés, corps façonnés: Autoportraits des femmes," 56–92, especially 61–69. Roca, a "plasticienne," speaks of the liveliness and life-likeness of wax, a living organic material that actually comes from the secretions of the bee's body, a ductile substance which therefore incarnates, infiltrates, and transforms, combining death, hope, and cure, holding spirit in its very material. Basseporte was to leave Biheron two silver candlesticks, which perhaps had extra significance?

36. *HARS pour l'année 1759* (Paris, 1765), 94.

37. Morand, *Catalogue des pièces d'Anatomie, instruments, machines qui composent l'arsenal de Chirurgie à Petersbourg* (Paris: L'Imprimerie royale, 1759).

38. *Journal de médecine, chirurgie et pharmacie* (1759): 277–279.

39. *Journal de Trévoux* (September 1759): 2297–2299.

40. La Condamine would give blow-by-blow accounts of these grotesque goings on in 1761. See Grimm et al., *Correspondance Littéraire*, nouvelle édition (Paris: Furne, 1829), III, 18–29. Dubourg, who went earlier, told the police not to persecute or prosecute the participants, for if anything sinister occurred it would out. See Grimm et al., *Correspondance Littéraire*, ed. Tourneux, edition of 1878, IV, 208–217, especially 213–214. Morand discussed this recurrence of "ridiculous ceremonies" starting in 1759 in *Opuscules de chirurgie* (Paris: Guillaume Desprez, 1768–1772), II (1772), 298–306.

41. We saw that Basseporte was very close with devout Jansenists Pluche and Duhamel de Monceau. Morand's library indicated that he had similar sympathies. See Ole Peter Grell and Andrew Cunningham, eds., *Medicine and Religion in Enlightenment Europe* (Aldershot: Ashgate, 2007), 105.

42. See Arnaud Orain, "The Second Jansenism and the Rise of French 18th Century Political Economy," *History of Political Economy* 46, no. 3 (2014): 463–490. Also see Brian E. Strayer, *Suffering Saints: Jansenists and Convulsionnaires in France, 1640–1799* (Brighton: Sussex Academic Press, 2008). Strayer sums up similar arguments by Dale Van Kley and Thomas E. Keiser regarding the Jansenist challenges to the status

quo. I use the word *frondeur* here to mean rebellious, insubordinate, as were the protests against the French Crown during the seventeenth-century civil wars called the Fronde. I develop that concept in my articles on theater newspapers and in my book, *Feminine and Opposition Journalism in Old Regime France: Le Journal des Dames* (Berkeley: University of California Press, 1987).

43. *Anatomie artificielle: Annonce de l'exposition publique de pièces d'anatomie artificielle, exécutées par la demoiselle Bihéron* (Paris: Le Prieur, 1761). By 1762, just a year later, Biheron's "astonishing, absolutely unique" exhibit was still open daily, but only in the afternoons, from four to six. See *Gazette de France* (19 April 1762): 138.

44. *L'Avant-coureur* (1761): 309.

45. *Gazette de France* (19 April 1762): 138.

46. See Gelbart, *Feminine and Opposition Journalism*, especially 110, and the rest of that chapter (95–132), in which Beaumer, a true radical feminist *avant la lettre*, championed women in general.

47. *Gazette d'Epidaure* 1 (13 May 1761): 142, and 2 (23 December 1761): 392.

48. Victor Fournel, *le Vieux Paris: fêtes, jeux et spectacles* (Tours: Mame, 1887), 327–328, called her an "originale et savante personne," not reducible to a "simple montreuse." See also, on this "partage du savoir," Daniel Raichvarg and Jean Jacques, *Savants et ignorants: une histoire de la vulgarisation des sciences* (Paris: Seuil, 1991) and Raichvarg's *Science et Spectacle: Figures d'un rencontre* (Nice: Z'Editions, 1993).

49. Grimm et al., *Correspondance Littéraire*, nouvelle édition (Paris: Furne, 1829), VII (April 1771), 221.

50. Bachaumont, *Mémoires secrets* (London: John Adamson, 1784), I, 291 (29 October 1763).

51. *Supplément à l'Histoire Naturelle* (Paris: Imprimerie Royale, 1777), IV, 578–580, planche VI. The image is "d'après une tête en cire qui etait faite par mlle Biheron dont on connait le grand talent pour le dessin et la représentation des sujets anatomiques."

52. Dalibard to Benjamin Franklin, 14 June 1768, Founders Online. They had, for example, missed Buffon and Daubenton.

53. Father Joseph Etienne Bertier (sometimes Berthier) (1702–1783) wrote to Franklin on 27 February 1769 expressing everyone's eagerness for a return visit.

54. Franklin to Samuel Cooper, 30 September 1769, Founders Online.

55. Bachaumont, *Mémoires secrets*, IV, 234–235 (10 May 1769) [says 1768 in error].

56. Alexis Delacoux, *Biographies des Sages Femmes Célèbres* (Paris: Trinquart, 1834), 35–36; Caroline Healey Dall, *Historical Pictures Retouched* (Boston: Walker and Wise, 1860), 155.

57. The Linnaean Correspondence, http://linnaeus.c18.net/, letter L4163, letter from Bjornstahl to Linnaeus, 15 April 1769.

58. *HARS pour l'année 1770* (Paris, 1773), 49–50.

59. *Éloges lus dans les séances publiques de l'Académie de chirurgie . . . par A. Louis* (Paris: Baillière et fils, 1859). He refers to Mme le Boursier du Coudray as "dame

Bourcier, sage-femme à Clermont," so as she had not yet begun her traveling teaching mission when living in that city, this model would have been a very crude mock-up.

60. See Nina Rattner Gelbart, *The King's Midwife: A History and Mystery of Mme du Coudray* (Berkeley: University of California Press, 1998).

61. M. Schweigheuser, *Tablettes chronologiques de l'histoire de la médecine puerpérale* (Strasbourg: Levrault, 1806), 88.

62. Portal, *Histoire de l'anatomie*, 6 vols. (Paris: Didot, 1770–1773), V, 2. See also Albrecht von Haller, *Bibliotheca anatomica* (Lugduni Batav: Haak & Socc, 1777), II, 665.

63. Delacoux, Prudhomme, and Dall all repeated that Gottingen had a uterus by Biheron on display at about this time.

64. *The Gentleman's Guide in His Tour through France*, 4th ed. (London: G. Kearsly, 1770), 57, 58.

65. Grimm et al., *Correspondance Littéraire*, nouvelle édition (Paris: Furne, 1829), VII, 221–222.

66. Diderot, *Correspondance*, ed. Roth (Paris: Editions de Minuit), X, 245, Diderot to Grimm, March 1771.

67. See Shane Agin's "Comment se font les enfants?" *Modern Language Notes* 117, no. 4 (2002): 722–736, and his "Sex Education in the Enlightened Nation," *Studies in Eighteenth-Century Culture* 37 (2008): 67–87.

68. Marie Leca-Tsiomis, "Une anecdote familiale de Diderot," *Cahiers Voltaire* 4 (2005): 202–206.

69. See Emma Spary, *Utopia's Garden: French Natural History from Old Regime to Revolution* (Chicago: University of Chicago Press, 2000), on elaborate patronage fights over rights to demonstrate in the Jardin du Roi.

70. Diderot, *Correspondance*, XI, 210–211, dated 19 October 1771.

71. See a discussion of this in Amy Wygant, *Medea, Magic and Modernity in France: Stages and Histories, 1553–1797* (Aldershot: Ashgate, 2007), 189–191.

72. Diderot, *Correspondance*, XII, 84, Diderot to Wilkes, 10 July 1772.

73. Letter from Franklin addressed to Miss Stevenson, dated only "Wednesday PM," Founders Online. This must have been before her marriage on 10 July 1770, as after that she gave up her maiden name and became Mrs William Hewson.

74. Antoine Laurent de Jussieu letter to his mother, BCMNHN, MS 3515, fol. 104, 24 February 1772.

75. *Gazetteer and New Daily Advertiser*, Tuesday, 10 March 1772. This notice was not in the classified ads but in a section titled "News."

76. Dubourg to Franklin, 31 May 1772, Founders Online.

77. Biheron to Franklin, 10 September 1772, Founders Online.

78. As seen in the previous chapter, A. L. de Jussieu encouraged his mother to try some of Basseporte's cures in Lyon that worked in Paris despite the skepticism of medical faculty doctors. BCMNHN, MS 3515, letters 17 (31 May 1766), 19 (30 June 1766), and 20 (23 July 1766).

79. Diderot, *Correspondance*, XII, 164, 13 November 1772. Angelique, after she became Mme de Vandeul, spoke with displeasure about this harsh, fanatical uncle who wanted Diderot to recant everything he had ever written, which of course the philosophe refused to do. See Vandeul, "Vie de Diderot," in Diderot, *Oeuvres choisies*, 37–38.

80. Dubourg to Franklin, 28 October 1772, Founders Online.

81. Franklin to Hunter, 30 October 1772, Founders Online.

82. Franklin to Dubourg, torn, November 1772, Founders Online.

83. This April 1773 letter from Franklin appears in Jacques Barbeu-Dubourg, ed., *Oeuvres de M. Franklin . . .* 2 vols. (Paris: Quillau, 1773) I: 328–329.

84. The ads for her exhibit appeared multiple times, in the *Gazetteer* on 7, 8, 14, 30 January, 1, 4, 9, 11, 24 February, and in the *Public Advertiser* on 20, 21, 26, 29 January, 2, 8 March (now exhibiting only noon to 3 p.m.), 22 March, 13 April (shorter, every day at noon), 28 April (that she will stay until 6 May), 6 May (that she will stay until 4 June), and finally on 25 May that she will in fact close up on 4 June.

85. See, for example, the *Public Advertiser*, Tuesday, 2 March 1773.

86. She did not compromise, as did her compatriot, the "ingenious Cosmopolita, late from Paris" who in early 1772 had brought "Mathematical Amusements" but kept changing and lengthening his act to entice the public with novelty and discount pricing (7 and 26 February, 4 March).

87. This summary is based on various websites about money in England and colonial Williamsburg, and on Jenny Uglow, *The Lunar Men* (New York: Farrar, Straus and Giroux, 2003), 416.

88. In a letter to Franklin on 26 June 1773 Biheron, by then back in Paris, wrote that she was sending over a young French surgeon to continue work with Hewson. Founders Online.

89. In 1998 when renovations were done on Franklin's house at 36 Craven Street, two-hundred-year-old bones, including at least ten skeletons, were discovered buried in the basement. See Dr. Brian Owen-Smith's discussions of the discovery of these bones, as for example at https://wshoms.co.uk/about-2/.

90. See Tania Kausmally, "William Hewson and the Craven Street Anatomy School," in Piers Mitchell, ed., *Anatomical Dissection in Enlightenment England and France* (Farnham: Ashgate, 2012).

91. Delacoux, *Biographies des Sages Femmes*, 35: "No doubt that the protégée who had made a special study of the uterus and its annexes in the state of gestation, was the first teacher (maître) of the protector: because in comparing the works of the one with the other we see a perfect analogy, and in comparing the dates we recognize that the work of the English author is 7 years after a uterus molded in wax by mademoiselle Biheron that adorns the anatomical museum of Gottingen." See also Caroline Healey Dall, *Historical Picture Retouched* (Boston: Walker and Wise, 1860), 114: "But here again the genius of woman came to the aid of Science." And see Matilda Gage, *Woman as*

Inventor (Fayetteville, N.Y.: F. A. Darling, 1883), 487, who also asserts that Hunter learned from Biheron, not vice versa.

92. G. C. Peachy, in *A Memoir of William and John Hunter* (Plymouth, Eng.: Printed by the author, 1924) confirms that some of Hunter's uterus images were made in 1750 but others not until 1772, just when Biheron visited him.

93. Franklin to Hunter, 30 October 1772, Founders Online.

94. William Hunter, *Two Introductory Lectures* (London: J. Johnson, 1784), 56.

95. Biheron to Franklin, 26 June 1773, Founders Online.

96. After the Treaty of Paris in 1783 Bertier congratulated Franklin for the definitive American victory in the Revolution, lauding him as the father of a new people. And he then sent him his book reconciling physics and Genesis.

97. Dubourg to Franklin, September, no year, beginning torn, Founders Online.

98. Biheron to Franklin, 10 October 1774, Founders Online.

99. See Diderot's *Mémoires pour Catherine II*, "Chronologie," xxxviii. The letter was dated 25 June 1776.

100. Franklin to Mary Hewson, 12 January 1777; Mary Hewson to Franklin, 11 March 1778; Ingenhousz to Franklin, 15 November 1776, Founders Online.

101. See numerous letters in the Franklin papers, now at Founders Online, showing their offers of help in these various capacities.

102. *Gazette de France* (November 1773): 418 and 424, and *L'Avant-coureur*, no. 49 (6 December 1773): 768–769.

103. Diderot, *Mémoires pour Catherine II*, "Sur la maison des jeunes filles," 84–85.

104. Ibid., "Sur le même sujet: la maison des jeunes filles," 86–91.

105. Ibid., "Sur les leçons d'anatomie," 193–194.

106. Diderot, *Correspondance*, XIV, 44–47, letter to Betzky, 15 June 1774.

107. Diderot, *Correspondance*, 83, letter to Catherine, 13 September 1774.

108. Mme Vandeull, "Vie de Diderot," in Diderot, *Oeuvres choisies*, 28–32.

109. Gauthier de Simpré, *Voyage en France de M le Comte Falckenstein* (Paris: Cailleau, 1778), I, 179–187. He reports that this exhibit of artificial anatomy in Saint Petersburg opened "about 3 years ago" (that would be 1775) and that Biheron's "figure" is the most "curious and important" part of it.

110. *Gazette de France* (2 December 1774): 426. It was now open Wednesdays only, from eleven to one.

111. Charles Joret, *D'Ansse de Villoison et l'Hellénisme en France* (Paris: Honoré Champion, 1910), 61, 88, 477–479, 485.

112. *HARS pour l'année* 1770, 49.

113. Grimm et al., *Correspondance Littéraire*, IX (April 1771), 276–277.

114. *Nouvelles à la main sur la comtesse Du Barry trouvée dans les papiers du Cte de* *** (Paris: Plon, 1861), 255–256. "Pouah! Je ne sais pourquoi je parle de cela!"

115. *État de médecine . . . . 1776* (Paris: Didot jeune, 1776), 230.

116. Buffon, "Supplément" of vol. IV of *Histoire Naturelle* (servant de suite a *l'Histoire naturelle de l'homme*), an engraving of a one-eyed child "d'après une tête en cire faite par Mlle Biheron. L'enfant ayant servi de modèle est née en octobre 1776 et n'a vécu que quelques heures."

117. Gauthier de Simpré, *Voyage en France de M le Comte Falckenstein.* See the discussion of Biheron, I, 179–187, which includes a long excerpt from the *Avant-coureur* of 1761, no 2.

118. Riballier and Cosson, *De l'Education physique,* 152–154. These words were written in 1775, judging from the dates provided.

119. Chastellux, *Voyage de M le Marquis de Chastellux dans l'Amérique Septentrionale* (Paris: Prault, 1788), I, 196. He was referring to the eccentric doctor Abraham Chovet whose wax anatomies went on display in Philadelphia around 1774.

120. Archives de l'École de Pharmacie, registre n. 48, p. 26. The described location indicates that she was still at the rue des Poules address.

121. R.-N. Dufriche Des Genettes wrote in 1793 that citoyenne Biheron was now very old but for more than thirty years she displayed her cabinet, so that would suggest, given that it opened in 1761, that it was still at least occasionally open after 1791. *Journal de physique,* 43 (1793): 83.

122. See AN MC ET LXXIII/1008, 16 March 1780, where she sets up payment in gold and silver now and the rest in *rentes,* to be inherited by her loyal domestic Marie Elisabeth Vibart if Biheron died first. She signs with a confident signature.

123. Claude-Denis Cochin, *Compte rendu à l'assemblée de charité de Saint-Jacques-du-Haut-Pas tenue le 7 août 1783, par M. Cochin, . . . exécuteur testamentaire de feu M. Cochin, . . . curé de ladite paroisse, de la situation de l'hospice pour les pauvres malades . . .* (Paris: Desprez, 1783).

124. See Louis-Sébastien Mercier, *Tableau de Paris* (Hambourg et Neuchatel, 1781) "Anatomie," I, 152–154, and "Raretés," II, 337. Mercier wrote that the commerce of anatomists with gravediggers was repulsive, that anatomy students were encouraged to climb cemetery walls and unearth coffins with their freshly buried cadavers, that there was a real danger of encountering one's father, brother, or friend buried properly the night before on the black marble dissection slab. Lofty professors should incur severe punishments for violating tombs, hashing the corpses to bits, and disposing of the pieces in the river, sewers or latrines. Biheron, in contrast, taught from her faithful imitations without the revulsion of handling real bones that seemed to shudder in one's hands.

125. In 1786 when the queen purchased Biheron's collection, her children were eight, five, and one. Was she really picturing anatomy lessons for them, or more probably for some adult courtiers instead? The purchase price was 9,000 livres, 6,000 paid immediately, and the other 3,000 to be paid over time in *rentes.* But the records are full of holes. The *Archives Parlementaires* of December 1789 shows her receiving 1,500 for the second half of that year, so perhaps she had been getting 1,500 twice a year since 1786? If the royal treasury was just trying to hit the requested 3,000, the payments

would have been done by then and the obligation over. But Biheron expected funds to continue, so the overall purchase price was therefore much higher. See *Archives Parlementaires*, XVI (6 June 1790): 125, another mention of her pension of 3,000. And again, more semi-annual pensions promised in *Archives Parlementaires*, XXVI, 34, *séance du vendredi 13 mai, 1791, article 26:* "A Marie-Anne-Marguerite Bihéron, tant pour compléter les pièces d'anatomie artificielle qui composent son cabinet . . . soit les démonstrations qu'elle est chargée d'en faire à la famille royale."

126. According to Boulinier, this inventory of 106 pieces was found in Biheron's home at her death, along with another by Thillaye that added 23 more pieces; the two together made possible the drafting of her inventaire après décès on 27 July 1795, AN MC ET LXXIII/1140.

127. See in *La Médecine Internationale et Illustrée, journal mensuel* (September 1909): 289–292, an article about Biheron by a Dr. Rondelet who writes that this note was found in an anatomy "carton," no doubt the same carton that will be referred to later by the committees of the National Convention.

128. A year earlier the minister Roland, in a letter of 23 December 1792, had agreed that her 3,000 annual pension be continued, and Delaporte and Fourcroy concurred in a letter of 22 January 1793, but then things got bogged down. See Archives et Manuscrits de la Bibliothèque de l'Académie de Médecine, SRM C195, dossier 4.

129. Bibliothèque de l'Académie de Médecine, "État des pièces d'anatomie artificielle qui composent le cabinet de Mlle Biheron, dressé par Vicq d'Azyr," SRM C195, dossier 4, pièce n 2, pièces 48 à 77. An initial list of 106 pieces dated 12 April has a later complement with 23 more models, "État dressé par Thillaye" (member of the commission des arts), for a total of 129 objects.

130. See *Procès-verbaux de la commission temporaire des arts II. 5 Nivôse an III–5 Nivôse an IV,* publiés et annotés par M. Louis Tuetey (1912), I, 59, 61, 233, 278; and *Procès-verbaux du Comité d'instruction publique de la Convention,* par M. J. Guillaume, IV, 452–453 and V, 411.

131. Marie-Jacques Barrois sold scientific books, publishing among other things the geographic works of Edmé Mentelle, the eulogist and dear friend of Basseporte who knew Biheron well.

132. AN MC ET LXXIII/1139, 4 Messidor an III (22 June 1795).

133. See AN MC ET LXXIII/1140, 27 July 1795, inventaire après décès de Mlle Biheron, and AN AJ/16/6563—documents from the revolutionary years IV and V concerning Mlle Biheron's cabinet.

134. AN AJ/16/6563, letter of 7 Ventôse an IV (26 February 1796).

135. *Magasin encyclopédique, ou Journal des sciences, des lettres et des arts,* seconde année, tome quatrième, rédigé par Aubin-Louis Millin, 1796, under "Nouvelles littéraires," 414–415.

136. *Magasin encyclopédique,* Prairial, an VII (roughly May–June 1799). "Son cabinet, auquel il a joint celui de Mlle Biheron, se voit rue Hautefeuille." On Bertrand-

Rival, as he called himself, see Jean-Marie Le Minor, "Le Cabinet de cires médicales du cyroplasticien J. F. Bertrand à Paris," *Histoire des Sciences Médicales* 33, no. 3 (1999): 275–286. Bertrand displayed mostly abnormal and diseased states to encourage righteous and virtuous living in visitors.

137. Fragonard had a different approach, injecting wax-like materials into the arteries and veins of actual cadavers. See Christophe Degueurce, "The Celebrated Ecorchés of Honoré Fragonard: Part I: The Classical Techniques of Preparation of Dry Anatomical Specimens in the 18th Century," *Clinical Anatomy* 23 (2010): 249–257, and "Part II:" 258–264. He used mutton tallow and pine resin diluted in essence of turpentine and essential oils, and a varnish of Venice turpentine, made from larch resin and known to repel insects. Some of these ingredients may well have been inspired by Biheron.

CHAPTER FIVE. CHEMIST AND EXPERIMENTALIST

1. AN MC ET LI/1248, 6 Nivôse, an XIV [27 December 1805]. "Ouverture et dépot du Testament mistique de Mme Geneviève Charlotte Darlus, veuve Louis Lazare Thiroux d'Arconville." The testament itself had been dictated two years before, on 28 and 31 December 1803.

2. Condorcet, "Éloge de Macquer," in *Oeuvres de Condorcet* (Paris: Firmin Didot frères, 1847), III, 138.

3. Two books have recently appeared on Mme d'Arconville. They treat her literary and moral works as well as her science. See Patrice Bret and Brigitte Van Tiggelen, eds., *Madame d'Arconville, 1720–1805: une femme de lettres et de sciences au siècle des Lumières* (Paris: Hermann, 2011), and the newer *Madame d'Arconville, moraliste et chimiste au siècle des lumières; Études et textes inédits,* ed. Marc André Bernier and Marie-Laure Girou-Swiderski (Oxford: Voltaire Foundation, 2016). On aspects of her science, see articles by Elisabeth Bardez, Margaret Carlyle, Brigitte Van Tiggelen, and myself in the first, and Bardez, Carlyle, and Sarah Benharrech in the second. See also the online article by Marie-Laure Girou-Swiderski, "Écrire à tout prix. La présidente Thiroux d'Arconville, polygraphe, 1720–1805," at http://aix1.uottawa.ca/~margirou /Perspectives/XVIIIe/arconvil.htm.

4. There is a portrait of him in 1733 by Hyacinth Rigaud for which he paid 600 livres. It is now at the Château de Cheverny.

5. This and much of the other biographical material in this chapter comes from d'Arconville's twelve-volume unpublished memoirs, *Pensées, réflexions et anecdotes,* referred to henceforth as *PRA*. This autobiographical manuscript is at the University of Ottawa, where Professor Marie-Laure Girou-Swiderski has been good enough to share some portions with me over the years, and to which I was granted complete digital access in 2019, for which I thank her. See "Histoire de ma littérature," *PRA,* V, 170.

6. "Histoire de mon enfance," *PRA,* III, 311–489.

7. Born 1712 in Bourges, he joined the *parlement* de Paris in 1732, became president

of the first Chambre des Enquêtes in 1748, and retired in 1758. He too was painted for 600 livres by Rigaud, in 1743, as his sister had been in 1742. His wife, Mlle Darlus before her marriage, had a youthful pastel done of her by Charles Antoine Coypel, but no painting by Rigaud.

8. Françoise Vaysse, "Mme Marie-Geneviève-Charlotte Thiroux d'Arconville: La Plume et la Cornue," *Bulletin de la Société de l'histoire du 3e arrondissement de Paris,* no. 51 (February 2008): 39–52. Mlle Vaysse's father wrote a history of Crosne, and her family has occupied the dove-house, which once belonged to Mme d'Arconville.

9. A long, minutely detailed 1748 death inventory of André Gillaume Darlus was generously given to me by Françoise Vaysse, showing the layout of the vast property.

10. Elisabeth Bardez, "Mme d'Arconville a-t-elle sa place dans la chimie du XVIIIe siècle?," in Bernier and Girou-Swiderski, eds., *Madame d'Arconville, moraliste et chimiste,* 161–182, especially 169–170.

11. [Hippolyte de La Porte], *Notices et observations à l'occasion de quelques femmes de la société du 18e siècle* (Paris: H. Fournier, 1835), 14–15. A portrait of her at the Château de Cheverny by Alexander Roslin, however, shows her with a great deal of rouge, perhaps put on just for that occasion.

12. *Mérope* tells of a woman who mourns a long-lost son, convinced he has been murdered, but who is eventually restored to her rightful throne by the very boy she thought gone forever when he returns and kills the usurper who had deposed her.

13. "Sur Moi," *PRA,* XI, 151.

14. "Sur Moi," *PRA,* XI, 160–161. Emphases hers.

15. "Histoire de ma littérature," *PRA,* V, 169–226, 182; "Mes souvenirs," *PRA,* IX, 335.

16. Carolyn Heilbrun, *Writing a Woman's Life* (New York: W. W. Norton, 1988), 116–117.

17. Such biographical details come from her grand-nephew Bodard and from Hippolyte de La Porte, who knew Bodard's work and added to it based on additional material from another relative, Mme Joubert (whose portrait is at the Getty in Los Angeles). See M. Bodard [Pierre Henri Hippolyte Bodard de la Jacopière], *Cours de botanique médicale comparée,* tome I (Paris: Mequignon, 1810), in the long footnote, xxvi–xxx, and [Hippolyte de La Porte] *Notices et observations à l'occasion de quelques femmes,* 13–39.

18. Hippolyte de La Porte, *Notices,* 16–17, 32.

19. "Sur ma mélancolie," *PRA,* VI, 3–18.

20. It was displayed in the Salon of 1763, and she probably acquired it shortly thereafter.

21. "Sur la mélancolie," *PRA,* I, 97–109, especially 98, 100–101, 108–109.

22. The many volumes of memoirs are filled with the term "faim canine" to describe her insatiable curiosity.

23. *Pensées et réflexions morales sur divers sujets* (Avignon [Paris], 1760), 149–150.

24. See "Sur le mariage," *PRA*, I, 51–52, and Hippolyte de La Porte, *Notices*, 20–22. See Elisabeth Badinter's controversial study of this phenomenon in her *L'Amour en Plus: Histoire de l'amour maternel (XVII–XXe siècle)* (Paris: Flammarion, 1980).

25. "Sur l'intérieur des ménages," *PRA*, VI, 19–38, 33.

26. See her early "Sur le mariage" in *Pensées et réflexions*, 92–97, and memoirs on the same subject half a century later, in her manuscript "Sur le mariage," *PRA*, I, 16–54, especially 45–47. See also Julie Chandler Hayes, "Réflexions sur le mariage: Mme d'Arconville et la tradition moraliste," in Bernier and Girou-Swiderski, eds., *Madame d'Arconville*, 121–136.

27. "Sur la liberté," *PRA*, I, 351.

28. "Mes voyages," *PRA*, V, 226–292. Her sister's marriage, in contrast and exceptionally, seems to have been lastingly happy. See "Par Mme d'Alleray à son mari," *PRA*, I, 411–415, and the ten-page poem (with its own pagination) that follows. D'Arconville refers frequently to the beauty of this particular and rare union.

29. "Du célibat," *PRA*, XI, 137–138. The emphases are hers. See also "Sur la solitude," I, 300ff.

30. "Sur la botanique," *PRA*, IX, 26. See also her preamble to *PRA*, V.

31. "Sur l'agriculture," *PRA*, II, 93.

32. Sarah Benharrech, "L'anti-Tournefort, ou la botanique d'une paresseuse," in Bernier and Girou-Swiderski, eds., *Madame d'Arconville*, 211–220. See also Yvonne Letouzey, *Le Jardin des Plantes à la croisée des chemins avec André Thouin, 1747–1824* (Paris: Edition du Muséum National d'Histoire Naturelle, 1989), 51.

33. BN MS letters to Macquer, now conveniently digitized on Gallica.

34. Félix Vicq d'Azyr, "Éloge de Macquer," in *Oeuvres de Vicq d'Azyr* (Paris: Duprat-Duverger, 1805), I, 300–305.

35. Condorcet, "Éloge de Macquer," in *Oeuvres de Condorcet* (Paris: Firmin Didot frères, 1847), III, 137.

36. Letter of 19 October 1775, as quoted in W. C. Ahlers, "Un chimiste du XVIIIe siècle: Pierre Joseph Macquer, 1718–1784" (PhD diss., Paris, La Sorbonne, 1969), 20–21. See also Vicq d'Azyr, "Éloge de Macquer," in *Oeuvres de Vicq d'Azyr*, I, 301–303, on Macquer's inclination to flee the "whirlwind" [*tourbillon*] of society.

37. Vicq d'Azyr, "Éloge de Poulletier de la Salle," in *Oeuvres de Vicq d'Azyr*, II, 1–18, especially 13–14.

38. Karl Feltgen, private communication. See also his article "Poulletier de la Salle et la découverte du cholesterol," *Groupe d'histoire des Hôpitaux de Rouen*, 26 October 1994. Poulletier isolated the crystals of cholesterol around 1758 but really clinched it in 1769. See also *Essai pour servir à l'histoire de la putréfaction. Par le traducteur des Leçons de chymie de M. Shaw* (Paris: Didot le Jeune, 1766), xxvi.

39. Poulletier's translation of Pemberton, *Pharmacopée du collège royal de médecine de Londres*, appeared in 1761, two years after d'Arconville's two scientific translations from English works.

40. *Essai pour servir,* v.

41. Ibid., xxvi–xxvii.

42. "Sur Moi," *PRA,* XI, 176. See also Benharrech, "L'anti-Tournefort," 216–218, and *PRA,* IX, "Sur la botanique," 28, as quoted in Bernier and Girou-Swiderski, eds., *Madame d'Arconville, moraliste et chimiste,* 154, note 26.

43. Malesherbes is discussed throughout the memoirs. See, for example, "Portrait de M de Malesherbes par M de Ségur," *PRA,* II, 398–401; "Sur la botanique," IX, 30–33; and "Des Abus," XI, 272–280.

44. See Christine Lehman, "Pierre-Joseph Macquer: Chemistry in the French Enlightenment," *OSIRIS* 29 (2014): 235–261.

45. See Mi Gyung Kim, *Affinity, That Elusive Dream: A Genealogy of the Chemical Revolution* (Cambridge: MIT Press, 2003), 201–215, 220.

46. See Walter Hamilton, *French Book-Plates* (London: G. Bell, 1896), 323–324. For discussions of the significance of such markers, see Norna Labouchere, *Ladies' Book-Plates: An Illustrated Handbook* (London: George Bell and Sons, 1895) and Germaine Meyer-Noirel, *L'Ex Libris* (Paris: Picard, 1989).

47. See Gérard Genette, *Seuils* (Paris: Éditions du Seuil, 1987), trans. as *Paratexts: Thresholds of Interpretation* (Cambridge: Cambridge University Press, 1997). He discusses paratexts as powerful zones that frame, influence, and even control the very readings of the text itself.

48. See her "Translator's Note" preceding Lord Halifax, *Avis d'un père à sa fille* (London [Paris], 1756). See also Marie-Pascale Pieretti, "Women Writers and Translation in 18th Century France," in *French Review,* 75, no. 3 (February 2002): 474–488, 479. Derrida famously said all translations are transformations, and d'Arconville's surely were.

49. See "Translating and Translations in the History of Science," the special issue of *Annals of Science* 73, no. 2 (2016), especially Patrice Bret, "The Letter, the Dictionary and the Laboratory: Translating Chemistry and Mineralogy in Eighteenth Century France," 122–142. The translating team that was created in Dijon later in the century by Macquer's friend Guyton de Morveau tested the laboratory findings in the original works, a precedent set by d'Arconville who likely served as a model for his enterprise.

50. Adeline Gargam, "La Chair, l'os et les éléments: 'l'heureuse fécondité' de la traduction scientifique au XVIIIe siècle: Le Cas de Marie-Geneviève Thiroux d'Arconville," in Guyonne Leduc, dir., *Les Rôles transfrontaliers joués par les femmes dans la construction de l'Europe* (Paris: L'Harmattan, 2012), 59–77.

51. *Traité d'ostéologie, traduit. de l'anglais de M. Monro, professeur d'anatomie et de la Société Royale d'Edinbourg, ou l'on a ajouté des planches en taille-douce qui représentent au naturel tous les os de l'adulte et du fœtus avec leurs explications, par M Sue, professeur et démonstrateur d'anatomie aux Écoles Royales de Chirurgie, de l'Académie Royale de Peinture et de Sculpture,* 2 vols. (Paris: Guillaume Cavalier, 1759).

52. Nina Rattner Gelbart, "Splendeur et squelettes; la "traduction" anatomique de

Madame Thiroux d'Arconville," in Bret and Van Tiggelen, eds., *Madame d'Arconville, 1720–1805*, 55–71.

53. Badinter, *Les Passions intellectuelles*, II, 252–254.

54. D'Arconville's explanations of these images in the *Traité d'ostéologie* are in I, i.

55. M. Bodard, *Cours de botanique*, note xxvi–xxx.

56. See Nina Rattner Gelbart, *The King's Midwife: A History and Mystery of Mme du Coudray* (Berkeley: University of California Press, 1999), 49, 74.

57. See Londa Schiebinger, *The Mind Has No Sex?* (Cambridge: Harvard University Press, 1989), 189–213, and her "Skelettestreit," *ISIS* 94 (2003): 307–313.

58. *Éléments d'anatomie à l'usage des peintres, des sculpteurs et des amateurs, ornés de 14 planches, en taille-douce, représentant au naturel tous les Os de l'adulte et ceux de l'enfant du premier âge, avec leur explication. Par M. Sue le fils* . . . (Paris: Méquignon, 1788), although it was widely known to have been written by his father of the same name, 46.

59. Gelbart, "Splendeur et squelettes," 58–67.

60. Ibid., 57–58, 67–70.

61. *Leçons de Chymie, propres à perfectionner la physique, le commerce, et les arts. Par M Pierre Shaw, premier médecin du Roi d'Angleterre, Traduites de l'Anglais* (Paris: Herissant, 1759). For her notes that contradict Shaw's findings, see, among others, 21, 26, 60, 143, 148, 280, 281, 349, 360, 419, and 443.

62. *Leçons de Chymie*, v.

63. Ibid., xlii.

64. In her memoirs d'Arconville made clear her distaste for war. She would also revisit this notion that not all discoveries were benign, and many downright dangerous. See, for example, "Sur les découvertes et la navigation," *PRA*, XI, 111–128.

65. *Leçons de Chymie*, xxiv. D'Arconville, like Ferrand, was well acquainted with Graffigny's 1747 *Lettres d'une Péruvienne*, which by 1759 had gone through numerous editions.

66. Ibid., xxxiii.

67. Steven Shapin and Simon Schaffer, *Leviathan and the Air-Pump: Hobbes, Boyle and the Experimental Life* (Princeton: Princeton University Press, 1985), made good use of this concept.

68. *Leçons de Chymie*, xciv.

69. *Journal de médecine* 10 (April 1759): 291–300. These remarks are on 292, 294, and 300. The *Journal des sçavans*, the *Journal encyclopédique*, and the *Mémoires de Trévoux* all agreed that the *Leçons de Chymie* and especially its long introduction were extremely edifying and deserved applause.

70. "Histoire de ma littérature," *PRA*, V, 189.

71. *Leçons de Chymie*, lxxxiv.

72. *Pensées et réflexions*, 73.

73. Ibid., 76.

74. Ibid., 84.

75. See Andrew Sparling, "Putrefaction in the Laboratory: How an 18th Century Experimentalist Refashioned Herself as an *Hommes de Lettres*," in QUER*ELLES* 10 (2005): 173–188. For the response, see Benharrech, "L'anti-Tournefort," 213–216, and Bernier and Girou-Swiderski, "Présentation," 25–29, in their *Madame d'Arconville, moraliste et chimiste.*

76. See *Pensées et réflexions*, 104–105, and *Mélanges*, I, 391–392. See also her section "Sur le mariage" in *Pensées et réflexions*, 92–97.

77. *Pensées et réflexions*, 160.

78. Ibid., 72.

79. *De l'Amitié* (1761), 83.

80. Ibid., 106–135.

81. BN MS fr 12306, fol. 203.

82. *Des Passions*. Par l'auteur du *Traité de l'Amitié* (London, 1764), vii–viii.

83. Ibid., 122.

84. Ibid., 2, 6.

85. Ibid, 19, 23, 29, 30, 32, 100, 106–107.

86. Joseph de la Porte and Jean-Francois de la Croix, "Lettre XXXIII," *Histoire littéraire des femmes françaises*, IV (Paris: Lacombe, 1769), 543.

87. Grimm et al., *Correspondance Littéraire* (Paris: Furne, 1829–31), IV (October 1764), 93–94.

88. The two treatises appeared as *Oeuvres Morales de Diderot, contenant son traité de l'Amitié et celui des Passions* (Francfort aux dépenses de la compagnie, 1770). They also appeared translated into German that same year, again falsely attributed to Diderot.

89. Elisabeth Bardez, a chemist, has addressed this claim in several articles. She recapitulates her thoughts in, most recently, "Mme d'Arconville a-t-elle sa place dans la chimie du XVIIIe siècle?," in Bernier and Girou-Swiderski, eds., *Madame d'Arconville* (2016), 161–182.

90. Newton, *Opticks*, 4th ed. (London: William Innys, 1730), Query #30, 350ff., with more mentions of putrefaction specifically in Query #31.

91. John Pringle, "Experiments upon Septic and Antiseptic Substances, with Remarks Relating to Their Use in the Theory of Medicine," *Proceedings of the Royal Society* (1750): xlvii.

92. *Essai pour servir*, iii.

93. Ibid., xviii.

94. Ibid., xxxv–xxxvi.

95. Macquer's approbation 10 October 1765, Privilege 14 November, Registered 10 December 1765.

96. *Essai*, i.

97. Ibid., vii, x, xi, xii.

98. Ibid., xxiii, xxviii.

99. Ibid., 77.

100. Ibid., 2, 5, 357, 367, 369.

101. Lissa Roberts, "Chemistry on Stage: G. F. Rouelle and the Theatricality of Eighteenth-Century Chemistry," in Bernadette Bensauade-Vincent and Christine Blondes, eds., *Science and Spectacle in the European Enlightenment* (Aldershot: Ashgate, 2008), 136.

102. Alain Corbin, *The Foul and the Fragrant* (Cambridge: Harvard University Press, 1986), 19. See also Sabine Barles, *La Ville délétère: médecins et ingénieurs dans l'espace urbain XVIII–XIX siècles* (Seyssel: Champ Vallon, 1999), 65–67, 92.

103. Some specific criticisms of Pringle are in the *Essai*, iv, xxi, 54, 62, 90 note a, 91 note b, 93, 99, 118 notes a and b, 184, 269, 453.

104. *Essai*, xxvi, 9, 11, 19, 21, 38, 60, 191, 200, 223–225, 268, 294, 455.

105. *Essai*, iii, xviii, xxiv.

106. *Essai*, xxxii.

107. See for some instances of this in the *Essai*, xii, xiv, xxxii, 37, 46, 54, 119, 176, 363.

108. *Essai*, 164, 303.

109. Ibid., 374.

110. Ibid., 21.

111. For some examples, see ibid., 9, 185, 191, 257, 268, 294.

112. Ibid., xxix.

113. Ibid., 542, 546–547.

114. Ibid., 374, 461–462.

115. "Histoire de ma littérature," *PRA*, V, 194.

116. See, for example, the detailed attention given to her as a "pioneer of the study of antisepsis" in Sir Henry Solomon Wellcome, *The Evolution of Antiseptic Surgery: An Historical Sketch of the Use of Antiseptics from Earliest Times* (London: Burroughs Wellcome, 1910), 52–57.

117. Macquer—though his name does not appear on the title page—*Dictionnaire de chymie* (Paris: Lacombe, 1766), II, 338–339.

118. *Journal oeconomique* (May 1766): 224–228. The unknown author of this review confirmed many of the findings in the *Essai*.

119. *Journal des sçavans* (October 1766): 683–687.

120. David Macbride, *Experimental Essays*, 2nd ed. (1767), 116 note.

121. Her husband had another brother just a year younger than he, Thiroux de Montregard, who was a regular at brothels. Not surprisingly, he does not figure in her reminiscences. I thank Kate Norberg for information on this less than savory sibling-in-law.

122. "Histoire de ma littérature," *PRA*, V, 201–202. See also "Mes souvenirs," IX, 327–366, especially 338–347.

123. See the review of this and all of d'Arconville's other works in Joseph de la Porte, *Histoire des femmes qui se sont rendues célèbres dans la littérature française* (Paris: Costard, 1771), IV, 542–599. The author knew who she was but respected her secret.

124. "Sur la peinture et la sculpture," *PRA*, II, 3–20. See especially 11–12, 19.

125. Diderot, *Correspondance,* ed. George Roth, VIII, 27–46, in long letter to his good friend Falconet in May 1769. Catherine seems to have settled for a copy of the statue, as there was a larger-than-life one in the Hermitage.

126. "De la philosophie et des philosophes," *PRA,* X, 353–441, especially 401–441. For her thoughts on Rousseau, see "Examen de l'ouvrage de M. J.-J. Rousseau," *PRA,* VIII, 22–66. Rousseau was in her circle during the 1750s and 1760s because a playing card in his possession—there were twenty-seven of these on which he famously scribbled notes for his *Rêveries du promeneur solitaire*—had on the back of it an invitation to visit Thiroux d'Espersenne and one of his uncles La Curne, to whom d'Arconville devoted a fond discussion in her memoirs. Oddly, Rousseau held onto this card for a long time, as d'Espersenne died in 1767 and the *Rêveries* were written a decade later.

127. BN MS fr 12305, fols. 19, 21–22, 23–24 for d'Arconville's letters, and BN MS fr 12306, fols. 89, 91, 131, 206 for those written by others to Macquer but concerning her.

128. *Vie de Marie de Médicis,* III, 515–516.

129. *Mélanges de littérature, de morale, et de physique.* Ed. Rossel, 7 vols. (Amsterdam: aux dépense de la Compagnie), I, "Avertissement de l'éditeur," i–iv.

130. *Mélanges* (1775), I, 417.

131. "Histoire de ma littérature," *PRA,* V, 197–199. Why she selected articles from the year 1720 of the TRS is unknown, perhaps because it was the year of her birth?

132. All of these take up pages 217–349 of the third volume of the *Mélanges.*

133. Macquer, *Dictionnaire de chymie,* 2e edition (1778), II, Tables des Auteurs, 847, lists Arconville (Madame la Presidente d') and then still in the table mentions articles involving her: I, "Fiel des animaux," 50 (his index comment refers to "a modern work, filled with research done with the greatest care, and called *Essai pour servir. . . . putréfaction*"); II, "Putréfaction," 303–304; sel commun, 389 (his index comment refers to her as "the exact and learned author of the *Essai . . . putréfaction*"), and sel neutres, 408.

134. Jean-Pierre Poirier, *Lavoisier, Chemist, Biologist, Economist,* trans. Rebecca Balinski (Philadephia: University of Pennsylvania Press, 1998), 52, 112.

135. Ibid., 113.

136. Ibid., 112–113.

137. Hippolyte de La Porte, *Notices,* 36–38.

138. Jean-Baptiste-Modeste Gence (1755–1840) was a prolific archivist and grammarian. For his admiration of d'Arconville, see his *Stances lyrique et morales à la savante et modeste Lettrée du dix-huitième siècle, Mme d'Arconville née à Paris en 1720 et morte en 1805* (Paris: Thomassin et Cie, n.d.).

139. "Histoire de ma littérature," *PRA,* V, 221.

140. Préface, *Vie de Marie de Médicis,* 3 vols. (Paris: Ruault, 1774), 1, ii–vi.

141. "Sur la politique," *PRA,* I, 186, and "Sur l'histoire" *PRA,* I, 226–233.

142. Gelbart, *King's Midwife,* 210. He was intelligent, sympathetic, helpful, and forced recalcitrant surgeons to attend her enlightened courses.

143. "Sur Moi," *PRA*, XI, 210–212, 221–225.

144. See "Portrait de M. de Malesherbes par M de Ségur," *PRA*, II, 398–401. The respected agronomist and physiocrat Louis-Paul Abeille said d'Arconville saved Malesherbes's manuscript attacking Buffon's *Histoire Naturelle*—a manuscript that without her prescience would have been lost forever. A copy was found "in the hands of a woman worthy in every respect of the title virtuoso. Connected to most of our best-known savants, she herself cultivated, and fruitfully, the sciences and the arts, [including] a good translation of a useful and voluminous treatise on a science hardly any person of her sex practiced." Abeille, ed., *Observations de Lamoignon de Malesherbes sur l'histoire générale et particulière de Buffon and Daubenton*, 2 vols. (Paris: C. Pougens, 1798), "Introduction."

145. "Sur Moi," *PRA*, XI, 148–230, 211–212. See also her description of the Revolution and the "brigands" who overwhelmed France in "Parallèle entre Charles I roi d'Angleterre et Louis XVI roi de France," *PRA*, XII, 211–276.

146. "Sur Moi," *PRA*, XI, 230.

147. L. Quarré-Reybourbon, *Pascal François Joseph Gosselin, Géographe Lillois* (Lille: L. Quarré, 1887), pp. 4, 8, 16, and note on 16. See also Gence, *Stances lyrique et morales*, which sheds light on many things including her closeness with Gosselin after the Terror, and a fictional conversation by Gence between her, Malesherbes, and the bibliophile Barbier that reiterates this, in *Annales de la littérature et des arts*, 34 (1829): 106–109. See also Hippolyte de La Porte, *Notices*, 26.

148. "Sur les projets," *PRA*, VI, 356–370, 368.

149. "Histoire de ma littérature," *PRA*, V, 221–224.

150. Her "embonpoint," she said, was the torment of her life. *PRA*, V, opening comments. On Jeanroy, who seems to have been a particularly caring medical authority, see Tessier, *Notice sur M. [Dieudonné] Jeanroy* . . . a nine-page pamphlet printed by Migneret, no date, and many other speeches made at his funeral. Gence's *Stances* refers to Jeanroy and Nauch as her last doctors, and Jacques Nauch contributed often to the *Journal du Galvanisme* and became president of the Société Galvanique in 1802.

151. "Sur la médecine," *PRA*, II, 108–116.

152. "Sur la médecine et la chirurgie," *PRA*, II, 42–56, 51. C.-N. Deslon (or d'Eslon) was one of Mesmer's more influential disciples.

153. "Sur la mort de Mme d'Alleray," *PRA*, II, 177–181.

154. "Sur la vieillesse et la mort," *PRA*, II, 171–176.

155. "Histoire de ma littérature," *PRA*, V, 169–225, 181.

156. Joseph de la Porte, *Histoire littéraire des femmes françaises; Oeuvres complètes de Diderot*, ed. J Assezat, IX (Paris, 1875), 453–455; see Bardez, "Mme d'Arconville a-t-elle sa place dans la chimie du XVIIIe siècle?," 179, note 96; John Berkenhout, "First Lines of the Theory and Practice of Philosophical Chemistry," Chart with no page number (London: T. Cadell, 1788); Abeille, ed., *Observations de Lamoignon de Malesherbes*, "Introduction"; William Higgins, *An Essay on the Theory and Practice of Bleaching*

(London: Printed for the Author, 1799), 3–4; Marguerite Ursule Fortunée Briquet *Dictionnaire historique, littéraire, et bibliographique des Françaises et des étrangères naturalisées en France* (Paris: Gillé, 1804), 13–14.

157. M. Bodard, *Cours de botanique*, note, xxvi–xxx.

158. Hippolyte de La Porte wrote the entry for Michaud's *Biographie universelle* on Thiroux d'Arconville (volume 45) and then a supplement the next year, 1829, in which he mentioned new information from Marie-Louise Joubert, a young relative of Poulletier de la Salle who knew d'Arconville and her sons well. See Hippolyte de La Porte's "Un dernier mot sur Madame Thiroux d'Arconville," in *Annales de la littérature et des arts* 34 (1829): 141–142, and 203–206. And then finally, see his 1835 *Notices et observations à l'occasion de quelques femmes*. Naturalist Georges Cuvier named her, in 1841, as the true author of the "magnificent edition" of Monro. See his *Histoire des sciences naturelles depuis leurs origines jusqu'à nos jours, chez tous les peuples,* III (Paris: Fortin, Masson et Cie, 1841), 233.

159. "Préambule," *PRA*, XII, 131, note.

160. "Réflexions sur les arts," *PRA*, X, 269, 278–281.

161. See Brigitte Van Tiggelen, "Entre anonymat et traduction: la carrière d'une femme en science," in Bret and Van Tiggelen, eds., *Madame d'Arconville (1720–1805)*, 93–111.

162. Bodard, *Cours de botanique*, xxvii, note. "N'ayant qu'elle seule pour diriger ses études, elle put dire avec raison qu'elle était elle-même son ouvrage."

EPILOGUE

1. Fourcroy, *Bibliothèque universelle des dames: Principes de Chimie* I (Paris: Rue et Hôtel Serpente, 1787), xix.

2. For arenas outside the traditional ones, see Katherine Park and Lorrain Daston, eds., *The Cambridge History of Science* 3: *Early Modern Science* (Cambridge: Cambridge University Press, 2006), Part II, "Personae and Sites of Natural Knowledge," 177–362.

3. *The Memoirs of Elisabeth Vigée-Le Brun* (London: Camden Press, 1989), 49.

4. As quoted in Lynn Hunt, *The French Revolution and Human Rights: A Brief Documentary History* (Boston and New York: Bedford/St Martins, 1996), 62.

5. Sydney Ross, "Scientist: The Story of a Word," *Annals of Science* 18, no. 2 (1962): 65–85.

6. See, for example, Heather Ellis, "Knowledge, Character and Professionalization in 19th Century British Science," *Journal of the History of Education Society* 43, no. 6 (2014): 777–792.

7. Ehrenreich and English, *For Her Own Good: Two Centuries of the Experts' Advice to Women,* 2nd ed. (New York: Anchor, 2005).

8. Evelyn Fox Keller, *A Feeling for the Organism* (San Francisco: W. H. Freeman, 1983).

9. See Marika Hedin, "A Prize for Grumpy Old Men? Reflections on the Lack of Female Nobel Laureates," *Gender and History* 26 (2014): 52–63.

10. Elisabeth Pain, "Action and Data for Women in Science: A French Example," *ScienceMag.org*, 13 February 2017.

Index

Italicized page numbers indicate illustrations and figures.

Smolny Institute for Girls, 197–198

Söderqvist, Thomas, 13, 270–271n40

Sommerville, Mary: *The Connexion of the Physical Sciences*, 263

Sonnerat, Pierre, 131, 145

La Specola museum (Florence), 209

Spinoza, Baruch, 252

spontaneous generation, 241, 243

Stahl, Georg Ernst, 227, 236

Stamp Act repeal, 195

Stevenson, Madame, 187, 188, 189

Strayer, Brian E., 305n42

Stuarts: Charles Edward Stuart (Young Pretender), Ferrand's hiding of, 6, 16, 18, 34–35, 37–38, 40, 43, 59; French support for, 18, 25, 34–35; James Francis Edward Stuart (Old Pretender), 23

Sue, Jean-Joseph, 167, 221–222, 225; works of: *Éléments d'anatomie à l'usage des peintres, des sculptuers et des amateurs*, 225–226

Talmont, Princesse de, 35, 37–38, 40, 41, 258

Tenon, Jacques-René, 200

Terrall, Mary, 273n14

Terror. *See* Reign of Terror

Thiroux de Crosne, Louis, 212, 244, 250–251

Thiroux de Gervillier, 244

Thiroux de Mondésir, 244, 251

Thiroux de Montregard, 257, 318n121

Thiroux d'Espersenne, 231, 244–245, 257, 319n126

thought experiments, 46, 279n132

Thouin, André, 105, 111, 112, 125, 140, 143–145, 149–150, 217

Thouret, Michel-Augustin, 206

time: description in clockmaking terms, 70; each woman scientist racing against, 261–262

translations and art of translating, 220–221, 315nn48–49

Treaty of Aix-la-Chapelle (1748), 34, 276n72

Treaty of Paris (1763), 76

Tronchin, Théodore, 24, 36

Tu Youyou, 264

Turgot, Anne Robert Jacques: d'Arconville and, 257; Clairaut and, 97; Commerson and, 111, 144–145; Condillac and, 25, 27; Ferrand and, 22, 23; Helvétius and, 25; Lalande and, 94; Lepaute and, 92; life of, 25; Louis XVI and, 142; Poivre and, 129; Rousseau and, 37; works of: "A Philosophical Review of the Successive Advances of the Human Mind," 25

Turrea, 297n115

Uglow, Jenny, 9

UNESCO, 265

Uranus (planet), 98

Vachier, Cleriade, 111, 133, 136, 138, 142, 144–145, 159, 292n29

Vaillant, Sebastien, 162

Vandermonde, Charles-Augustin, 177, 229

Van Kley, Dale K., 269n23

van Spaandonck, Albert, 293n41

van Spaendonck, Gerard, 147, 300n171, 300n173

Vassé, Madame la Comtesse de: after Ferrand's death, 54; bequests from Ferrand, 42–43; Condillac and, 55; Condillac's *Traité des sensations* dedicated to, 19, 44, 45, 46–48; at Couvent Saint Joseph, 21; early life of, 21; Helvétius and, 25, 26, 54; hiding of Charles Edward (Edouard) Stuart and, 37, 54, 59; illness and death of, 55, 280n140, 280n142; Mably and, 26, 40, 55; as partner of Ferrand, 16, 17, 21, 26, 42, 59–60; related to Jaucourt family, 24; will and bequests by, 55, 273n20

Vaucanson, Jacques de, 183

Vauquelin, Louis-Nicolas, 257

Vaysse, Françoise, 313n8

Venus (planet), 95; transit of, 10, 76–77, 81, 85, 88–89, 91, 261, 285n52

Véron, Pierre-Antoine, 111, 291n14

Versailles, Trianon at, 121, 142

Vibart, Marie Elisabeth, 205, 208, 310n122